Lecture Notes in Mathematics

Edited by A. Dold and B. Eckmann

1066

Numerical Analysis

Proceedings of the 10th Biennial Conference held at Dundee, Scotland, June 28 – July 1, 1983

Edited by David F. Griffiths

Springer-Verlag
Berlin Heidelberg New York Tokyo 1984

Editor

David F. Griffiths
Department of Mathematical Sciences, University of Dundee
Dundee DD1 4HN, Scotland

AMS Subject Classification (1980): 65-06, 65D10, 65F99, 65K10, 65L05, 65M25, 65M99, 65N30, 65R20, 65S05

ISBN 3-540-13344-5 Springer-Verlag Berlin Heidelberg New York Tokyo
ISBN 0-387-13344-5 Springer-Verlag New York Heidelberg Berlin Tokyo

Printing and binding: Beltz Offsetdruck, Hemsbach/Bergstr.
2146/3140-543210

Preface

The Tenth Dundee Biennial Conference on Numerical Analysis, held at the University of Dundee, Scotland on the four days 28 June - 1 July, 1983, attracted over 200 participants from 25 countries. The organizers were again fortunate in gaining the services of 16 eminent numerical analysts covering a broad spectrum of the subject and it is their papers which appear in these notes. Unfortunately Professor Dupont's contribution was not available at the time of going to press. In addition to the invited talks, short contributions were solicited and 69 of these were presented at the conference in three parallel sessions. A complete list of these submitted papers, together with authors' addresses, is also given here.

I would like to take this opportunity of thanking Professor Dr L Collatz who, as after dinner speaker, kept the audience greatly amused with anecdotes, some true, some with only a grain of truth and others apocryphal, concerning many well-known mathematicians. It is also a pleasure to thank all the speakers, the session chairmen and the members of the Mathematical Sciences Department of this University for their contributions and assistance with the successful outcome of this conference. I am particularly indebted to Mrs Dorothy Hargreaves for attending to the considerable task of typing the various documents associated with the conference and for coping so admirably with many of the organizational details.

Financial support for this conference was obtained from the European Research Office of the United States Army. This support is gratefully acknowledged.

Dundee, January 1984. D F Griffiths

INVITED SPEAKERS

R H Bartels:	Department of Computer Science, University of Waterloo, Waterloo, Ontario, Canada N2L 3G1.
A Bjorck:	Department of Mathematics, Linkoping University, S-581 83 Linkoping, Sweden.
C de Boor:	Mathematics Research Center, University of Wisconsin-Madison, Madison, Wisconsin 53706, USA.
H Brunner:	Institut de Mathematiques, University of Fribourg, CH 1700 Fribourg, Switzerland.
T Dupont:	Department of Mathematics, University of Chicago, 5734 University Avenue, Chicago, Illinois 60637, USA.
D M Gay:	Bell Laboratories, 600 Mountain Ave., Murray Hill, NJ 07974, USA.
P W Hemker:	Mathematical Centre, Kruislaan 413, 1098 SJ Amsterdam, The Netherlands.
M J D Powell:	Department of Applied Mathematics and Theoretical Physics, University of Cambridge, Silver Street, Cambridge CB3 9EW, England.
P A Raviart:	Analyse Numérique, Université Pierre et Marie Curie (Paris VI), 4 Place Jussieu, 75230 Paris, Cedex 05, France.
L F Shampine:	Numerical Mathematics Division 1642, Sandia National Laboratories, Albuquerque, New Mexico 87185, USA.
A Spence:	School of Mathematics, University of Bath, Clavertown Down, Bath, England.
H J Stetter:	Institut fur Numerische Mathematik, Technische Hochschule Wien, A-1040 Wien, Gusshausstr 27-29, Austria.
Ph L Toint:	Department of Mathematics, Facultes Universitaires de Namur, Rempart de la Vierge 8, B-5000 Namur, Belgium.
G A Watson:	Department of Mathematical Sciences, University of Dundee, Dundee DD1 4HN, Scotland.
M F Wheeler:	Mathematical Sciences Department, Rice University, Houston, Texas 77001, USA.
J R Whiteman:	Institute for Numerical Mathematics, Brunel University, Kingston Lane, Uxbridge, Middlesex UB8 3PM, England.

SUBMITTED PAPERS

C A Addison and P M Hanson, Department of Computer Science, University of Victoria, P O Box 1700, Victoriam B.C., Canada V8W 2Y2.
An investigation into the stability properties of second derivative methods with perfect square iteration matrices.

J M Aitchison, Computing Laboratory, University of Oxford, 19 Parks Road, Oxford OX1 3PL, England.
The calculation of free surface flows using a conformal transformation and boundary element techniques.

S Amini and D T Wilton, Department of Mathematics, Statistics and Computing, Plymouth Polytechnic, Drake Circus, Plymouth, Devon PL4 8AA, England.
A comparison between boundary element methods for the acoustic radiation problem.

H Arndt, Institut fur Angewandte Mathematik, Wegelerstr. 6, D-5300 Bonn, W Germany.
On step size control for Volterra integral equations.

U Ascher, Department of Mathematics, University of British Columbia, Vancouver, Canada.
Collocation methods for singular perturbation problems.

M L Baart, NRIMS-CSIR, P O Box 395, Pretoria 0001, South Africa.
Recursive calculation of curved finite element stiffness matrices.

K Barrett, Mathematics Department, Coventry Polytechnic, Priory Street, Coventry CV1 5FB, England.
A multigrid finite element method for general regions.

L Brutman, Department of Mathematics and Computer Science, University of Haifa, Mount Carmel, Haifa 31999, Israel.
Generalized alternating polynomials, some properties and numerical applications.

P Camino and J M Sanz-Serna, Departmento de Ecuaciones Funcionales, Universidad de Valladolid, Facultad de Ciencias, Valladolid, Spain.
ODE solvers for trajectory problems and applications to PDEs.

C Carter, Department of Mathematics, Trent University, Peterborough, Ontario, Canada K9J 7B8.
Spectral solution of stiff piecewise-linear differential equations occurring in heat transfer through a network.

M M Cecchi, Instituto di Elaborazione del Informasione, Via S Maria 46, 56100 Pisa, Italy.
Spline approximation for linear two point boundary value problems.

A R Conn and N I M Gould, Department of Combinatorics and Optimization, University of Waterloo, Waterloo, Ontario, Canada N2L 3G1.
The calculation of directions of negative curvature for non-linear minimization problems.

G J Cooper, School of Mathematics and Physical Sciences, University of Sussex, Brighton BN1 9QH, England.
Algebraic stability for explicit Runge-Kutta methods.

I C Demetriou, DAMTP, University of Cambridge, Silver Street, Cambridge, England.
Piecewise monotonic least squares data fitting.

D B Duncan, Chalk River Nuclear Laboratories, Atomic Energy of Canada, Chalk River, Ontario, Canada KOJ 1J0.
Numerical scheme for a groundwater flow problem with radionuclide seepage.

C M Elliott, Mathematics Department, Imperial College, Queens Gate, London SW7 2BX, England.
Ordinary differential inclusions and equations with discontinuous right hand sides.

R England, IIMAS, Universidad Nacional Autónoma México, Apdo Postal 20-726, Delegacion de Alvaro Obregón, 01000 Mexico, D.F.
Implicit multistage methods for ordinary differential equations and their efficient implementation.

Th Fawzy* and I Refat+, *Mathematics Department, Suez Canal University, Ismailia, Egypt, +P O Box 8878, Salmiya, Kuwait.
On the Lacunary interpolation with splines.

Th Fawzy* and I Refat+, *Mathematics Department, Suez Canal University, Ismailia, Egypt, +P O Box 8878, Salmiya, Kuwait.
On the approximate solution of the D E $y'' = f(x,y,y')$.

R Fletcher, Department of Mathematical Sciences, University of Dundee, Dundee DD1 4HN, Scotland.
Expected conditioning.

R Fletcher and S P J Matthews, Department of Mathematical Sciences, University of Dundee, Dundee DD1 4HN, Scotland.
Stable modification of explicit LU factors for simplex updates.

M Friedman and G Erez, Department of Physics, Nuclear Research Center-Negev, Beer-Sheva, P O Box 9001, Israel.
An improved Newtons' method.

C W Gear and O Osterby, Computer Science Department, University of Aarhus, Ny Munkegate, DK8000 Aarhus C, Denmark.
Methods for ordinary differential equations with discontinuities.

R Ge, DAMTP, University of Cambridge, Silver Street, Cambridge CB3 9EW, England.
A filled function method for finding the global minimizer.

H Gerritsen, Delft Hydraulics Laboratory, P O Box 177, 2600 MH Delft, The Netherlands.
Accurate boundary treatment in shallow water flow computations.

J K Gibson, Mathematics Department, Birkbeck College, Malet Street, London WC1, England.
The double secant method for real polynomial equations.

I Gladwell* and C A Addison+, *Mathematics Department, University of Manchester, Manchester M13 9PL, England, +Department of Computer Science, University of Victoria, Victoria, B.C., Canada V8W 2Y2.
Implementation of second derivative methods for implicit second order systems.

K Hebeker, Fachbereich 17 Mathematik/Informatik, Universtat GHS Paderborn, Warburger Str 100, D-4790 Paderborn, W Germany.
A boundary element method for computing 3-D viscous incompressible fluid flows.

G W Hedstrom, Lawrence Livermore Laboratory, University of California, P O Box 808, Livermore, California 94450, USA.
A moving-grid method for tracking shocks.

B M Herbst, Department of Mathematical Sciences, University of Dundee, Dundee DD1 4HN. Scotland.
Stability and the nonlinear Schrodinger equation.

N J Higham, Half Edge Lane, Eccles, Manchester M30 9BA, England.
Upper bounds for the condition number of a triangular matrix.

N Houbak, Numerical Institute, Technical University of Denmark, Building 303, DK-2800 Lyngby, Denmark.
SIL - a simulation language.

P J van der Houwen and B P Sommeijer, Mathematical Centre, Kruislaan 413, 1098 SJ Amsterdam, The Netherlands.
Linear multistep methods with minimized truncation error for periodic initial value problems.

W H Hundsdorfer, Department of Mathematics, University of Leiden, Wassenaarseweg 80, Postbus 951, 2300 Ra Leiden, The Netherlands.
B-Stability for semi-implicit methods.

A Iserles, DAMTP, University of Cambridge, Silver Street, Cambridge CB3 9EW, England.
Order stars and finite differences for parabolic PDE's.

K Jonasson, Department of Mathematical Sciences, University of Dundee, Dundee DD1 4HN.
Numerical solution of continuous Chebyshev approximation problems.

P E Koch, Institute of Information, University of Oslo, P O Box 1080, Blindern, Oslo 3, Norway.
A collocation method for singularly perturbed two-point boundary value problems with a turning point.

D P Laurie, NRIMS-CSIR, P O Box 395, Pretoria 0001, South Africa.
Practical error estimates in numerical integration.

P E Manneback, Division of Information and Computing, National Physical Laboratory, Teddington, Middlesex, TW11 OLW, England.
Solution of weighted linear least squares problems with simple structure.

V S Manoranjan, Department of Mathematical Sciences, University of Dundee, Dundee DD1 4HN, Scotland.
Bifurcation studies in reaction-diffusion.

J C Mason, Mathematics Branch, Royal Military College of Science, Shrivenham, Swindon, England.
L_p approximation by real and complex Chebyshev series.

E J W Ter Maten and G L G Sleijpen, Department of Mathematics, University of Utrecht, Budapestlaan 6, P O Box 80.010, 3508 TA Utrecht, The Netherlands.
Stability analysis of high accuracy methods for fourth order parabolic equations.

G Moore, School of Mathematical Sciences, National Institute for Higher Education, Dublin 9, Republic of Ireland.
Divided differences and defect correction for the finite element method.

F D Van Niekerk, Department of Mathematics, University of Pretoria, Pretoria, South Africa.
A global-local finite element method in space-time for a convection-diffusion problem.

J Nocedal and M Overton, Courant Institute of Mathematical Sciences, New York University, 251 Mercer Street, New York, NY 10012, USA.
Numerical methods for solving inverse eigenvalue problems.

F A De Oliveira, Departmento de Matematica, Universidade de Coimbra, 3000-Coimbra, Portugal.
Two-point-boundary-value problems with interval analysis.

G Opfer, Institut fur Angewandte Mathematik, University of Hamburg, Bundesstr. 55 D 2000 Hamburg 13, Germany.
Richardson's iteration for nonsymmetric matrices.

F Patricio, Departmento de Matematica, Universidade de Coimbra, 3000-Coimbra, Portugal.
A class of methods for stiff ordinary differential equations.

P W Pedersen, Department of Mathematics, Technical University of Denmark, DK 2800 Lyngby, Denmark.
Computing square roots with 15 decimal accuracy using only 4 multiplications - and some algorithms.

H J J te Riele and P Schroevers, Mathematical Centre, Kruislaan 413, 1098 SJ Amsterdam, The Netherlands.
A comparison of numerical methods for the linear generalized Abel integral equation.

R A Sack and R Brown, Department of Mathematics, University of Salford, Salford M5 4WT, England.
Vector analogues of Aitken's δ^2-process.

L L Schumaker, Department of Mathematics, University of Texas at Austin, Austin, Texas 78712, USA.
Finding the zeros of splines.

B W Scotney, Department of Mathematics, University of Reading, Whiteknights, Reading RG6 2AX, England.
Optimal error estimates for Petrov-Galerkin methods.

A Sharma, Department of Mathematics, University of Alberta, Edmonton, Alberta, Canada.
An extension of a theorem of J L Walsh.

S Sigurdsson, Faculty of Engineering and Science, University of Iceland, Hjardarhagi 2-6, Reykjavik, Iceland.
Construction of generalized Adams methods for stiff ODEs.

G L G Sleijpen and E J W Ter Maten, Department of Mathematics, University of Utrecht, Budapestlaan 6, P O Box 80.010, 3508 TA Utrecht, The Netherlands.
Stability analysis of hopscotch methods for fourth order parabolic equations.

D M Sloan, Department of Mathematics, University of Strathclyde, 26 Richmond Street, Glasgow G1 1XH., Scotland.
Stability of a boundary condition for a fourth order hyperbolic difference scheme.

M N Spijker, Subfacultad der Wiskunde, University of Leiden, Wassenaarseweg 80, Postbus 9512, 2300 Leiden, The Netherlands.
The existence of solutions to the algebraic equations in implicit Runge-Kutta methods.

J Steinberg, Technion, Israel Institute of Technology, Haifa, Israel.
Computation of stress intensity factors.

R Thatcher, Department of Mathematics, UMIST, P O Box 88, Sackville Street, Manchester M60 1QD., England.
Variational methods for solving free boundary problems.

P G Thomsen and P Norsett, Institute for Numerical Analysis, Technical University of Denmark, DK-2800 Lyngby, Denmark.
Control of error in codes for stiff ODEs.

E F Toro and M J O'Carroll, Department of Applied Mathematical Studies, University of Leeds, Leeds LS2 9JT., England.
A Kantarovic computational method for free surface gravity flows.

M van Veldhuizen, Wiskundig Seminarium, Vrije Universiteit, De Boelelaan 1081, Postbus 7161, 1007 MC Amsterdam, The Netherlands.
D-stability and Kaps-Rentrop methods.

A J Wathen and M J Baines, Department of Mathematics, University of Reading, Whiteknights, Reading, Berks., England.
The structure of the moving finite element.

B Werner and A Spence, Department of Mathematics, University of Hamburg, Bundesstr. 55, D-2000 Hamburg 13, W Germany.
Computation of symmetry-breaking bifurcation points.

R A Williamson, DAMTP, University of Cambridge, Silver Street, Cambridge CB3 9EW.
The stability and accuracy of semi-discretised schemes in the numerical solution of hyperbolic differential equations.

T J Ypma, Department of Applied Mathematics, University of Witwatersrand, Johannesburg 2001, S. Africa.
Local convergence in inexact Newton methods.

Y Yuan, DAMTP, University of Cambridge, Silver Street, Cambridge CB3 9EW, England.
An example of only linear convergence of trust region algorithms for nonsmooth optimization.

Z Zlatev, R Berkowicz and L P Prahm, Air Pollution Laboratory, Riso National Laboratory, DK-4000 Roskilde, Denmark.
Selfadaptive time-integration in a code for numerical treatment of an air pollution model.

CONTENTS

Splines in Interactive Computer Graphics

Richard H. Bartels

ABSTRACT

Computer graphics, particularly interactive computer graphics, is not, as the name might imply, concerned with drawing graphs, but rather with the broadest issues of manipulating, transforming, and displaying information in visual format. It is interactive in so far as operations can be carried out in real time – which requires algorithms of high computational efficiency and low complexity.

Splines are a valuable tool in graphics, but they are often applied in a way not used by the mathematician. This difference raises computational issues which the numerical analyst might otherwise never see. This talk will provide a brief introduction to such issues and follow with a study of two current developments.

We begin with a review of the graphics environment, mentioning the modelling and display process and pointing out some of the costly issues. The novel use of splines in interactive graphics comes through the construction of surfaces as weighted averages of selected points, called "control vertices" in which B-splines are taken as the weighting functions. Some examples will illustrate the characteristics of this use of B-splines.

With this background we consider two recent developments. The first is the control-vertex recurrence of Riesenfeld, Cohen, and Lyche; the second is Barsky's work on geometric vs. mathematical continuity, and his introduction of Beta-splines. We will close with some results on current research concerned with a synthesis of these two developments.

1. Introduction

The Computer Graphics Laboratory at the University of Waterloo has embarked on a programme to investigate techniques of potential use for the next generation of computer-aided design systems. The terms of reference are *interactive* and *surface modeling*. The context is not that of fitting curves or surfaces to data or objects which exist – plotting or approximation is not of interest. The context is that of (1) providing a mathematically naive industrial draftsman, sitting before a screen and control panel, with an easy way of creating the mathematical description of an object which does not yet exist, (2) displaying and manipulating images of that object on the screen, (3) modifying the object, and (4) ultimately generating machine-tool commands which will provide a means for producing the object. In design systems of this type splines have been very important in the past, and they are undergoing interesting developments for the future.

In this presentation we will look at a typical visual display environment in computer graphics to set the stage and provide a motivation for some of the things to be mentioned subsequently. There will be a brief, informal, intuitive review of the classical construction of B-splines, concentrating on simple knots, to provide a paradigm for some new developments. The use of B-splines in computer graphics to construct curves and surfaces is distinctly different from the use of B-splines in approximation and interpolation. This use will be presented, along with some of the reasons it is particularly appropriate from the point of view of computational efficiency and human interface. The work of Lyche, Riesenfeld, and Cohen for subdividing curves and surfaces, as a means of modification and display, will be outlined.

This will end one of the thrusts of the presentation. A second will cover the concept of geometric (as opposed to mathematical) continuity which was explored by Barsky, and which provides a generalisation of B-splines to functions which can serve as generators for "tensed" and "biased" splines. Some examples of curves produced by these functions, called Beta-splines, will be given.

The third portion of this presentation will cover recent progress in expanding the notion of geometric continuity to a more general context and in adapting the classical B-spline construction methods to the production of Beta-splines. The goal of this work is to develop subdivision recurrences for Beta-splines.

The primary references for the spline material in this presentation are [2,5,8] and [9]. Good backgrounds on the graphics environment are to be found in [1,6] and [7].

2. The Graphics Display Environment

The following is not the only example of a visual display environment for a design system, but it is typical and will serve to motivate the later discussion. Figure 1 gives an overview of a *display pipeline.* In it, mathematical objects are defined separately as *templates*, each in its own *local coordinate system*, each arranged according to some canonical scaling and orientation. These objects a placed together in a *world coordinate system* using rotations, translations and scalings, all of which are rigid transformations. As such, they preserve the character of the objects; in particular, polynomials remain polynomials. The result of these *modelling transformations* is a composite scene to be displayed. Each time a new view of the scene is to be taken, *viewing transformations* must be carried out. In industrial design systems these may be no more complicated than a single orthographic projection. In graphic art systems however, a camera model is often used: an eye-point is specified, as are viewing direction, upward orientation, angle of view, aperture, depth of field, and image plane. The *viewing frustum* which results is often mapped to a canonical viewing configuration by further rigid transformations. This canonical configuration is then subjected to a distortion by a *perspective transformation.* Magically, straight lines are preserved, but polynomials become rational functions. Under this distortion, the viewing frustum is changed into a canonical *clipping box.* Up until now, objects in the scene which are not within the viewing volume may have been involved in the computations. Algorithms are now invoked to trim away all portions of the scene outside of this volume and project only what remains onto the image plane. The projected objects must then be discretised, if the display is a *raster device*, or be approximated in outline by simple, known curves, if the display is a *calligraphic device.*

Profound implications are hidden in the above. To do all of this exactly, transformations would have to be applied to infinite numbers of points or to functional descriptions of surfaces; extensive root-finding techniques would be required to determine the curves of intersection between pairs of surfaces, as well as between surfaces and the sides of the viewing frustum; information would also have to be extracted to determine which surfaces are obscured by which – requiring that additional root-finding operations be performed to trace the silhouettes of objects with respect to the eye-point – and finally, discretisation processes akin to differential-equation solvers would have to be applied to paint an image on the display. For the purposes of industrial design, a wire-frame rendering of each object in the scene may suffice, but for other purposes a more realistic rendering of the objects may be required. This may involve something as elaborate as using a mathematical model of the optical characteristics of some collection of materials together with computations requiring normal vectors to the various surfaces and the illuminant details of a number of light sources. If all of this is to be interactive, then it must take place as fast as the refresh rate of the display – that is: typically within a sixtieth to a thirtieth of a second. (It is small wonder that the Cray computer corporation has started to advertise in the graphics trade literature.) Finally, if machine-tool descriptions are needed, while they don't need to be performed with the speed of display computations, they will still involve determining how to track contours on a surface while holding a prescribed orientation to the normal – which could be a nontrivial problem in control theory.

The only salvation for owners of small computers lies in determining how little one can do, how efficiently and how approximately, without the disturbing the final result within visual or machine-tool tolerances. In graphics this is accomplished by "creative cheating", which frequently involves applying all of the above processes only to a small numbers of representers of the objects in question. For example, it is still usual to deal mostly with polyhedral bodies, plus a few additional primitive forms such as spheres and cylinders, and carry out the transformations of the pipeline merely on vertices, centres, radii, etc., using the images of these few representers at the bottom of the pipeline to recreate approximate pictures. One feature of the unique way in which spline surfaces are used in graphics derives from the fact it provides an efficient way for the surfaces to be approximated to arbitrary accuracy by polyhedral surfaces. It is to these polyhedra rather than to the splines themselves that the above computations in the pipeline can be applied.

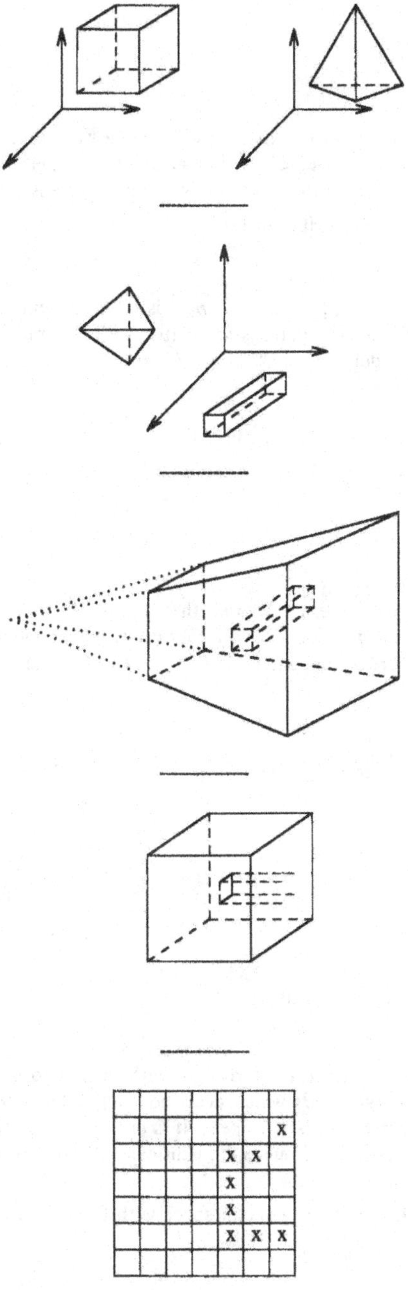

Figure 1. The graphics display pipeline.

3. Splines

For any $k > 0$, let $\mathbf{P}^k$ be the space of polynomials of order k with real coefficients:

$$p(u) = c_0 + c_1 u + \cdots + c_{k-1} u^{k-1} .$$

For any $m > 0$ let $\mathbf{U}^m$ be a sequence of m *knots*:

$$\mathbf{U}^m = \{ u_0, \ldots, u_m \} ,$$

where

$$u_0 \le u_1 \le \cdots \le u_m .$$

Each knot u_i has *multiplicity* μ_i, the count of the knots in $\mathbf{U}^m$ having value equal to u_i. This includes u_i itself, By convention, $\mu_i \le k$ for all i; hence, $u_i < u_{i+k}$ for all $i = 0, \ldots, m-k$. As a further convention, let $a < u_0$ and $b > u_m$, and agree that nothing outside of $[a, b)$ is of any interest.

The set of the M *distinct*, consecutive values in $\mathbf{U}^m$

$$u_{i_0} < \cdots < u_{i_M}$$

are the *breakpoints* of $\mathbf{U}^m$. ($\{ i_0, \ldots, i_m \} \subseteq \{0, \ldots, m\}$ is any conveniently chosen subsequence which picks out the breakpoints.) Each breakpoint is associated with the multiplicity of its corresponding knots, and the *breakpoint intervals* defined by $\mathbf{U}^m \bigcup [a, b)$ are the half-open intervals

$$\mathbf{I}_0 = [a, u_{i_0})$$

$$\mathbf{I}_j = [u_{i_{j-1}}, u_{i_j}) \quad \text{for } j = 1, \ldots, M .$$

and

$$\mathbf{I}_{M+1} = [u_{i_M}, b) .$$

Formally:

Definition: Assuming that $k \ge 1$, that $m \ge k-1$, and that $1 \le \mu_i \le k$ for all $i = 0, \ldots, m$, then $\mathbf{S}(\mathbf{P}^k, \mathbf{U}^m, a, b)$, the set of all splines of order k on $[a,b)$ with the knot sequence $\mathbf{U}^m \subset [a,b)$, breakpoints $\{ u_{i_0}, u_{i_1}, \ldots, u_{i_M} \} \subseteq \mathbf{U}^m$, and breakpoint intervals $\mathbf{I}_0, \ldots, \mathbf{I}_{M+1}$ is the set of all functions of the form $s(u)$ satisfying:

$$s(u) \in \mathbf{P}^k \quad \text{for each } \mathbf{I}_j \ (j = 0, \ldots, M+1)$$

and for any breakpoint u_{i_j} associated with multiplicity μ_{i_j} $(j = 0, \ldots, M)$, if

$$s(u) \equiv p_j \in \mathbf{P}^k \quad \text{on } \mathbf{I}_j$$

and

$$s(u) \equiv p_{j+1} \in \mathbf{P}^k \quad \text{on } \mathbf{I}_{j+1}$$

then

$$D_u^{(l)} p_j(u_{i_j}) = D_u^{(l)} p_{j+1}(u_{i_j}) \quad \text{for } l = 0, \ldots, k-1-\mu_{i_j} .$$

In the definition, $D_u^{(l)} p(u_i)$ stands for the lth derivative with respect to u of $p(u)$ evaluated at u_i. We denote p_j, p_{j+1} as the *segment polynomials* which describe $s(u)$ in the interval $\mathbf{I}_j$ and $\mathbf{I}_{j+1}$ respectively. Notice that any issue of end conditions is left open. It is only necessary that a spline be a polynomial in the intervals $\mathbf{I}_0$ and $\mathbf{I}_{M+1}$. Imposing various end conditions will serve merely to isolate subsets of splines, and this issue will be left aside.

It is easily verified that $\mathbf{S}(\mathbf{P}^k, \mathbf{U}^m, a, b)$ is a vector space with $\mathbf{P}^k$ as a subspace.

4. The One-Sided Basis

The dimension of $\mathbf{S}(\mathbf{P}^k, \mathbf{U}^m, a, b)$ is

$$k + \mu_{i_0} + \cdots + \mu_{i_M} .$$

This can be made plausible by considering any $s(u) \in \mathbf{S}(\mathbf{P}^k, \mathbf{U}^m, a, b)$ as u moves from a rightwards to b. On the breakpoint interval $\mathbf{I}_0 = [a, u_{i_0})$, $s(u) = p_0(u)$ is a polynomial of order k; hence it can be represented as a linear combination of

$$(u-a)^0, (u-a)^1, \ldots, (u-a)^{k-1} . \tag{4.1}$$

In the next interval to the right $u_{i_0} \le u < u_{i_1}$, $s(u)$ changes to

$$\begin{aligned} s(u) &= p_0(u) + (p_1(u) - p_0(u)) \\ &= p_0(u) + \Delta_{1_1}(u) , \end{aligned}$$

where $\Delta_1(u)$ "touches zero with $C^{k-1-\mu_{i_0}}$ continuity" at $u = u_{i_0}$. It proves true that $\Delta_1(u)$ can be represented as a linear combination of the truncated power functions

$$(u - u_{i_0})_+^{k-1}, (u - u_{i_0})_+^{k-2}, \ldots, (u - u_{i_0})_+^{k-\mu_{i_0}} . \tag{4.2}$$

Consequently $s(u)$ can be represented on the interval $a \le u < u_{i_1}$ as a linear combination of the functions (4.1) together with those of (4.2).

The same arguments apply as u crosses u_{i_j} and $p_j(u)$ changes into

$$p_{j+1}(u) = p_j(u) + \Delta_{j+1}(u)$$

for each $j = 0, \ldots, M$.

After further considerations:

Theorem: The functions

$$(u-a)^0, (u-a)^1, \ldots, (u-a)^{k-1} ,$$

together with

$$\begin{array}{ll} (u - u_{i_0})_+^{k-1}, (u - u_{i_0})_+^{k-2}, \ldots, (u - u_{i_0})_+^{k-\mu_{i_0}} & \text{for } u_{i_0} \\ (u - u_{i_1})_+^{k-1}, (u - u_{i_1})_+^{k-2}, \ldots, (u - u_{i_1})_+^{k-\mu_{i_1}} & \text{for } u_{i_1} \\ \bullet\bullet\bullet \quad\quad \bullet\bullet\bullet & \bullet\bullet\bullet \\ (u - u_{i_M})_+^{k-1}, (u - u_{i_M})_+^{k-2}, \ldots, (u - u_{i_M})_+^{k-\mu_{i_M}} & \text{for } u_{i_M} . \end{array} \tag{4.3}$$

form a basis for $\mathbf{S}(\mathbf{P}^k, \mathbf{U}^m, a, b)$.

There are precisely $k + \mu_{i_0} + \cdots + \mu_{i_M}$ functions in (4.3). This is the *one-sided basis* of $\mathbf{S}(\mathbf{P}^k, \mathbf{U}^m, a, b)$.

5. Linear Combinations and Cancellation

Computing the coefficients in the representation of a spline with respect to this basis is often an ill-conditioned problem. This arises, roughly, as follows. Most splines with which we would want to deal in practice have moderate values throughout the interval $[a,b)$. The one-sided basis functions, on the other hand, blow up as u increases. Hence, if this basis is used to express "reasonable" spline curves and surfaces, the coefficients required for this could be expected to flip-flop between large positive and negative values in order to force *numerical cancellation* of the basis-function values as u increases.

A second shortcoming, from the point of view of graphics, is that the one-sided basis functions do not have *compact support*; they are all nonzero on half of the real line. If a curve or surface is represented in terms of the one-sided basis and some change is made to the representation to provide an adjustment of shape, then the change has an influence over the entire curve or surface. A complete recomputation of the curve or surface is necessary; no local updates are possible. The continual need engage in costly recomputations will all but rule out interactive graphical design. In graphics it is as significant that the B-splines have compact support as that they provide well-conditioned representations.

The key to constructing a desirable basis from the less desirable (but conceptually simple) one-sided basis is to recognise (1) that a basis with compact support will be an answer to the numerical objections above as well as (2) being desirable from the point of view of computational efficiency. Compact support can be achieved by a process of *symbolic cancellation*, before any numerical computations are begun. It is through this door that divided differences enter.

To illustrate, let $k=4$ (cubic splines), and consider

$$u_i < u_{i+1} < u_{i+2} < u_{i+3} < u_{i+4} \ .$$

We have

$$\frac{(u-u_{i+1})_+^3-(u-u_i)_+^3}{(u_{i+1}-u_i)} = \begin{cases} 0 & u<u_i \\ -(u-u_i)^3 & u_i \le u < u_{i+1} \\ -3u^2+3u(u_{i+1}+u_i) - (u_{i+1}^2+u_{i+1}u_i+u_i^2) & u_{i+1} \le u \ . \end{cases}$$

which is a "nicer" function than either $(u-u_i)_+^3$ or $(u-u_{i+1})_+^3$ in that it grows only quadratically for $u \to \infty$. Denote the result by

$$[u_i, u_{i+1}:t](u-t)_+^3 = \frac{(u-u_{i+1})_+^3 - (u-u_i)_+^3}{u_{i+1}-u_i} \ .$$

Since this function is a linear combination of $(u-u_i)_+^3$ and $(u-u_{i+1})_+^3$, we may substitute it for one of these truncated power functions, e.g. for $(u-u_i)_+^3$. We may carry out a similar operation for the pairs

$$\{u_{i+1}, u_{i+2}\} \ , \ \{u_{i+2}, u_{i+3}\} \ , \text{ and } \{u_{i+3}, u_{i+4}\} \ ,$$

to produce quadratic-growing substitutes for

$$(u-u_{i+1})_+^3 \ , \ (u-u_{i+2})_+^3 \ , \text{ and } (u-u_{i+3})_+^3 \ .$$

Going one stage further, let

$$[u_i, u_{i+1}, u_{i+2}:t](u-t)_+^3 = \frac{[u_{i+1}, u_{i+2}:t](u-t)_+^3 - [u_i, u_{i+1}:t](u-t)_+^3}{u_{i+2}-u_i} \ ,$$

This function grows only linearly for $u \to \infty$. It may be used to replace $[u_i, u_{i+1}:t](u-t)_+^3$, and the differencing and replacement may be repeated pairwise, again. Ultimately we arrive at a function which "grows as zero" for $u \to \infty$.

When multiple knots appear, the construct

$$[u_i, u_{i+1}:t](u-t)_+^3 = \frac{(u-u_{i+1})_+^3 - (u-u_i)_+^3}{u_{i+1}-u_i}$$

is regarded as being taken in the limit as $u_i \to u_{i+1}$. This provides the convention that the derivatives of the one-sided power function $(u-t)_+^r$ with respect to t for fixed u:

$$D_t^{(l)}(u-t)_+^r \quad \text{for } l,r=0,1,2,3,\cdots$$

are to be understood in the left-handed sense. The derivatives of $(u-t)_+^r$ with respect to u for fixed t are similarly to be understood in the right-handed sense. So long as we hold to this, the truncated power functions behave under differentiation just as the ordinary power functions do. This brings us to

---

Definition: For any values $z_i \leq \cdots \leq z_{i+l}$ the l-th *divided difference* with respect to the variable x of any function f depending upon x (and possibly other variables) is given
for $l=0$ by

$$[z_i:x]f(x) = f(z_i)$$

and for $l>0$ by

$$[z_i,\ldots,z_{i+l}:x]f(x) =$$

if $z_{i+l} > z_i$ **then**

$$\frac{[z_{i+1},\ldots,z_{i+l}:x]f(x) - [z_i,\ldots,z_{i+l-1}:x]f(x)}{z_{i+l} - z_i}$$

else

$$\frac{1}{l!}D_x^{(l)}f(x)\,|_{x=z_i}$$

When this general definition is applied to $(u-t)_+^r$, with t in the role of x and u replacing z, compact-support substitutes for the truncated power functions result, even for the case of multiple knots, and we have

---

Definition: Assuming that $i+k \leq m$, the *B-spline of order k associated with the knots* $u_i,\ldots,u_{i+k}$ is given by

$$B_{i,k}(u) = (-1)^k(u_{i+k}-u_i)[u_i,\ldots,u_{i+k}:t](u-t)_+^{k-1}$$

This function has support on $[u_i,u_{i+k})$. The factor $(-1)^k$ ensures nonnegativity, and the factor $(u_{i+k}-u_i)$ provides normalisation. These three observations will be stressed more formally somewhat below.

This does not, as such, provide a full replacement for the one-sided basis. If it had been our intention, for example, to construct a compact-support basis for $\mathbf{S}(\mathbf{P}^k,\mathbf{U}^m,a,b)=\mathbf{S}(\mathbf{P}^4,\{0,1,2,3,4\},(-1),5)$, then we would not have enough power functions to carry the differencing process to completion. The one-sided basis for the splines in question would be

$$\begin{array}{lllll} (u-(-1))^0, & (u-(-1))^1, & (u-(-1))^2, & (u-(-1))^3, & \\ (u-0)_+^3, & (u-1)_+^3, & (u-2)_+^3, & (u-3)_+^3, & (u-4)_+^3 \ , \end{array}$$

and this provides us with only the one B-spline $(-1)^4(4-0)[0,1,2,3,4:t](u-t)_+^3$. There is nothing to the right of $u_4=4$ to use as a "partner" for u_1,u_2,u_3 and u_4 to produce a compact-support function which could substitute for $(u-u_1)_+^3$, for example. Furthermore, there is nothing which can form a compact-support substitute for any of the functions $(u-(-1))^j$, $j=0,\ldots,3$. To compensate for this, introduce additional knots:

$$u_{-4} \leq u_{-3} \leq u_{-2} \leq u_{-1} \leq a \equiv (-1)$$

and

$$b \equiv 5 \leq u_5 \leq u_6 \leq u_7 \leq u_8 \ .$$

Then the differencing process will be able to produce B-splines with supports in the intervals

$$[u_{-4},u_0)\,,\,[u_{-3},u_1)\,,\,[u_{-2},u_2)\,,\,[u_{-1},u_3)\,,\,[u_0,u_4)\,,$$
$$[u_1,u_5)\,,\,[u_2,u_6)\,,\,[u_3,u_7)\,,\,[u_4,u_8)\quad .$$

These basis functions can represent any spline in $\mathbf{S}(\mathbf{P}^4,\{0,1,2,3,4\},(-1),5)$, but the representation is only good on the interval $[a,b)=[(-1),5)$. Since we are agreed that nothing outside of this interval is of interest, this is good enough. That this is true in general is the Curry-Schoenberg result:

Construction: To the sequence of knots in $\mathbf{U}^m$ add arbitrarily chosen knots

$$u_{-k}\leq\cdots\leq u_{-1}\leq a$$

and

$$b\leq u_{m+1}\leq\cdots\leq u_{m+k}\quad .$$

Then let

$$B_{i,k}(u)\;=\;(-1)^k(u_{i+k}-u_i)[u_i,\ldots,u_{i+k}](u-t)_+^{k-1}$$

for $i=-1,\ldots,m$.

Theorem: Any $B_{-1,k}(u),\ldots,B_{m,k}(u)$ constructed in this fashion is a basis for $\mathbf{S}(\mathbf{P}^k,\mathbf{U}^m,a,b)$.

It is usual in graphics to ignore the possibilities

$$u_{-1}<a\quad\text{and}\quad b<u_{m+1}$$

and to fix

$$u_{-1}=a\;,\;u_{m+1}=b\quad .$$

6. B-spline Properties

The differences of products of functions is governed by the *Leibniz rule*:

Theorem: For any $z_i\leq\cdots\leq z_{i+l}$ and any appropriately differentiable functions, $f_1(x)$ and $f_2(x)$:

$$[z_i,\ldots,z_{i+l}:x]\{f_1(x)f_2(x)\}$$
$$=\sum_{r=0}^{l}\{[z_i,\ldots,z_{i+r}:x]f_1(x)\}\{[z_{i+r},\ldots,z_{i+l}:x]f_2(x)\}\quad .$$

This has relevance, since

$$B_{i,k}(u)\;=\;(-1)^k(u_{i+k}-u_i)[u_i,\ldots,u_{i+k}:t](u-t)_+^{k-1}\quad .$$

And

$$(u-t)_+^{k-1}\;=\;(u-t)\cdot(u-t)_+^{k-2}\;,$$

for $k\geq 2$. That is, that $B_{i,k}(u)$ is constructed by differencing a product. Therefore,

$$[u_i,\ldots,u_{i+k}:t](u-t)_+^{k-1}$$
$$=\sum_{j=i}^{i+k}\{[u_i,\ldots,u_j:t](u-t)\}\cdot\{[u_j,\ldots,u_{j+k}:t](u-t)_+^{k-2}\}$$

It is found that

$$[u_i,\ldots,u_j:t](u-t)$$

is zero save for two of the terms in the summation, and these two terms can be rearranged to give

$$B_{i,k}(u) = (-1)^{k-1}(u_{i+k}-u)[u_{i+1},\ldots,u_{i+k};t](u-t)_+^{k-2} + (-1)^{k-1}(u-u_i)[u_i,\ldots,u_{i+k-1};t](u-t)_+^{k-2} . \tag{6.1}$$

It is easy to recognise two lower-order B-splines in this expression. In other words, the B-splines satisfy a *recurrence relation.*

Theorem: For any $i \in \{-k, -k+1, \ldots, m\}$

$$B_{i,1}(u) = \begin{cases} 1 & u_i \le u < u_{i+1} \\ 0 & \text{otherwise} \end{cases}$$

and

$$B_{i,r}(u) = (u_{i+r}-u)Q_{i+1,r-1}(u) + (u-u_i)Q_{i,r-1}(u) ,$$

for $r = 2, 3, \ldots, k$,

where

$$Q_{i,r}(u) = \begin{cases} B_{i,r}(u) \,/\, (u_{i+r}-u_i) & u_i \le u < u_{i+r} \\ 0 & \text{otherwise} \end{cases} .$$

From this recurrence can be established the three properties of *nonnegativity*, *compact support*, and *partition of unity*:

Theorem:

$$B_{i,k}(u) > 0 \quad \text{for } u \in (u_i, u_{i+k}) ,$$

$$B_{i,k}(u) = 0 \quad \text{for } u \notin [u_i, u_{i+k}) ,$$

and

$$\sum_{i=-k}^{m} B_{i,k}(u) = 1 \quad \text{for each fixed } u .$$

(At $u = u_i$ we have: $B_{i,k}(u_i) = 1$ for $k = 1$ and $B_{i,k}(u_i) = 0$ for $k > 1$.) The proofs are inductive in each case, starting at the step functions $B_{i,1}(u)$, for which the properties obviously hold, and marching upward in k by means of the recurrence.

7. Parametric Curves and Surfaces

The above properties are the crux of the utility which B-splines offer to graphics. In graphics, the space $\mathbf{S}(\mathbf{P}^k, \mathbf{U}^m, a, b)$ is not at issue; it is the B-splines themselves which are important. The three properties above allow B-splines to be used as *weight functions* to produce curves and surfaces in *parametric* form.

The *parametric representation of a curve* is a mapping of some interval

$$a \le u < b$$

onto a continuum of points

$$\mathbf{Q}(u) = [X(u), Y(u)]$$

or

$$Q(u) = [X(u), Y(u), Z(u)] \ .$$

A surface is provided by a similar sort of mapping

$$Q(u,v) = [X(u,v), Y(u,v), Z(u,v)]$$

for

$$a \leq u < b \ \text{ and } \ c \leq v < d \ .$$

Consider the special cases of the above provided by

$$Q(u) = \sum_{i=-k}^{m} V_{i+k} B_{i,k}(u) \tag{7.1}$$

where

$$V_0, \ldots, V_{m+k}$$

is a collection of points in the plane (or in space), and

$$Q(u,v) = \sum_{i=-k}^{m} \sum_{j=-k}^{n} V_{i+k,j+k} B_{i,k}(u) B_{j,k}(v) \tag{7.2}$$

where

$$V_{0,0}, \ldots, V_{m+k,n+k}$$

are points in space. Note that (7.1) provides a curve in the plane or in space, and (7.2) gives a surface.

The points V are known as *control vertices*, and the curves and surfaces $Q(u), Q(u,v)$ are weighted averages of these points. In the case of cubics, any point on the curve, $Q(u)$ for each u, rests in the *convex hull* of only 4 adjacent V_i, because of the compact support of the B-splines. In like manner, each point of a surface, $Q(u,v)$ for each u,v. lies in the convex hull of only 16 adjacent $V_{i,j}$. It is this fact which causes the graphics community to refer to the nonnegativity, compact support and partition of unity properties of B-splines collectively as the *convex hull property*.

This method of constructing curves and surfaces has the desirable features that (1) only a small number of points are needed to determine quite extensive curves and surfaces, (2) it is easy for a mathematically unschooled person to "sculpt" curves and surfaces by placing control vertices and moving them around using a graphics display, (3) these representations of curves and surfaces can be computed efficiently, (4) when any change to a curve and surface involving only a change in a single control vertex is made, these representations can be efficiently updated – the changes induced in the curve or surface are local to the vertex in question.

We close with three pictorial examples.

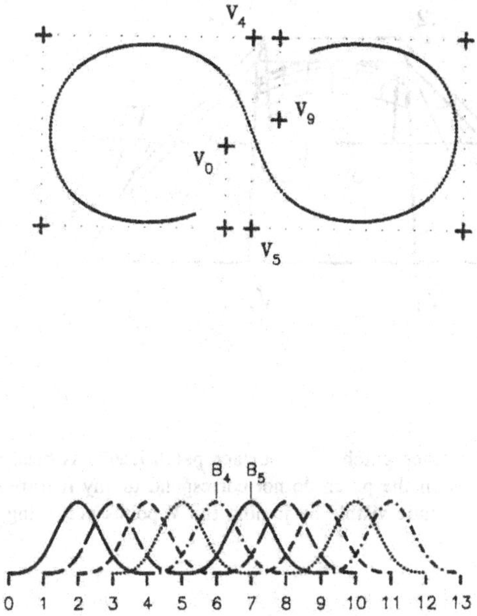

Figure 2. A B-spline generated curve with the B-splines shown for purposes of reference. The separate, pure-polynomial *segments* of the curve have been plotted in alternating solid and dotted lines to add insight. The *joints* between the segments are the images of the knots. Additionally, the lines between successive control vertices have been put in to show the *control graph*. It defines the curve and may be considered a rough approximation to it.

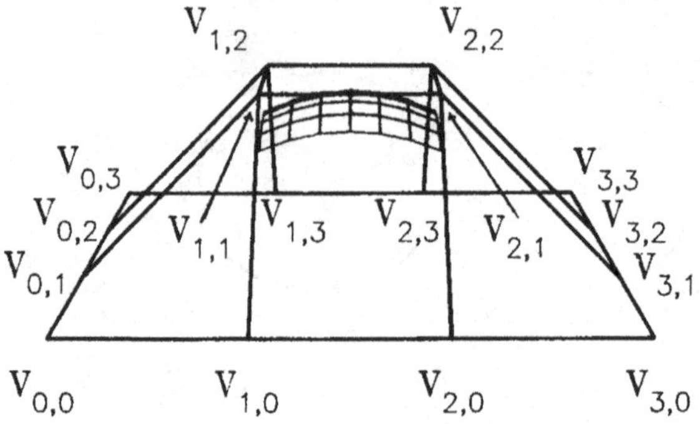

Figure 3. A B-spline generated surface patch. The surface patch itself has been represented as a wire frame mesh. The grid lines on the patch do not correspond to any feature of the splines. Again, the control graph has been made visible by joining the **V** points according to their index adjacencies in *i* and *j*.

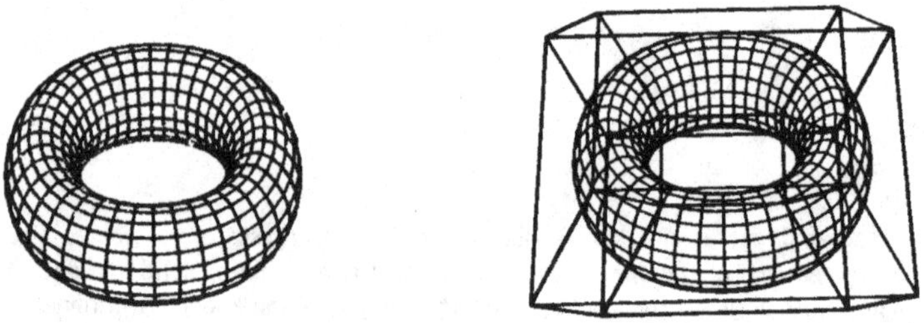

Figure 4. A B-spline generated "torus". (Actually not a torus. Control vertices have been chosen in such a way as to force periodicity in two directions. The cross section of the object is a closed cubic curve rather than a circle.) Again, a wire-frame rendering of the surface has been used, the grid lines do not correspond to any feature of the splines. The control graph has been drawn in.

8. The Oslo Algorithm

Suppose we have constructed a curve

$$Q(u) = \sum_{i=-k}^{m} V_{i+k} B_{i,k}(u)$$

or a surface

$$Q(u,\bar{v}) = \sum_{i=-k}^{m} \sum_{j=-k}^{n} \mathbf{V}_{i+k,j+k} B_{i,k}(u) B_{j,k}(\bar{v})$$

using control vertices, and we now wish to express the same curve or surface in terms of these control vertices plus some additional ones. There are two different reasons for wanting to do this. The obvious one is that we may wish to "fine tune" the curve or surface by adding some control vertices near a section which is being adjusted. The less obvious reason is that we may wish to increase the number of control vertices as an intermediate step in displaying the curve or surface. It is a consequence of the convex hull property that this is a reasonable thing to do. To explain: consider any curve constructed from control vertices using cubic B-splines, so that each segment of the curve lies in the convex hull of no more than 4 control vertices. Let d stand for the minimum of the distances from the curve to any of the control vertices. Suppose the same curve can also be represented by double the number of control vertices (or a hundred times that number, or a thousand). It seems plausible that, if the number of control points is increased enough, and if they are distributed uniformly enough about the curve, they cannot all lie outside the minimum distance d. Each segment of the curve must still lie in the convex hull of only 4 vertices, so presumably there can be a large number of vertices in any region about the curve only at the cost of having them close to the curve. This is, indeed, the case. Similar results apply to surfaces constructed from products of B-splines. This suggests that vertices can be added to such an extent that the control graph "converges visually" to the curve or surface; i.e. the graph "clamps down on" the curve or surface as each of its facets is subdivided by additional vertices until each of the resulting subfacets span no more than a pixel or so. The graph rather than the curve or surface can then be subjected to all of the transformations needed to bring a presentable image onto a display device. Since each facet is a polygon (in fact, a quadrilateral for surfaces and a line segment for curves), and since polygons remain polygons throughout the display pipeline (as contrasted with parametric cubics, which are mapped into rational functions by perspective transformations), this adds a significant amount of computational simplicity through the length of the pipeline.

Using curves as the subject of our further discussion, we begin by noting the connection between control vertices and knots. Each new control vertex which we might like to add needs to be weighted by some new B-spline; each new B-spline which we might like to construct needs some knot at which to become nonzero. Thus, we can approach the problem in two ways. We want

$$Q(u) = \sum_{i=-k}^{m} \mathbf{V}_{i+k} B_{i,k}(u) \equiv \sum_{j=-k}^{n} \mathbf{W}_{j+k} N_{j,k}(u) \ , \tag{8.1}$$

where

$$n > m \ .$$

We can attempt to find the vertices **W** explicitly, or we can attempt to find the new B-splines, $N_{j,k}(u)$, and use them to determine the **W** implicitly. The latter is the route which is taken. More deviously, everything can be accomplished merely by adding new knots to the existing knot sequence. This process will implicitly define the new control vertices. This approach is the one developed in [8] and is known as the "Oslo algorithm".

Add knots

$$s_1 \leq \cdots \leq s_f$$

to the existing sequence

$$u_{-k} \leq \cdots \leq u_{m+k}$$

to obtain a new sequence

$$\{ u_{-k}, \ldots, u_{m+k} \} \cup \{ s_1, \ldots, s_f \} \equiv \{ w_{-k}, \ldots, w_{n+k} \} \ .$$

We will denote the multiplicity of each w_j by ν_j.

Definition: The knot sequence

$$\{ w_{-k}, \ldots, w_{n+k} \}$$

formed in the above fashion will be called a *refinement* of the knot sequence

$$\{ u_{-k}, \ldots, u_{m+k} \} \quad ,$$

provided that

$$v_j \leq k \quad \text{for all } j = -k, \ldots, n+k$$

and provided that

$$a < s_1 \text{ and } s_f < b \quad .$$

The restrictions guarantee that the process of refinement does not lead to a knot sequence which would be illegal for constructing splines of order k and that the spline space we develop with the refined mesh is compatible with the original spline space in the sense that $B_{i,k}(u) \in \mathbf{S}(\mathbf{P}^k, \mathbf{W}^n, a, b)$ for all i,k.

The $N_{j,k}(u)$, of course, are given by

$$N_{j,k}(u) = (-1)^k (w_{j+k} - w_j)[w_j, \ldots, w_{j+k} : t](u-t)_+^{k-1} \tag{8.2}$$
$$\text{for } j = -k, \ldots, n \quad .$$

The next major goal to be reached is the following

Theorem: For each $j = -k, \ldots, n$

$$\mathbf{W}_j = \sum_{i=1}^{m} \mathbf{V}_i \alpha_{i,k}(j) \quad , \tag{8.3}$$

for some collection of numbers $\alpha_{i,k}(j)$.

The proof of this resides in the "add-a-knot" result of the standard B-spline literature. The first step lies in finding a representation for the polynomial $(u-t)^{k-1}$, for any $u \in [a,b)$ and any t, in terms of the basis $N_{j,k}$. In fact, the representation is good for a slightly larger interval than $[a,b)$:

Theorem: For any t and any $u \in [w_{-1}, w_{n+1})$

$$(u-t)^{k-1} = \sum_{j=-k}^{n} \psi_{j,k}(t) N_{j,k}(u) \quad ,$$

where

$$\psi_{j,k}(t) = 1 \text{ if } k = 1 \quad ,$$

and

$$\psi_{j,k}(t) = \prod_{r=1}^{k-1} (w_{j+r} - t) \text{ if } k > 1 \quad .$$

This is known as Marsden's Lemma.

When t is restricted to the discrete set $\{w_{-1}, \ldots, w_{n+1}\}$, then the above representation of $(u-t)^{k-1}$ can be "chopped off" to obtain the representation of $(u-t)_+^{k-1}$ as follows:

Definition: Let

$$\phi_{j,k}(t) = (w_j + \varepsilon_j - t)_+^0 \psi_{j,k}(t) \quad ,$$

where the numbers ε_j satisfy

$$w_j \leq w_j + \varepsilon_j < w_{j+k} .$$

Theorem: For any $t \in \{w_{-k}, \ldots, w_{n+1}\}$ and any $u \in [w_{-1}, w_{n+1})$,

$$(u-t)_+^{k-1} = \sum_{j=-k}^{n} \phi_{j,k}(t) N_{j,k}(u) .$$

The proof of this result is given in [8]. The particular form of the $\phi_{j,k}(t)$, including the ε's, has been chosen to allow differentiation (left-handed) with respect to t, hence the divided-difference operator with respect to t can be applied to obtain

$$\begin{aligned} B_{i,k}(u) &= (-1)^k (u_{i+k} - u_i)[u_i, \ldots, u_{i+k} : t](u-t)_+^{k-1} \\ &= (-1)^k (u_{i+k} - u_i)[u_i, \ldots, u_{i+k} : t] \sum_{j=-k}^{n} \phi_{j,k}(t) N_{j,k}(u) \\ &= \sum_{j=-k}^{n} \left\{ (-1)^k (u_{i+k} - u_i)[u_i, \ldots, u_{i+k} : t]\, \phi_{j,k}(t) \right\} N_{j,k}(u) . \end{aligned}$$

We put these together in the obvious way:

Definition:

$$\alpha_{i,k}(j) = (-1)^k (u_{i+k} - u_i)[u_i, \ldots, u_{i+k} : t]\phi_{j,k}(t) .$$

Theorem:

$$B_{i,k}(u) = \sum_{j=-k}^{n} \alpha_{i,k}(j) N_{j,k}(u) .$$

This completes the add-a-knot result, since it specifies how anything represented in terms of the functions $B_{i,k}(u)$ can be represented in terms of the $N_{j,k}(u)$, the B-spline basis functions for the added-knot mesh.

The functions

$$\phi_{j,k}(t) = (w_j - t)_+^0 \prod_{r=1}^{k-1} (w_{j+r} - t)$$

are similar to the truncated power functions

$$(u-t)_+^{k-1} = (u-t)_+^0 \prod_{r=1}^{k-1} (u-t) .$$

The variable u has been restricted in a peculiar way to the discrete values in the sequence of knots $\mathbf{W}^n$. As a result, $\alpha_{i,k}(j)$ looks like a "discretised" version of $B_{i,k}(u)$. The alpha's are called *discrete B-splines*.

The final step in establishing the result (8.3) comes in observing in (8.1) that

$$\begin{aligned} \sum_{j=-k}^{n} \mathbf{W}_{j+k} N_{j,k}(u) &= \sum_{i=-k}^{m} \mathbf{V}_{i+k} B_{i,k}(u) \\ &= \sum_{i=-k}^{m} \mathbf{V}_{i+k} \sum_{j=-k}^{n} \alpha_{i,k}(j) N_{j,k}(u) \\ &= \sum_{j=-k}^{n} \left[\sum_{i=-k}^{m} \mathbf{V}_{i+k} \alpha_{i,k}(j) \right] N_{j,k}(u) . \end{aligned}$$

We appeal to the linear independence of the functions $N_{j,k}(u)$ in order to equate coefficients on the left and the right.

As might be hoped, the discrete B-splines satisfy a recurrence, from which nonnegativity, compact support, and partition of unity can be established.

Theorem:

$$\alpha_{i,1}(j) = \begin{cases} 1 & u_i \le w_j < u_{i+1} \\ & \text{and } u_i < u_{i+1} \\ 0 & \text{otherwise} \end{cases}$$

and

$$\alpha_{i,r}(j) = (w_{j+r-1} - u_i)\gamma_{i,r-1}(j) + (u_{i+r} - w_{j+r-1})\gamma_{i+1,r-1}(j) \quad ,$$

for $r = 2,3,\ldots,k$, where

$$\gamma_{i,r}(j) = \begin{cases} \alpha_{i,r}(j)/(u_{i+r} - u_i) & \text{if } u_{i+r} > u_i \\ 0 & \text{otherwise} \end{cases}$$

(The $\gamma_{i,r}(j)$ correspond to the Q-spline functions in the continuous case.)

Theorem:

$$\alpha_{i,k}(j) \ge 0$$

$$\sum_{i=-k}^{m} \alpha_{i,k}(j) = 1$$

and

For each j let δ be such that $u_\delta \le w_j < u_{\delta+1}$.

Then $\alpha_{i,k}(j) = 0$ for $i \notin \{\delta-k+1,\ldots,\delta\}$.

The Oslo algorithm itself is simply the application of these results directly to the control vertices. Briefly, recall that

$$\mathbf{W}_j = \sum_{i=1}^{m} \mathbf{V}_i \alpha_{i,k}(j) \quad .$$

Equivalently,

$$\mathbf{W}_j = \sum_{i=\delta-k+1}^{\delta} \mathbf{V}_i \alpha_{i,k}(j) \quad ,$$

from the compact-support property of the discrete B-splines. This means that the **W** "depend locally" on the **V** in the sense that adding knots in a certain region of u will only change the control vertices being weighted by the B-splines whose nonzero intervals are touched by these new knots. Obviously, this has implications of computational efficiency. In the graphics environment it is possible to update spline curves and surfaces locally, rather than needing to recompute them globally, whenever control vertices are moved or added to.

Note that the recurrence for the alpha's can be applied to produce

$$\mathbf{W}_j = \sum_{i=\delta-k+1}^{\delta} \mathbf{V}_i \alpha_{i,k}(j)$$

$$= \sum_{i=\delta-k+1}^{\delta} \mathbf{V}_i \left[(w_{j+r-1} - u_i)\gamma_{i,r-1}(j) + (u_{i+r} - w_{j+r-1})\gamma_{i+1,r-1}(j) \right] .$$

This develops into a recurrence for the control vertices themselves. The third item in the theorem immediately above guarantees that

$$\gamma_{\delta-k+1,k-1}(j) = \gamma_{\delta+1,k-1}(j) = 0 ,$$

which permits us to collect terms to obtain

$$\mathbf{W}_j = \sum_{i=\delta-k+2}^{\delta} \mathbf{V}_{i,2}(j)\,\alpha_{i,k-1}(j) ,$$

where

$$\mathbf{V}_{i,2}(j) = \left[(w_{j+k-1} - u_i)\mathbf{V}_i + (u_{i+k-1} - w_{j+k-1})\mathbf{V}_{i-1} \right] / (u_{i+k-1} - u_i) .$$

This may be repeated to yield

Control Vertex Recurrence:

Let

$$\mathbf{V}_{i,1}(j) = \mathbf{V}_i ,$$

and

$$\mathbf{V}_{i,r}(j) = \left[(w_{j+k-r+1} - u_i)\mathbf{V}_{i,r-1}(j) + (u_{i+k-r+1} - w_{j+k-r+1})\mathbf{V}_{i-1,r-1}(j) \right] / (u_{i+k-r+1} - u_i)$$

for $r = 2, \ldots, k$, where ratios with zero denominators may be taken as zero.

Then

$$\mathbf{W}_j = \mathbf{V}_{\delta,k}(j) .$$

This permits the direct computation of the **W** vertices from the **V** vertices using only the $\{ w_j \}$ and the $\{ u_i \}$ knots. This recurrence is exceedingly efficient.

Strictly speaking, if the display process is carried out by means of this refinement recurrence, using facets of the control graph rather than patches of the spline curve or surface, the entire graphics environment could work purely in terms of the control vertices – a course graph "at the top of the pipeline" for the purposes of human interface and design, refined to a fine graph "at the bottom" for the purposes of display. The splines which are supposedly underlying this process could be ignored entirely!

We close with one example of this process. The B-splines which weight the seven control vertices in the figure below are defined on the uniform knots $u_i = i$. The beginning and ending control vertices are repeated once. The coordinates x_i, y_i of the $\mathbf{V}_i$ are given below:

0.4568	1.3369	
0.4568	1.3369	
0.4122	0.2562	$\mathbf{V}_1$
1.3482	0.3788	$\mathbf{V}_2$
1.4100	1.5153	
3.2199	1.4930	
2.8746	0.3565	
1.9387	0.6685	
1.9387	0.6685	

We have flagged the second and third control vertices because, if we introduce a new knot at $u = 4.5$, the new control vertices $\mathbf{W}_j$ prove to be:

0.4568	1.3369	
0.4568	1.3369	
0.4196	0.4363	V_1'
0.8802	0.3175	P
1.3585	0.5682	V_2'
1.4100	1.5153	
3.2199	1.4930	
2.8746	0.3565	
1.9387	0.6685	
1.9387	0.6685	

and the corresponding figure is:

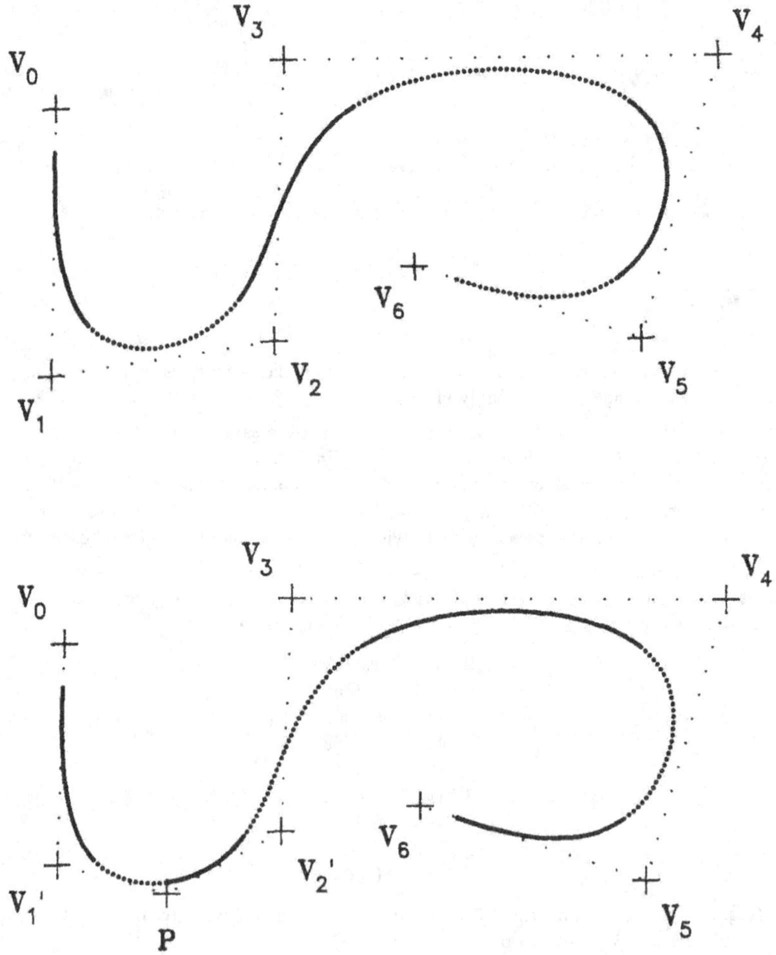

Figure 5. An example of the Oslo algorithm at work.

In particular, notice that the three new control vertices lie closer to the curve than did the two vertices which they replace.

9. Geometric Continuity

We shift gears now to consider an issue which is particularly important in the graphics context, that of the selection of the parameter mapping function. Up until now we have assumed that the interval $[a,b)$ is mapped directly onto a curve $\mathbf{Q}(u)$. In graphics it is often the case that each breakpoint interval $\mathbf{I}_j = [u_{i_j}, u_{i_{j+1}})$ is mapped to the unit interval

$$s \;=\; s(u) \;=\; \frac{u - u_{i_j}}{u_{i_{j+1}} - u_{i_j}} \tag{9.1}$$

and the unit interval is then mapped to an individual curve segment

$$\mathbf{Q}(s) \;=\; \mathbf{Q}(s(u)) \quad .$$

This is done particularly when breakpoint intervals are of equal length (or when there are only a small number of possible lengths), for then the number of different segment polynomials is limited, and code can be written especially tuned to each distinct segment polynomial.

Why should we assume that the mapping (9.1) is the best one to use? The mapping given by

$$\bar{s} \;=\; \bar{s}(u) \;=\; \left[\frac{u - u_{i_j}}{u_{i_{j+1}} - u_{i_j}} \right]^2$$

might be preferred for some reason.

Unfortunately, if one is allowed to reparameterise curves and surfaces at will, some unpleasant and counter-intuitive effects can arise. An example of what is possible is given in Figure 6 below. The curve shown has a mathematically continuous first derivative by virtue of a cleverly-chosen reparameterisation. It obviously does not have a continuous tangent vector.

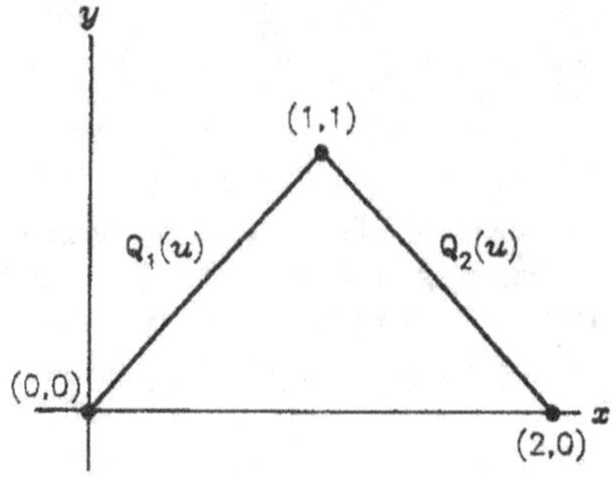

Figure 6. An example of a parametric curve which has a mathematically continuous derivative but does not have a continuously-varying tangent vector.

The parameterisations which have been chosen to make this happen are

$$\mathbf{Q}(u) = [X(u), Y(u)] = \begin{cases} [2u - u^2, 2u - u^2] & \text{for } 0 \le u < 1 \\ [u^2 - 2u + 2, 2u - u^2] & \text{for } 1 \le u < 2 \end{cases}$$

Consequently,

$$\mathbf{Q}^{(1)}(u) = [X^{(1)}(u), Y^{(1)}(u)] = \begin{cases} [2 - 2u, 2 - 2u] & \text{for } 0 \le u < 1 \\ [2u - 2, 2 - 2u] & \text{for } 1 \le u < 2 \end{cases}$$

and so

$$\mathbf{Q}^{(1)}(1) = \begin{cases} [0,0] & \text{from the left} \\ [0,0] & \text{from the right} \end{cases}$$

It is also easy to give examples of (1) a geometrically smooth curve whose tangent vector is mathematically discontinuous, (2) a curve which suffers an abrupt change in curvature but whose second derivative is mathematically continuous, and (3) a curve which is everywhere reasonably curved but whose second derivative changes abruptly. See [2] for these.

What is needed to avoid these anomalies is a definition of continuity which allows for a reasonable change of parametrisation between segments without permitting visual smoothness to be destroyed. Suppose

$$u \le u^+ \ ; \ s \ge s^-$$

$$u = u(s) \ , \ \text{monotone}$$

and

$$u^+ \text{ corresponds to } s^- \ ,$$

giving the same point on $\mathbf{Q}$.

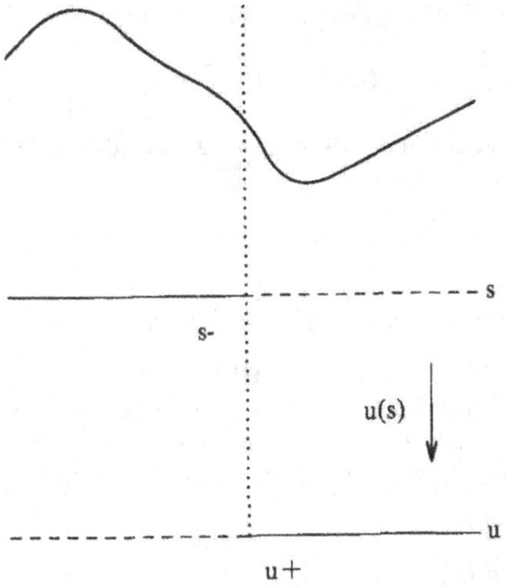

Figure 7. A diagram showing a typical reparameterisation.

This presents the situation in which the parameter mapping on the right is viewed as a reference, and the mapping on the left is viewed as newly introduced. We want the derivative of the curve on the reference side of the join, taken with respect to its parametric variable, to be consistent with the derivative on the other side, taken with respect to the new parametric variable. Moreover, the derivative of the curve should be consistent on both sides of the join, if we were to keep the reference parameterisation throughout, which is just what is required in the usual spline case. That is, we want

$$D_s^{(1)}\mathbf{Q}(u^+) \;=\; D_s^{(1)}\mathbf{Q}(s^-) \;,$$

or (by the chain rule)

$$[D_s^{(1)}u^+]\,D_u^{(1)}\mathbf{Q}(u^+) \;=\; D_s^{(1)}\mathbf{Q}(s^-) \;.$$

That is:

$$[\beta^{(1)}]\,\mathbf{Q}^{(1)}(u^+) \;=\; \mathbf{Q}^{(1)}(s^-) \;. \tag{9.2}$$

Similarly:

$$[D_s^{(1)}u^+]^2 D_u^{(2)}\mathbf{Q}(u^+) + [D_s^{(2)}u^+]\,D_u^{(1)}\mathbf{Q}(u^+) \;=\; D_s^{(2)}\mathbf{Q}(s^-) \;.$$

That is:

$$[\beta^{(1)}]^2\mathbf{Q}^{(2)}(u^+) + [\beta^{(2)}]\,\mathbf{Q}^{(1)}(u^+) \;=\; \mathbf{Q}^{(2)}(s^-) \;, \tag{9.3}$$

and so on. The β's are merely the derivatives of the function which relates the u-parameterisation to the s-parameterisation, the derivatives being evaluated at the joint.

Equations (9.2) and (9.3) are referred to as the conditions for *first geometric continuity*, or G^1 continuity, and *second geometric continuity*, or G^2 continuity, respectively. They were explored by Barsky in [2] on geometric grounds alone, using considerations involving the arc-length parameterisation of a curve, the unit tangent vector, and the curvature vector. In his investigations he dropped emphasis on the underlying parameter mappings and considered the equations (9.2) and (9.3) in their own right, regarding the β's as free (nonnegative) parameters which could be chosen. Clearly the choice of

$$\beta^{(1)} = 1 \text{ and } \beta^{(2)} = 0$$

correspond to the familiar conditions of mathematical continuity. The chain-rule approach above indicates that the notions of geometric continuity extend beyond those considered by Barsky, up to any order.

In the particular case of spline-generated curves, the conditions suggest that, since

$$D_u^{(l)}\mathbf{Q}(u) = \sum_{i=-k}^{m} \mathbf{V}_{i+k}\,[D_u^{(l)}B_{i,k}(u)] \quad ,$$

an easy way of ensuring that conditions (9.2) and (9.3) should hold for $\mathbf{Q}(u)$ is to insist that corresponding conditions hold for $B_{i,k}(u)$, namely

$$[\beta^{(1)}]\,B_{i,k}^{(1)}(u_j^-) = B_{i,k}^{(1)}(u_j^+)$$

and

$$[\beta^{(1)}]^2 B_{i,k}^{(2)}(u_j^-) + [\beta^{(2)}]\,B_{i,k}^{(1)}(u_j^-) = B_{i,k}^{(2)}(u_j^+)$$

at every knot u_j in the interval of support for $B_{i,k}(u)$.

This suggests that we construct new versions of B-splines – geometrically continuous splines, of compact support, suitable for use as weighting functions. Such functions are known as *Beta-splines*. For example, consider quadratic splines and G^1 continuity, using the knots u_0, u_1, u_2, u_3, so that

$$B_{i,k}(u) = \begin{cases} p_1(u) & \text{on } [u_0, u_1) \\ p_2(u) & \text{on } [u_1, u_2) \\ p_3(u) & \text{on } [u_2, u_3) \\ 0 & \text{elsewhere} \end{cases}$$

The conditions of G^1 continuity for this case are

$$\begin{aligned} 0 &= p_1(u_0) \\ p_1(u_1) &= p_2(u_1) \\ p_2(u_2) &= p_3(u_2) \\ p_3(u_3) &= 0 \\ 0 &= p_1^{(1)}(u_0) \\ \beta^{(1)}p_1^{(1)}(u_1) &= p_2^{(1)}(u_1) \\ \beta^{(1)}p_2^{(1)}(u_2) &= p_3^{(1)}(u_2) \\ p_3^{(1)}(u_3) &= 0 \end{aligned} \qquad (9.4)$$

Also, to normalise the result so that it may be used as a weighting function

$$1 = p_2(u_1) + p_3(u_2) \quad .$$

(Note, for future reference, that the value of $\beta^{(1)}$ does not depend upon its associated knot. The same value is used at u_1 and u_2. The result will be what is known as a *uniformly-shaped* quadratic Beta -spline.)

The result of "hand-crafting" such Beta-splines in the case of G^2 continuity and cubics is given below in the next sequence of figures. Note that geometric continuity is not continuity in the ordinary sense. The Beta-splines have obvious discontinuities in tangent, for example, for $\beta^{(1)} \neq 1$. But these discontinuities cancel out when the Beta-splines are used with control vertices to construct parametric curves. The same holds true when Cartesian products of Beta-splines are used to construct surfaces. In the sequence of pictures below giving examples of curves constructed from the beta splines, note how variations in the values of the β's serve to "tense" the curve. Barsky has explored the relationship between Beta-splines and the usual splines under tension, and some of this is to be found in [3]. Beta-splines are receiving enthusiastic interest in the graphics community, because they offer an obvious tool to the graphics designer in helping him "tune" the shape of a curve or surface. Until recently, Beta-spline weighting functions had to be constructed *ab initio* from the defining equations of geometric continuity.

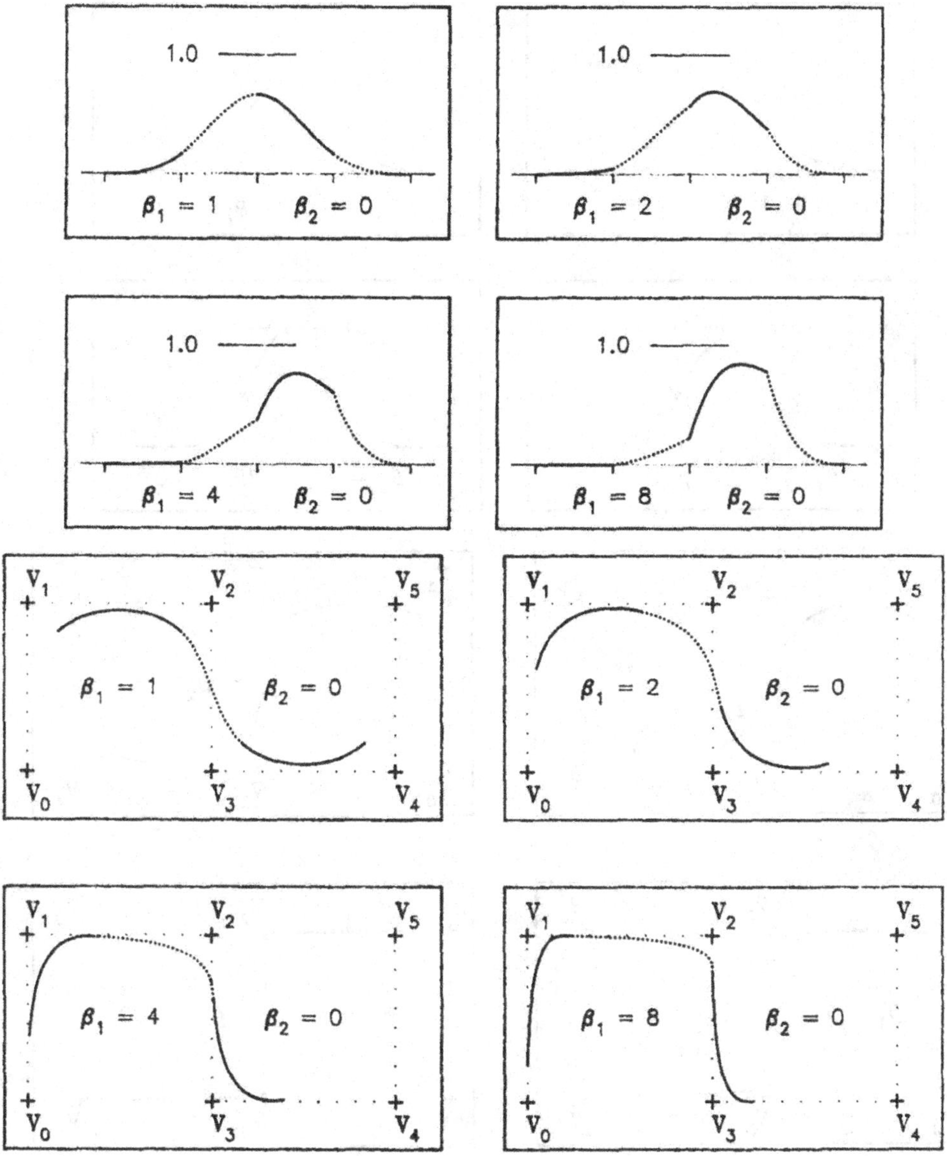
1.0
$\beta_1 = 1$ $\beta_2 = 0$
1.0
$\beta_1 = 2$ $\beta_2 = 0$
1.0
$\beta_1 = 4$ $\beta_2 = 0$
1.0
$\beta_1 = 8$ $\beta_2 = 0$
V_1 V_2 V_5
$\beta_1 = 1$ $\beta_2 = 0$
V_0 V_3 V_4
V_1 V_2 V_5
$\beta_1 = 2$ $\beta_2 = 0$
V_0 V_3 V_4
V_1 V_2 V_5
$\beta_1 = 4$ $\beta_2 = 0$
V_0 V_3 V_4
V_1 V_2 V_5
$\beta_1 = 8$ $\beta_2 = 0$
V_0 V_3 V_4

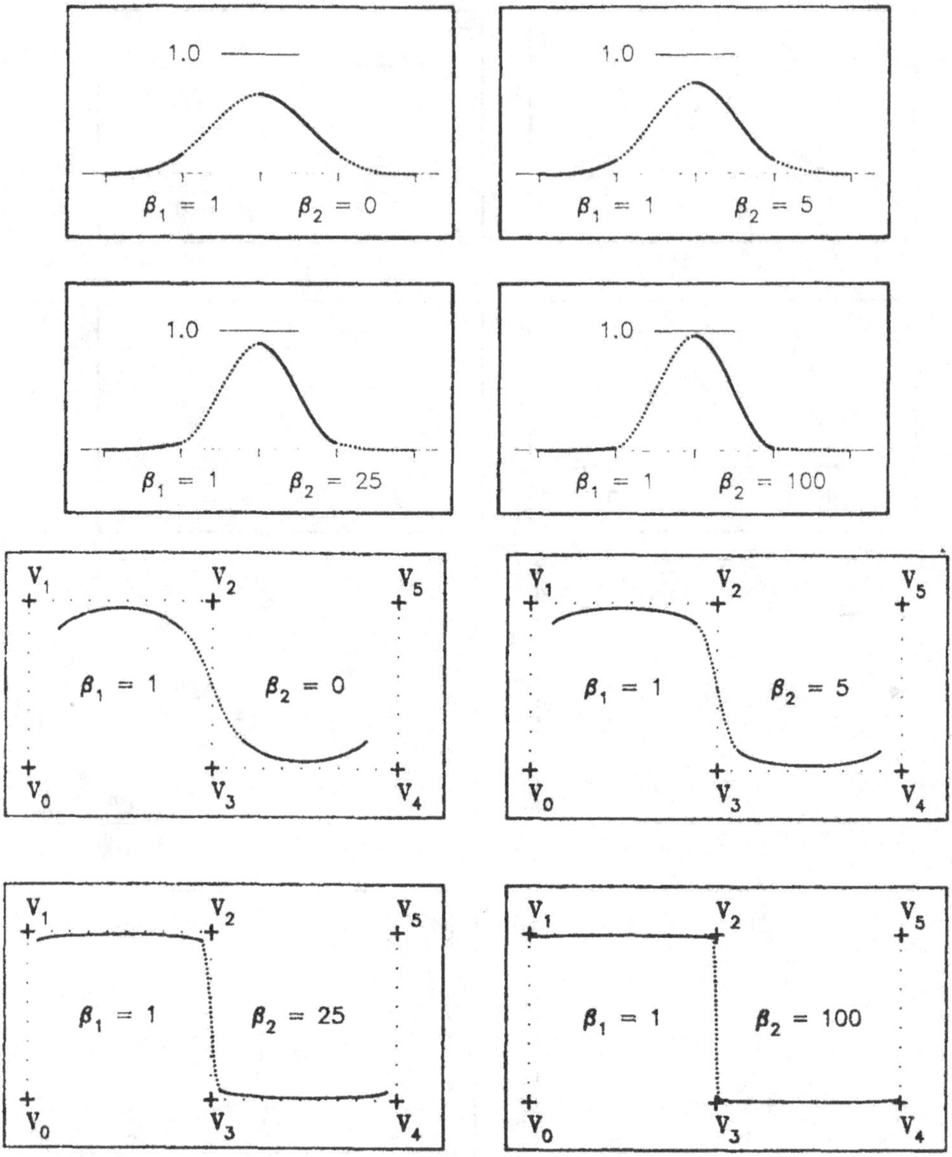
1.0
β1 = 1
β2 = 0
1.0
β1 = 1
β2 = 5
1.0
β1 = 1
β2 = 25
1.0
β1 = 1
β2 = 100
V1 V2 V5
β1 = 1
β2 = 0
V0 V3 V4
V1 V2 V5
β1 = 1
β2 = 5
V0 V3 V4
V1 V2 V5
β1 = 1
β2 = 25
V0 V3 V4
V1 V2 V5
β1 = 1
β2 = 100
V0 V3 V4

10. Extensions

We close by giving an overview of some work in progress.

Geometrically continuous splines on an interval $[a,b)$ with respect to a sequence of knots $\mathbf{U}^m$ form a vector space. This is true even when different values of the β's are associated with different knots; i.e.

$$\beta_1^{(1)} p_1^{(1)}(u_1) = p_2^{(1)}(u_{1)}$$

$$\beta_2^{(1)} p_2^{(1)}(u_2) = p_3^{(1)}(u_{2)} \quad .$$

If these were used instead of the corresponding conditions in (9.4), the result would be a *discretely-shaped* quadratic Beta-spline. (The term *continuously-shaped* might also be appropriate, but it has been used in [4] by Barsky and Beatty for a quite different function.) It should be clear that the ideal paradigm to be followed would be (1) find a one-sided basis for discretely-shaped geometric splines, (2) establish a differencing operation to obtain the discretely-shaped Beta-splines as a compact support basis, (3) extract a recurrence from the differencing operation, (4) establish the convex hull property from the recurrence, (5) use the recurrence to represent the one-sided basis in terms of the Beta-splines, (6) obtain the add-a-knot algorithm, an analogue of the discrete splines, and thereby the analogue of the Oslo algorithm. There are many interesting problems here. The remaining content of this presentation deals with the progress we have made on items (1) and (2) above. A remark about item (6), is in order, however. If knots are added to a sequence along with associated β values other than the ones corresponding to mathematical continuity, then the geometrically continuous splines on the refined mesh are not consistent with those on the coarse mesh. There is inclusion of spline spaces under knot refinement only within limited circumstances. This will probably make the discovery of an analogue to the Oslo algorithm quite challenging.

It is by no means clear what should correspond to the feature of a multiple knot in the case of geometric splines. Thus, all of the following discussion assumes simple knots.

The β-associated conditions of geometric continuity can be introduced into the simple truncated power functions by augmenting them with other truncated power functions of lower order. An example for the quadratic, G^1 case should suffice to give the idea.

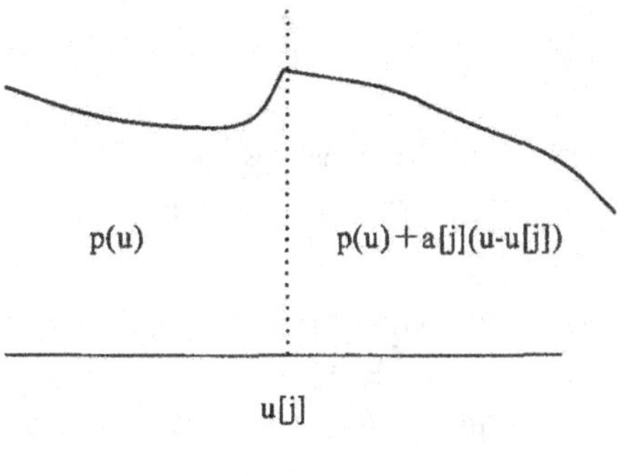

Figure 8. The G-1 jump achieved by a power-function.

With respect to Figure 8, note that, if we expect to have

$$[\beta_j^{(1)}]p^{(1)}(u_j^-) = p^{(1)}(u_j^+) + a_j ,$$

then we can set

$$a_j = [\beta_j^{(1)} - 1]p^{(1)}(u_j)$$

and, as a result, the function

$$p(u) + a_j(u - u_j)_+^1$$

will satisfy the conditions for G^1 continuity at u_j. For cubics and G^2 continuity, we need to consider

$$p(u) + a_j(u - u_j)_+^1 + b_j(u - u_j)_+^2$$

for a similar result.

In the light of this, the one-sided basis for quadratic, G^1 splines proves to consist of functions of the form

$$g_i(u) = (u - u_i)_+^2 + \sum_{j=i+1}^{m} a_{i,j}(u - u_i)_+^1$$

where the $a_{i,j}$ can be determined by

for $i := 0$ **step** 1 **until** $m-1$ **do**
begin
  $s := 0;$
  for $j := i+1$ **step** 1 **until** m **do**
  begin
    $a_{i,j} := (\beta_j^{(1)} - 1)[2(u_j - u_i) + s];$
    $s := s + a_{i,j}$
  end
end

The process for higher orders of geometric continuity is similar.

It is easy to collect terms in like powers of u; e.g. in the quadratic case:

$$g_i(u) = u^2 + [a_{i,i+1} + a_{i,i+2} + a_{i,i+3} - 2u_i]u + \cdots$$

$$g_{i+1}(u) = u^2 + [a_{i+1,i+2} + a_{i+1,i+3} - 2u_{i+1}]u + \cdots$$

$$g_{i+2}(u) = u^2 + [a_{i+2,i+3} - 2u_{i+2}]u + \cdots$$

$$g_{i+3}(u) = u^2 + [-2u_{i+3}]u + \cdots$$

Higher orders are just as easy to deal with. As yet, however, a symbolic cancellation process has not been discovered which works for all orders. In the quadratic case, for instance, we find that

$$g_i(u) = (u - u_i)_+^2 + \sum_{j=i+1}^{m} a_{i_j}(u - u_j)_+^1$$

$$\Delta g_i(u) = \frac{g_{i+1}(u) - g_i(u)}{\beta_{i+1}^{(1)}\beta_{i+2}^{(1)}(u_{i+1} - u_i)}$$

$$\Delta^2 p_i(u) = \frac{\beta_{i+1}^{(1)}\beta_{i+2}^{(1)}(\Delta g_{i+1}(u) - \Delta g_i(u))}{(\beta_{i+1}^{(1)}u_{i+2} + [1-\beta_{i+1}^{(1)}]u_{i+1} - u_i)} ,$$

but this becomes vastly more complicated for higher orders. However, a computational differencing process can be arranged quite easily for all orders. It is, trivially, the process illustrated here for quadratics:

Collect terms in the functions g_i as above to obtain

$$g_j(u) = u^2 + A_j u + B_j \quad \text{for } j = i, i+1, \cdots \quad .$$

Define

$$\Delta g_i(u) = \frac{(g_{i+1}(u) - g_i(u))}{(A_{i+1} - A_i)}$$

$$= u + \frac{(B_{i+1} - B_i)}{(A_{i+1} - A_i)}$$

$$= u + C_i$$

Then

$$\Delta^2 g_i(u) = \frac{(\Delta g_{i+1}(u) - \Delta g_i(u))}{(C_{i+1} - C_i)} = 1$$

And

$$\Delta^3 g_i(u) = B_{i,3}(u) = -(\Delta^2 g_{i+1}(u) - \Delta^2 g_i(u))$$

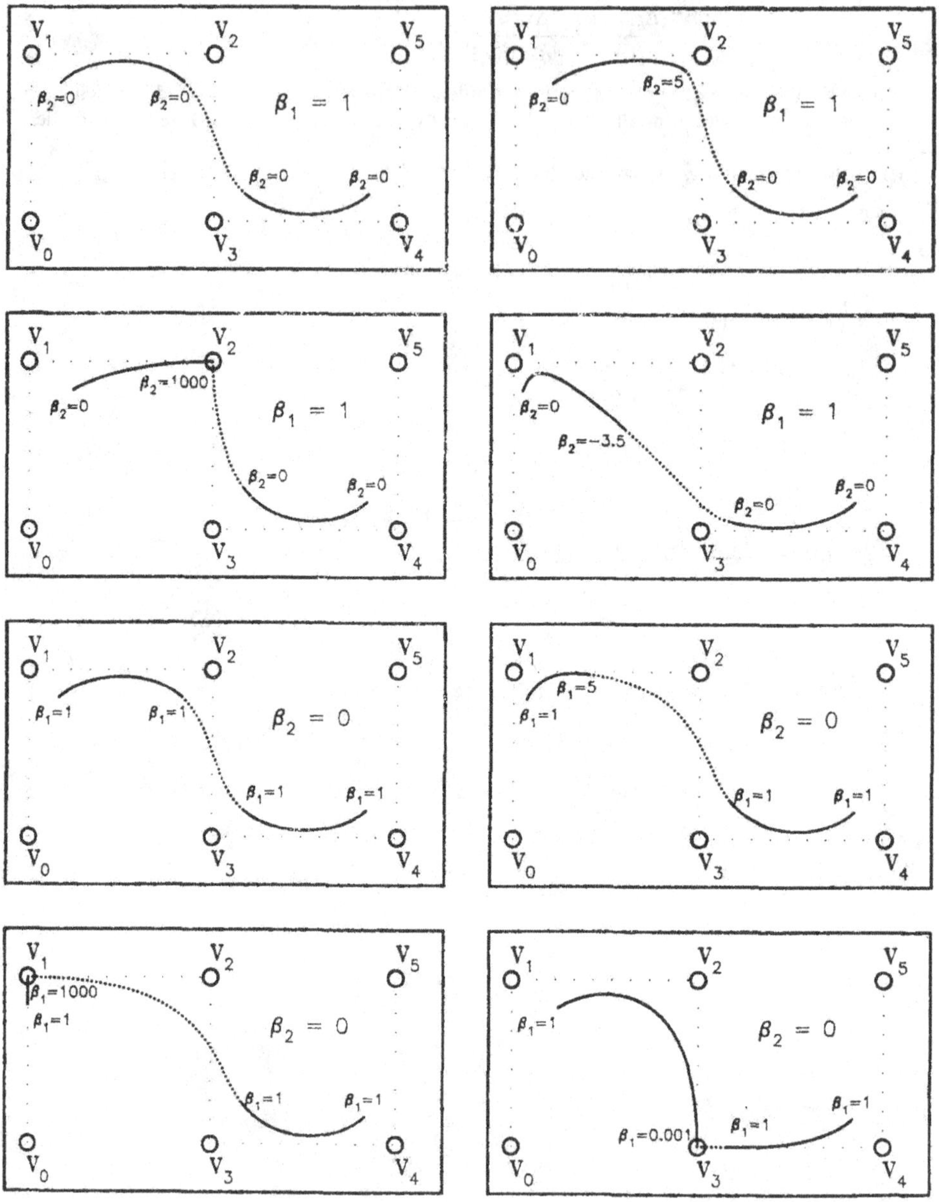
V1 V2 V5
β2=0 β2=0
β1 = 1
β2=0 β2=0
V0 V3 V4
V1 V2 V5
β2=0 β2=5
β1 = 1
β2=0 β2=0
V0 V3 V4
V1 V2 V5
β2=0 β2=1000
β1 = 1
β2=0 β2=0
V0 V3 V4
V1 V2 V5
β2=0 β2=−3.5
β1 = 1
β2=0 β2=0
V0 V3 V4
V1 V2 V5
β1=1 β1=1
β2 = 0
β1=1 β1=1
V0 V3 V4
V1 V2 V5
β1=5 β1=1
β2 = 0
β1=1 β1=1
V0 V3 V4
V1 V2 V5
β1=1000 β1=1
β2 = 0
β1=1 β1=1
V0 V3 V4
V1 V2 V5
β1=1
β2 = 0
β1=0.001 β1=1 β1=1
V0 V3 V4

The functions found in this way are nonnegative, have compact support, and can be verified symbolically to have the partition of unity property. A general proof of these results has not been found.

The corresponding process has been carried out for the cubic case, in particular, and the final drawings of this presentation were all computed using Beta-splines defined by computational differencing.

Currently, work is progressing on the construction of a recurrence for discretely-shaped Beta-splines, starting from first principles and working up from the step functions.

11. Acknowledgements

The author expresses his appreciation for the support he received in preparing this document from the sponsors of the 1983 Dundee Conference on Numerical Analysis. Some of the newer material derives from joint research by the author and John C. Beatty, and was done at the Computer Graphics Laboratory of the University of Waterloo. Funding has been received from the Natural Sciences and Engineering Research Council of Canada, from the National Science Foundation of the United States, from the Ontario Board of Industrial Leadership and Development, from the National Research Council of Canada, and from the Laidlaw Corporation. Appreciation is also expressed to the Mathematics Faculty Computing Facility of the University of Waterloo for its help in photo-printing.

12. References

[1] J. A. Adams and D. F. Rogers (1976), *Mathematical Elements for Computer Graphics*, McGraw-Hill.

[2] B. A. Barsky (1982), The Beta-spline: A Curve and Surface Representation for Computer Graphics and Computer Aided Geometric Design [submitted for publication].

[3] B. A. Barsky (1982), Exponential and Polynomial Methods for Applying Tension to an Interpolating Spline Curve [submitted for publication].

[4] B. A. Barsky and J. C. Beatty (1983), Local Control of Bias and Tension in Beta-splines, *Computer Graphics - SIGGRAPH '83 Conference Proceedings* **17**, 193-218.

[5] C. de Boor (1978), *A Practical Guide to Splines*, Applied Mathematical Sciences Volume 27, Springer-Verlag.

[6] I. D. Faux and M. J. Pratt (1979), *Computational Geometry for Design and Manufacture*, John Wiley & Sons.

[7] J. D. Foley and A. van Dam (1982), *Fundamentals of Interactive Computer Graphics*, Addison Wesley.

[8] R. Riesenfeld, E. Cohen, and T. Lyche (1980), Discrete B-splines and Subdivision Techniques in Computer-Aided Geometric Design and Computer Graphics, *Computer Graphics and Image Processing* **14**(2), October, 87-111.

[9] L. L. Schumaker (1981), *Spline Functions: Basic Theory*, John Wiley & Sons.

SOME METHODS FOR SEPARATING STIFF COMPONENTS IN INITIAL VALUE PROBLEMS

Å. Björck

ABSTRACT

When solving a stiff differential system by an implicit method, factorizing the Jacobian and solving the resulting linear equations often dominate the cost. We develop some methods related to a block Schur factorization of the Jacobian for separating the stiff components. These methods use block versions of the QR or LR algorithm or, for sparse Jacobians, orthogonal iteration to derive an approximate Jacobian. The technique is practical only for systems with relatively few stiff components.

1. INTRODUCTION

We consider here the stiff initial value problem

$$(1.1) \qquad \dot{y} = f(t,y), \; y(t_0) = y_0, \quad y \in R^n.$$

Some problems can be partitioned to have the singular perturbation form

$$(1.2) \qquad \begin{cases} \varepsilon \dot{y}_1 = f_1(t,y_1,y_2), & y_1 \in R^k \\ \dot{y}_2 = f_2(t,y_1,y_2), & y_2 \in R^{n-k} \end{cases}$$

where $\varepsilon \ll 1$, possibly after a diagonal scaling of t and y, see Mao Zu-fan [17]. The structure in (1.2) can be taken advantage of in at least two different ways. Dahlquist and Söderlind [6] recommend using two different methods for each subsystem. Alternatively one can use the same implicit method for both subsystems and exploit the structure only in the iterative solution of the resulting nonlinear system of equations as done e.g. by Robertson [12].

Recently, research has been carried out into generalizing the second approach above to systems, which cannot readily be partitioned as in (1.2), but where the dimension of the stiff subspace is small. We mention papers by Enright and Kamel [8], Watkins and HansonSmith [15] and Björck [2]. An advantage of this approach is that only the efficiency and not the accuracy is sensitive to the correct partitioning. Also such techniques could fairly easily be incorporated into existing software, since only the linear system solver is affected. For a description of a systematized collection of ode solvers incorporating different options for treating the linear systems see Hindmarsh [10].

2. BASIC TECHNIQUES

When an implicit method is used to solve (1.1) then in the Newton iterations we have to solve linear systems with the coefficient matrix

$$W = \frac{1}{h\gamma} I - J , \tag{2.1}$$

where h is the steplength, γ a number characteristic of the method being used and J is the Jacobian matrix

$$(J)_{ij} = \frac{\partial f_i}{\partial y_j}(t,y), \qquad 1 \leq i,j \leq n. \tag{2.2}$$

Factorizing W and solving the resulting linear equations often dominate the cost.

An eigenvalue λ_i of J with negative real part is called stiff if $|h\lambda_i| << 1$. (Note that the definition depends on the steplength h, see discussion in Söderlind [14].) The system (1.1) is called separably stiff at t, if λ_i, i = 1,...,k are stiff eigenvalues and

$$\min_{1\leq i\leq k} |\lambda_i| << \max_{k<i\leq n} |\lambda_i| .$$

For this to be a useful concept it is necessary that k << n. We assume in the following that the eigenvalues are ordered after decreasing magnitude and refer to

$$\mu = |\lambda_{k+1}|/|\lambda_k| \tag{2.3}$$

as the relative separation.

A problem of singular perturbation type (1.2) is separably stiff under mild restrictions. The Jacobian is of the form

$$J = \begin{pmatrix} \varepsilon^{-1}J_{11} & \varepsilon^{-1}J_{12} \\ J_{21} & J_{22} \end{pmatrix} ,$$

where we can assume that the matrices J_{ij} are reasonably scaled. To exhibit the eigenvalues of J we compute the block LR factorization

$$(2.4) \qquad J = \begin{pmatrix} I & 0 \\ \varepsilon J_{21}J_{11}^{-1} & I \end{pmatrix} \begin{pmatrix} \varepsilon^{-1}J_{11} & \varepsilon^{-1}J_{12} \\ 0 & J_S \end{pmatrix} \equiv LR ,$$

where $J_S = J_{22} - J_{21}J_{11}^{-1}J_{12}$ is the Schur complement. We then form

$$J_1 = RL = L^{-1}JL = \begin{pmatrix} \varepsilon^{-1}J_{11} + O(1) & \varepsilon^{-1}J_{12} \\ \varepsilon J_S J_{21} J_{11} & J_S \end{pmatrix},$$

i.e. we perform one step of a "block LR algorithm". From the theorem below by G.W. Stewart [13] it now follows that

$$\lambda_i = \begin{cases} \varepsilon^{-1}\mu_i + O(1), & i = 1,\dots,k \\ \nu_i + O(\varepsilon), & i = k+1,\dots,n \end{cases},$$

where μ_i and ν_i are the eigenvalues of J_{11} and J_S respectively.

Theorem 1. (G.W. Stewart) Let $B \in C^{n\times n}$ be partitioned in the form

$$B = \begin{pmatrix} B_{11} & B_{12} \\ B_{21} & B_{22} \end{pmatrix},$$

where $B_{11} \in C^{k\times k}$. Define δ by

$$\delta = \inf\{||PB_{11} - B_{22}P||: P \in C^{(n-k)\times k}, ||P||=1\}.$$

If

$$(2.5) \qquad \gamma = ||B_{21}||\ ||B_{12}||/\delta < \delta/4,$$

then the eigenvalues of B are the disjoint union

$$\lambda(B) = \lambda(B_{11} + E_{11}) \cup \lambda(B_{22} + E_{22}),$$

and $||E_{ii}|| < 2\gamma$, $i = 1,2$.

For a separably stiff system assume that we know a similarity transformation of J to block upper triangular form

$$(2.6) \qquad J = L \begin{pmatrix} T_{11} & T_{12} \\ 0 & T_{22} \end{pmatrix} L^{-1}.$$

where T_{11} is k by k and $\lambda_i(T_{11}) = \lambda_i$, $i = 1,\dots,k$. If we partition

$$L = (\underbrace{L_\alpha}_{k} \; L_\beta), \qquad L^{-1} = M = \begin{pmatrix} M_\alpha^T \\ M_\beta^T \end{pmatrix} \}k$$

then

$$J L_\alpha = L_\alpha T_{11} ,$$

Thus, L_α spans the dominant invariant subspace of J, i.e. the space spanned by the stiff right eigenvectors of J. Similarly M_β spans the subdominant invariant subspace of J^T, corresponding to the nonstiff left eigenvectors of J. We denote these two spaces by $D_k(J)$ and $S_{n-k}(J^T)$ respectively.

If we use an approximate Jacobian $\tilde{J}$ in (2.1), then the asymptotic convergence in the Newton iterations will be linear with rate $\rho(C)$, where

$$C = (\frac{1}{h\gamma}I - \tilde{J})^{-1}(J - \tilde{J}). \tag{2.7}$$

From the factorization (2.6) we get the approximate Jacobian

$$\tilde{J} = L \begin{pmatrix} T_{11} & T_{12} \\ 0 & 0 \end{pmatrix} L^{-1} = L_\alpha M_\alpha^T J. \tag{2.8}$$

Note that $\tilde{J}$ has rank k and that $L_\alpha M_\alpha^T$ is a projection matrix projecting onto $D_k(J)$. With $\tilde{J}$ as in (2.8) it can be shown that

$$\rho(C) = |h\gamma|\rho(T_{22}) = |h\gamma\lambda_{k+1}| ,$$

so the asymptotic rate of convergence in the Newton iterations will depend only on the largest nonstiff eigenvalue.

If we substitute $\tilde{J}$ from (2.8) in (2.1) we get

$$\tilde{W} = L \begin{pmatrix} \frac{1}{h\gamma}I_k - T_{11} & -T_{12} \\ 0 & \frac{1}{h\gamma}I_{n-k} \end{pmatrix} L^{-1} \quad . \tag{2.9}$$

Thus, in the Newton iterations we now only have to factor a k by k matrix instead of the n by n matrix W. If the factorization (2.6) can be cheaply computed, this can lead to great savings. We note that using the Woodbury formula we can derive the expression

$$\tilde{W}^{-1} = (\frac{1}{h\gamma}I - L_\alpha M_\alpha^T J)^{-1} = h\gamma[I + L_\alpha(\frac{1}{h\gamma}I_k - T_{11})^{-1}M_\alpha^T J], \tag{2.10}$$

where

$$T_{11} = M_\alpha^T J L_\alpha,$$

which can be used instead of (2.9).

An important special case is when the factorization (2.6) is an orthogonal similarity, i.e. when $L = Q$ is orthogonal. Then, $M = Q^T$, and the approximate Jacobian (2.8) become

$$J = Q \begin{pmatrix} T_{11} & T_{12} \\ 0 & 0 \end{pmatrix} Q^T = Q_\alpha Q_\alpha^T J,$$

where $Q_\alpha Q_\alpha^T$ now is the orthogonal projection matrix onto $D_k(J)$, see Björck [2].

For the practical use of the approximate Jacobians introduced here, we need an efficient method to compute an approximate basis for $D_k(J)$. In the next section we give some algorithms for doing this.

3. COMPUTATION OF THE DOMINANT INVARIANT SUBSPACE

The simplest and oldest method for computing the dominant invariant subspace $D_k(J)$ is simultaneous iteration. In its general form this method starts with an n by k matrix P_0, whose columns form a basis for a subspace $\mathcal{P}$ of dimension k. A sequence of matrices $\{P_s\}$ is then generated from

(3.1) $$J P_{s-1} = P_s R_s \quad , \quad s = 1,2,\ldots,$$

where R_s is a nonsingular upper triangular matrix chosen to make the columns of P_s sufficiently linearly independent. Usually R_s is chosen so that either P_s becomes lower unit trapezoidal or to make P_s have orthogonal columns. Here we only consider the second choice and also assume that P_0 is orthogonal. Then (3.1) is called orthogonal iteration.

From the relation

$$J^s P_0 = P_s (R_s R_{s-1} \ldots R_1)$$

it follows that the columns of P_s form a basis for the subspace $J^s\mathcal{P}$. The fundamental fact about orthogonal iteration is that if $|\lambda_k| > |\lambda_{k+1}|$, then under mild restrictions on $\mathcal{P}$ the subspaces $J^s\mathcal{P}$ tend to the dominant invariant subspace $D_k(J)$. The rate of convergence is almost linear with rate $\mu = |\lambda_{k+1}/\lambda_k|$, see Stewart [12]. More precisely one can show, see Golub and Van Loan [9], that

(3.2) $$\sin \theta_{max}\{D_k(J), J^s\mathcal{P}\} < \frac{c}{d}|\lambda_{k+1}/\lambda_k|^s,$$

if

(3.3) $$d = \sin \theta_{min}\{S_{n-k}(J),\mathcal{P}\} > 0 .$$

Here $\theta_i\{U,V\}$ denotes the principal angles between the subspaces U and V. The constant c is independent of P, but depends on J in a complicated way.

We next describe variants of the QR and LR algorithms, which are appropriate when we only want to compute the invariant subspace $D_k(J)$ rather than a complete set of eigenvectors. In the basic QR algorithm one takes $J_1 = J$ and computes

$$(3.4) \qquad Q_s^T J_s = R_s, \quad J_{s+1} = R_s Q_s, \quad s = 1,2\ldots$$

where Q_s is orthogonal. In the block QR algorithm we propose here we let R_s be of 2 by 2 block upper triangular form

$$(3.5) \qquad R_s = \begin{pmatrix} R_{11}^{(s)} & R_{12}^{(s)} \\ 0 & R_{22}^{(s)} \end{pmatrix},$$

where $R_{11}^{(s)}$ is k by k. If $R_{11}^{(s)}$ is chosen to be upper triangular then R_s can be computed by performing the first k-1 steps of the QR factorization of J_s.

It has long been known that simultaneous iteration is closely related to the QR and LR algorithms. Buurema [3] first exploited this idea and Parlett and Poole [8] used it to develop a rigorous geometric theory for the QR, LR and power iterations. To exhibit the basic equivalence between orthogonal iteration and the block QR algorithm we put

$$(3.6) \qquad \tilde{P}_0 = I, \text{ and } \tilde{P}_s = Q_1 Q_2 \ldots Q_s, \quad s > 0.$$

Then the basic relations for the block QR algorithm are

$$(3.7) \qquad J\tilde{P}_{s-1} = \tilde{P}_{s-1} J_s = \tilde{P}_s R_s, \quad s = 1,2,\ldots.$$

Now, let $I_n = (E_1, E_2)$, E_1 $n\times k$, be a partition of the unit matrix I_n and multiply (3.7) from the left with E_1 to get

$$(3.8) \qquad J(\tilde{P}_{s-1}E_1) = \tilde{P}_s R_s E_1 = (\tilde{P}_s E_1) R_{11}^{(s)}.$$

Here the last step follows from the block upper triangular form of R_s. It follows that $\tilde{P}_s E_1$, i.e. the <u>first</u> k columns of $\tilde{P}_s$ in the block QR algorithm span the same subspaces as generated by orthogonal iteration starting from the n by k matrix E_1. Thus the convergence theory is the same. Similarly the <u>last</u> n-k columns of $\tilde{P}_s$ span the orthogonal complement of these subspaces, which are the same subspaces as generated by inverse orthogonal iteration with J^T starting from E_2.

If we make the further assumption that $R_{11}^{(s)}$ is upper triangular, then the relationship (3.8) holds with $E_1 = \binom{I_q}{0}$ and $R_{11}^{(s)}$ $q \times q$, $q = 1,\dots,k$. The block QR algorithm then gives a nested sequence of orthogonal basis for the dominant invariant subspaces $D_q(J)$, $q \leq k$.

Similarly we can develop a block LR algorithm by taking $J_1 = J$ and computing

$$(3.9) \qquad J_s = L_s R_s, \quad L_{s+1} = R_s L_s, \quad s = 1,2,\dots$$

where R_s has the block upper triangular form (3.5) and L_s is lower triangular. Again by performing the first k-1 steps of the LR factorization of J_s we get (3.5) with $R_{11}^{(s)}$ upper triangular. If we define

$$(3.10) \qquad N_0 = I, \; N_s = L_1 L_2 \dots L_s, \quad s > 0,$$

then $N_s E_1$ span the same subspaces as generated by orthogonal iteration starting from E_1. Also $(N_s^{-1})^T E_2$ span the same subspaces as inverse orthogonal iteration with J^T starting from E_2.

We remark that the LR factorization in (3.9) may not always exist and this can make the LR algorithm fail. To overcome this one can introduce column pivoting. Even then it is possible for N_s to converge to a noninvertible matrix, see Wilkinson [16] p. 501. This is however unlikely to cause trouble in the application we have in mind here.

4. IMPLEMENTATION OF THE METHOD

In the block QR or block LR algorithm we have for the computed matrix

$$(4.1) \qquad J_s = \begin{pmatrix} J_{11}^{(s)} & J_{12}^{(s)} \\ J_{21}^{(s)} & J_{22}^{(s)} \end{pmatrix}$$

that under mild restrictions

$$\|J_{21}^{(s)}\| < c_1 \mu^s, \quad \mu = |\lambda_{k+1}/\lambda_k| < 1 .$$

Therefore, by Theorem 1 the eigenvalues of $J_{11}^{(s)}$ and $J_{22}^{(s)}$ will converge to the stiff and nonstiff eigenvalues respectively. When $J_{21}^{(s)}$ is sufficiently small we use, in analogy with (2.8), the approximate Jacobian

$$(4.2) \qquad \tilde{J} = N_{s-1} \begin{pmatrix} J_{11}^{(s)} & J_{12}^{(s)} \\ 0 & 0 \end{pmatrix} N_{s-1}^{-1} .$$

For example, for problems of singular perturbation type we could use (4.2) with $N_{s-1} = L$ as given by (2.4), $s = 2$.

With $\tilde{J}$ as in (4.2) we only need to factor the k by k matrix $\frac{1}{h\gamma} I_k - J_{11}^{(s)}$ in the Newton iterations instead of the n by n matrix J.

The block QR and block LR algorithms as described in section 3 require $4n^2k$ and $2n^2k$ flops (floating point operations) per iteration step, which is more than for orthogonal iteration even when the matrix J is dense. The work per iteration can however be reduced by a factor of k if J is initially reduced to the compact form

$$(4.3) \qquad \begin{pmatrix} H_{11} & J_{12} \\ (0\ b) & J_{22} \end{pmatrix} ,$$

where H_{11} is a k by k Hessenberg matrix and b a vector. This reduction can be acheived by performing the first k-1 steps in the reduction of J to full Hessenberg form. It can be shown, see Björck [2], that the form (4.3) is preserved by the block QR and block LR algorithms. Since the cost for the reduction is the same as for one iteration with the unreduced matrix, the total work will be less already when ≥ 2 iterations are required. An added advantage with using the reduced form (4.3) is that the factorization of $\frac{1}{h\gamma} I_k - H_{11}^{(s)}$ requires only $k^2/2$ flops, cf. Enright and Kamel [8].

We now show that the initial reduction to the compact form (4.3) cannot always be recommended. We note from (3.2) and (3.3) that even when $|\lambda_{k+1}/\lambda_k|$ is small, initial covergence can be slow if

$$d = \sin \theta_{max} \{S_{n-k}(J), \text{ span } (E_1)\}$$

is extremely small. As an example consider the Jacobian

$$(4.4) \qquad \underset{n\times n}{J} = \begin{bmatrix} -2/\varepsilon & 1/\varepsilon & & & & \\ 1 & -2 & 1 & & & \\ & 1 & -2 & 1 & & \\ & & \ddots & \ddots & \ddots & \\ & & & 1 & -2 & 1 \\ & & & & 1/\varepsilon & -2/\varepsilon \end{bmatrix}, \quad \varepsilon \ll 1 ,$$

which has k = 2 stiff eigenvalues well separated from the rest, $\mu \approx \varepsilon$. It can be verified that independent of ε more than n iterations are required in the block QR algorithm to separate the second stiff component. For this example there is a simple fix. If we permute rows and columns in J to put the second large diagonal element into position (2.2) then we get rapid convergence in the block QR and block LR algorithms. In general a good strategy is to permute rows and columns of J so that the k most negative diagonal elements of J are placed in the block J_{11}, before applying the block

QR or block LR algorithm. Unfortunately, we may not be able to do this and simultaneously acheive block Hessenberg form. Again this is illustrated by the matrix J in (4.4) - its block Hessenberg form is obviously destroyed by the permutation. If we try to reduce the permuted J to block Hessenberg form, then we will end up again with the initial matrix J! This is no coincidence but a consequence of the uniqueness of the block Hessenberg form, which we state in the Lemma below.

Lemma (cf. Wikinson [16] p. 352) If $JQ_1 = Q_1H_1$ and $JQ_2 = Q_2H_2$, where Q_1 and Q_2 are orthogonal and have the same first column and H_1 and H_2 are of block upper Hessenberg form then

$$Q_2 = Q_1D_1, \quad H_2 = D^{-1}H_1D$$

where

$$(4.5) \qquad D_1 = \mathrm{diag}(\pm 1, \pm 1, \ldots, \pm 1), \quad D = \begin{pmatrix} D_1 & 0 \\ 0 & \bar{Q} \end{pmatrix}$$

and $\bar{Q}$ is orthogonal.

The lemma is proved following Wilkinson [16] with obvious modifications. Consider now a symmetric permutation $\bar{J} = P^TJP$ of the Jacobian J, and note that P can be written $P = I_{1,j}\bar{P}$, where $\bar{P}$ is a permutation matrix only affecting components 2,...,n and $I_{1,j}$ interchanges components 1 and j. Let $\bar{J}$ be reduced to block upper Hessenberg form by the orthogonal matrix

$$\bar{Q} = \begin{pmatrix} 1 & 0 \\ 0 & Q_1 \end{pmatrix} \cdots\cdots \begin{pmatrix} I_k & 0 \\ 0 & Q_k \end{pmatrix}.$$

Then we have

$$H = \bar{Q}^T\bar{J}\bar{Q} = \bar{Q}^TP^TJP\bar{Q} = Q^TJQ, \quad Q = P\bar{Q} = I_{1,j}(\bar{P}\bar{Q}).$$

It is seen that the first column Qe_1 och Q is independent of $\bar{P}$ and only depends on $I_{1,j}$. From the lemma it then follows that the block upper Hessenberg matrix H is uniquely determined by $I_{1,j}$ up to a block diagonal similarity with D given by (4.5).

Enright and Kamel [8] suggested an algorithm, which reduces J to block upper triangular form in k-1 steps and at each step performs a symmetric permutation designed to make the norm of the unreduced part J_{22} as small as possible. By the same argument as above it follows that $||J_{22}||$ cannot be reduced by any choice of the permutations. This was first proved by Söderlind [14].

We now consider the use of orthogonal iteration, which for this purpose was first suggested by Watkins and HansonSmith [15]. A great advantage of orthogonal iteration is that the intitial matrix P_0 can be freely chosen, so that known information about $D_k(J)$ may be used. Also, this is the only possible algorithm when J is only implicitly available through a finite difference approximation

$$(4.6) \qquad J\,v = \{f(t, y+\delta v) - f(t, y)\}/\delta \, .$$

To get the approximate Jacobian we first perform s steps of orthogonal iteration (3.1) , put $Q_\alpha = P_s$ and compute

$$T_{11} = Q_\alpha^T(JQ_\alpha) \, .$$

We next transform T_{11} to Hessenberg form, update Q_α

$$W^T T_{11} W = H_{11}, \quad Q_\alpha = Q_\alpha W \, ,$$

and perform an LU factorization of the k by k matrix $\frac{1}{h\gamma}I_k - H_{11}$.

We now solve the linear systems in the Newton iterations using (2.10)

$$W^{-1}v = h\gamma\{I + Q_\alpha(\frac{1}{h\gamma}I_k - H_{11})^{-1}Q_\alpha^T J\}\,v \, .$$

This is a slightly more efficient scheme than that based on (2.9) suggested in Björck [2]. The leading terms in the operations counts and the corresponding typical frequences of operations are given below, where it is assumed that the operation J times a vector requires n_J flops.

Operation	flops	no/ time step
new Jacobian	$(s+1)k(n_J + nk)$	1/50
change of h	$k^2/2$	1/5
solve	$n_J + 2nk$	2

Table 1. Operation counts for method of orthogonal iteration

As remarked before the only purpose of the orthogonalization of $J\,P_{s-1}$ in (3.1) is to maintain sufficient linear independence among the columns of P_s. This orthogonalization might be skipped during some iterations, which if $n_J \ll nk$ will give a significant reduction in the operation count for computing a new Jacobian.

5. A NUMERICAL EXAMPLE

To illustrate the given methods for separating stiff components we give here some numerical results for a Jacobian generated by

(5.1) $$J = U\,\mathrm{diag}\{\lambda_i\}U^{-1}, \quad U = I - 2uv^T/u^Tv,$$

where

$$u = (r_1, r_2, \ldots, r_n)^T, \quad v = (r_{n+1}, \ldots, r_{2n})^T$$

and r_i are random numbers uniformly distributed on [0,1]. We have used the eigenvalues

(5.2) $$\lambda_i = \begin{cases} -(n+1-i)\varepsilon^{-1}, & i = 1,\ldots,k \\ -(n+1-i), & i = k+1,\ldots,n \end{cases}$$

giving a separation between stiff and nonstiff eigenvalues equal to $\mu = \varepsilon(n-k)/(n+1-k)$. In the example below we took

$$n = 10, \quad k = 3, \quad \varepsilon = 10^{-2} \text{ and } 10^{-3},$$

and applied the block LR algorithm (3.9) without pivoting. In table 2 we give the quantities

$$n_{ij}^{(s)} = ||J_{ij}^{(s)}||_2, \quad i,j = 1,2.$$

Note that $J_{i,j}^{(1)} = J_{ij}$ is the initial matrix.

s	$\mu = 10^{-3}$		$\mu = 10^{-2}$	
1	1.283 +4	1.744 +3	1.280 +3	1.739 +2
	3.792 +3	1.005 +4	3.781 +2	1.004 +3
2	1.067 +4	2.498 +3	1.001 +3	2.487 +2
	7.612 +2	1.374 +2	6.074 +2	1.230 +2
3	1.083 +4	2.501 +3	1.082 +3	2,494 +2
	4.529 −1	9.707 +0	4.497 +0	9.910 +0
4	1.085 +4	2.500 +3	1.085 +3	2.493 +2
	2.855 −4	9.687 +0	2.838 −2	9.688 +0

Table 2. Norms of subblocks in J_s computed by block LR algorithm.

We note that for both $\varepsilon = 10^{-3}$ and 10^{-2} the stiff eigenvalues are sufficiently well separated after two iterations in the block LR algorithm. Also the linear convergence of $||J_{22}^{(s)}||$ to zero with approximate rate equal to ε is clearly observed.

We remark here that it suffices to choose k to be an upper bound of the number of stiff eigenvalues. The actual number of stiff eigenvalues can then be found by inspecting the iterates J_s for the largest appropriate partitioning. However, for this to be feasible the upper bound k must satisfy $k << n$.

6. CONCLUDING REMARKS

We have developed several algorithms for separating the stiff components of a general, separably stiff system. Clearly the block QR algorithm can only be recommended for systems with a dense Jacobian, since sparse matrices tend to fill up rapidly under orthogonal similarity transformations.

In general, large systems have a sparse Jacobian and for such systems the block LR algorithm is to be preferred. The less satisfactory stability properties of this algorithm compared to block QR algorithm should not matter here. An implementation of the block LR algorithm could build on available software for sparse LR factorization e.g. Duff [7]. A possible way to further limit fill-in in the block LR algorithm is to use a "drop tolerance" which omits all fill-in elements below a specified magnitude as used by Carver and MacEwen [4].

As remarked above, for large systems where no explicit Jacobian is available and a difference approximation has to be used to compute the product of J and a vector, orthogonal iteration is the only possible choice among the algorithms considered here. Some experience with this method applied to small systems is given in Watkins and HansonSmith [14].

It should be stressed that the methods given here are not general tools for integrating stiff systems, but may be viable options for separably stiff systems where the number of stiff components is relatively small. Curtis [5] has made a detailed study of spectral properties for some large stiff ODE systems, which have occurred in practical problems. For these systems the concept of a relatively small number of stiff components was shown not to be valid in most of the time interval. However, it is still hoped that the methods suggested here are useful extensions of methods previously suggested for systems of singular perturbation form (1.2).

REFERENCES

1. Alefeld, P. and Lambert, J.D. "Correction in the dominant space: a numerical technique for a certain class of stiff initial value problems", Math. Comput. 31, 922-938 (1977).

2. Björck, Å. "A block QR algorithm for partitioning stiff differential systems", BIT 23:3 (1983).

3. Buurema, H.J. "A geometric proof of convergence for the QR method", Thesis, Rijksuniversiteit te Groningen (1970).

4. Carver, M.B. and MacEwen, S.R. "On the use of sparse matrix approximation to the Jacobian in integrating large sets of ordinary differential equations", SIAM J. Sci. Stat. Comput., 2, 51-64 (1981).

5. Curtis, A.R. "Jacobian matrix properties and their impact on choice of software for stiff ode systems", Report CSS 134, Computer Science and Systems Division, A.E.R.E Harwell, Oxfordshire (1983).

6. Dahlquist, G. and Söderlind, G. "Some problems related to stiff nonlinear differential systems", Computing Methods in Applied Sciences and Engineering V (eds. R. Glowinski and J.L. Lions), North-Holland Publ. Co., 57-74 (1982).

7. Duff, I.S. "MA28 - A set of Fortran subroutines for sparse nonsymmetric linear equations. AERE Report R.8730, Harwell (1979).

8. Enright, W.H. and Kamel, M.S. "Automatic partitioning of stiff systems and exploiting the resultant structure" ACM Trans. Math. Software <u>5</u>, 374-385 (1979).

9. Golub, G.H. and Van Loan, C. "<u>Applied Matrix Computation</u>" in preparation.

10. Hindmarsh, A.C. "Stiff system problems and solutions at LLNL", Report UCRL-87406, Lawrence Livermore National Laboratory, (1982).

11. Parlett, B.N. and Poole, W.G. "A geometric theory for the QR, LU and power iterations", SIAM J. Numer. Anal. <u>10</u>, 389-412 (1973).

12. Robertson, H.H. "Some factors affecting the efficiency of stiff integration routines", in <u>Numerical Software - Needs and Availability</u>, (ed. D.A.H. Jacobs), Academic Press, 279-301 (1978).

13. G.W. Stewart "Simultaneous iteration for computing invariant subspaces of non-Hermitian matrices", Numer. Math. <u>25</u>, 123-136 (1976).

14. Söderlind, G. "On the efficient solution of nonlinear equations in numerical methods for stiff differential systems", Report TRITA-NA-8114, The Royal Inst. of Tech., Stockholm (1981).

15. Watkins, D.S. and HansonSmith, R.W. "The numerical solution of separably stiff systems by precise partitioning", Report Dept. Pure and Appl. Math., Washington State Univ., Pullman (1982).

16. Wilkinson, J.H. "The algebraic eigenvalue problem", Clarendon Press, Oxford (1965).

17. Zu-fan, M. "Partitioning a stiff ordinary differential system by a scaling technique", Report TRITA-NA-8210, The Royal Inst. of Tech., Stockholm (1982)

APPROXIMATION ORDER FROM SMOOTH BIVARIATE PP FUNCTIONS

C. de Boor

This talk concerns approximation from

$$S \ := \ \pi^{\rho}_{k,\delta} \ := \ \pi_{k,\delta} \cap C^{\rho} ,$$

the space of pp (:= piecewise polynomial) functions of degree $\leq k$ on some partition or subdivision δ and constrained to have continuous derivatives of all orders $< \rho$. We measure the distance in the sup-norm on some compact domain G :

$$\mathrm{dist}(f,S) := \inf_{s \in S} \|f - s\| , \text{ with } \|g\| := \sup_{x \in G} |g(x)| .$$

For a smooth function f , we would expect to get

$$\mathrm{dist}(f,S) \sim \mathrm{const}_f \, |\delta|^m ,$$

with $|\delta|$ the mesh size, and m an exponent which, for $\rho = -1$, is $k+1$, and, for $\rho = k$, is 0 , and which we expect to decrease when we increase ρ . We are interested in the precise relationship between ρ and m . Information of particular interest includes the maximal ρ (if any) for which $m = k+1$ (i.e., for which there is <u>full</u> approximation power), and the maximal ρ for which m is still positive (i.e., for which there is <u>some</u> approximation power). In the univariate situation, all of this is well understood: $m = k+1$ as long as $\rho < k$, and $m = 0$ otherwise. But, already for bivariate functions, the situation is much more subtle and, in any case, not completely understood.

As a start, we consider only the situation in which the various partitions are all obtained from a fixed partition, by scaling. Precisely, we consider the function $h \longmapsto \mathrm{dist}(f,S_h)$, with $S_h := \sigma_h(S)$ and $\sigma_h f: x \longmapsto f(x/h)$. We say that m is the **approximation order** from S , or, more precisely, from the **scale** (S_h) in case

$(\mathrm{i})_m$ for all smooth functions, $\mathrm{dist}(f,S_h) = O(h^m)$,

$(\mathrm{ii})_m$ for some smooth function, $\mathrm{dist}(f,S_h) \neq o(h^m)$.

Because of the simple nature of such a scale, it is not hard to prove [3] that

(1) $$(i)_m \implies \pi_{m-1} \subseteq S .$$

This provides the obvious upper bound $k+1$ for m.

The converse of (1) clearly does not hold; take, e.g., $S = \pi_{m-1}$. What needs to be added to the right side of (1) to achieve equivalence with the left?

One would expect the approximation power of pp functions to come from their local flexibility, their piecewise polynomial character. The only way I know for making this precise, i.e., the only way I know for proving $(i)_m$ is with the aid of a local, bounded **quasi-interpolant** of order m. This is a linear, bounded map $Q: C \longrightarrow S$ which reproduces polynomials of degree $< m$, and is **local**, i.e.,

$$\text{for some } r \text{ and all } x : \quad f_{|B_r(x)} = 0 \implies (Qf)(x) = 0 ,$$

with $B_r(x)$ the ball of radius r around x. For such a Q,

$$|(f - Qf)(x)| \le \|1-Q\| \operatorname{dist}_{B_r(x)}(f,\pi_{m-1}) ,$$

and therefore, for the scaled map $Q_h := \sigma_h Q \sigma_h^{-1}$,

$$\|f - Q_h f\| = O(h^m)$$

for all smooth f.

I know of no way of getting such a quasi-interpolant except via a **local, stable partition of unity** in S, i.e., a collection (ϕ_j) in S for which

$$\sup_j \operatorname{diam} \operatorname{supp} \phi_j < \infty , \quad \|\Sigma |\phi_j|\| < \infty , \quad \Sigma \phi_j = 1 .$$

Note that having such a local, stable partition of unity in S implies that $m > 0$ since then

$$Qf := \Sigma f(\tau_j)\phi_j \quad , \text{ with } \tau_j \in \operatorname{supp} \phi_j , \text{ all } j ,$$

is a suitable quasi-interpolant of order 1. Note also that the stability condition is trivial in case the ϕ_j's are nonnegative. Thus the only hard requirement is the localness.

This sets up another adversary relationship involving the smoothness requirement ρ : The larger ρ is, the harder it is to come up with a local partition of unity in S since it gets harder to construct any compactly supported element in S.

It is traditional in finite elements to construct compactly supported smooth pp functions by trial and error. It has been found recently that interesting compactly supported smooth pp functions can be found as **multivariate B-splines**, i.e., as **shadows** of higher-dimensional polyhedral bodies. To explain, pick such a body B in $\mathbb{R}^n$ and choose an affine map $P:\mathbb{R}^n \longrightarrow \mathbb{R}^p$. Then the distribution M, defined on $\mathbb{R}^p$ by the rule

(2) $$M: \phi \longmapsto \int_B \phi \circ P \,, \quad \text{all} \quad \phi \in C_0^\infty(\mathbb{R}^p) \,,$$

has support in $P(B)$. If $P(B)$ is p-dimensional, then $M \in L_\infty$ and, in that case, M is pp of degree $n-p$, on the subdivision given by the images of (p-1)-dimensional faces of B, and the smoothness across such an interface is determined by the highest-dimensional face of B mapped by P into it. See [1] for an introduction to, and [7] for the most recent survey of, the fast developing area of multivariate B-splines.

The shadow of a cube or box is called a **box spline.** A particularly simple example is the standard linear finite element on a hexagonal (or, three-direction) mesh which is normally referred to as a pyramid, but, looked at properly, is clearly the shadow of a three-cube. These box splines are studied in [3] and [4]. They were introduced in [2] for the purpose of studying the questions of concern in this talk, in a very simple setting: $p = 2$, i.e., we are dealing with bivariate functions, and $\delta = \Sigma$:= a square subdivision, i.e., a **two-direction** mesh, or $\delta = \Delta$:= the triangulation obtained from Σ by drawing in all north-east diagonals, i.e., a **three-direction** mesh. In either case, denote by S_{loc} the span of the translates of the various box splines belonging to S. With this, one can show [4] that

(3) $$(i)_m \implies \pi_{m-1} \subseteq S_{loc} \,.$$

Since S_{loc} decreases when we increase the smoothness demand ρ, this makes the relationship between m and ρ somewhat more explicit. In particular, it allows the conclusion that the approximation order is 0 in case ρ is so large that there are no functions in S with compact support. According to [2], this happens unless

(4) $$\rho < \begin{cases} (k-2)/2 \,, & \text{if} \quad \delta = \Sigma \\ (2k-2)/3 \,, & \text{if} \quad \delta = \Delta \end{cases} .$$

Further, even if ρ satisfies (4), the approximation order from (S_h) may be less than $k+1$ (the optimal one according to (1)), unless $\rho = -1$. Precisely [2],

$$m = k-\rho$$

for the two-direction mesh. For the three-direction mesh, the precise approximation order has not yet been found. According to [4], if ρ

satisfies (4), i.e., if $\rho < \rho(k) := \lfloor (2k-2)/3 \rfloor$, then

$$m \in [\rho(k)+2, m(k)] \quad \text{with} \quad m(k) := \min\{2(k-\rho), k+1\} .$$

In particular, $m = \rho(k)+2$ when $\rho = \rho(k)$ and $k \equiv 1(3)$. But the interval increases as ρ decreases. Yet, [3] expresses the hope that m always stays within 1 of the upper bound $m(k)$. In [8], Jia proves that m always stays within 2 of that upper bound.

The conclusion of (3) is so much stronger than that of (1) that, for a while, we hoped that (3) could be reversed. In fact, [2] shows that

$$(i)_m \iff \pi_{m-1} \subseteq S_{loc}$$

in case of the two-direction mesh. This equivalence also holds for the three-direction mesh for all cases that can be settled by inspection, i.e., for $k = 0$, for $\rho < 1$, and for $\rho = 1$ and $k = 1, 2$. It breaks down, though, for the first nontrivial case, i.e., for $\rho = 1$ and $k = 3$: In this case, $m = 3$ even though $\pi_3 \subseteq S_{loc}$, as is shown in [5]. In [8], the same technique is used to show that, for $\rho = 1$ and $k = 4$, $m = 4$ even though $\pi_4 \subseteq S_{loc}$. This means that yet something else has to be added to the conclusion in (3) to make it equivalent to $(i)_m$.

An alternative is to strengthen $(i)_m$, as was recently done by Dahmen and Micchelli [6]. Fix and Strang (see, e.g., [10]) analysed carefully the approximation power of the space

$$T := \text{span}\ \{\phi(\cdot-q) : \phi\in\Phi,\ q\in\mathbb{Z}^p\}$$

generated by the $\mathbb{Z}^p$-translates of a finite collection Φ of compactly supported functions on $\mathbb{R}^p$. Their goal is to determine, for given m, necessary and sufficient conditions on Φ to guarantee, for every $f \in C^{(\infty)}(\mathbb{R}^p)$ and every h, the existence of sequences $w_\phi^h : \mathbb{Z}^p \longrightarrow \mathbb{R}$ such that

$$\| f - \sum_{\phi\in\Phi} \sum_{q\in\mathbb{Z}^p} w_\phi^h(q)\ \phi(\cdot/h - q) \| \le \text{const}\ h^m\ \|D^m f\|$$

while

$$\|w_\phi^h\| \le \text{const}\ \|f\| , \quad \text{all } \phi .$$

(Strictly speaking, they deal only with the L_2-norm; see [6] for the corresponding statements in the L_∞-norm which we are concerned with here.) They show that this holds if and only if there is a finite linear combination M of the ϕ's and their translates so that the quasi-interpolant of the form

$$Qf := \sum_{q\in\mathbb{Z}^p} f(q)M(\cdot-q)$$

carries π_{m-1} faithfully into itself. They call the largest m for which this holds the <u>**controlled approximation order**</u> of the corresponding scale (T_h). Dahmen and Micchelli [6] have added to this the observation that having controlled approximation order m is equivalent to having a quasi-interpolant of order m of the form

$$Qf := \Sigma_q \lambda f(\cdot+q)M(\cdot-q)$$

with λ a suitable compactly supported bounded linear functional and M a finite linear combination of the ϕ's and their translates.

Since [4] shows the approximation order from S to be the same as that from S_{loc}, it is possible to apply these ideas to the situation under discussion, and this is what Dahmen and Micchelli do.

Since [3] uses a quasi-interpolant to show that the approximation order from S is $m(k) = 2k - 2\rho$ in case $\rho = \rho(k)$ and $k \equiv 1(3)$ (and on the three-direction mesh), the controlled approximation order is the same in this case. For the other two cases, $k \not\equiv 1(3)$, of highest smoothness, i.e., $\rho = \rho(k)$, [6] shows that the controlled approximation order is $m(k) - 1$.

I had originally intended to devote this talk to the relationship between approximation order and controlled approximation order, for it is not at all clear that the two are the same (neat though it would be if they were). In particular, I had begun calculations, based on the Strang-Fix characterization of controlled approximation order in terms of the Fourier transform of M, with the hope of showing that the controlled approximation order from S_{loc} is only 5 in case of C^1-quintics on a three-direction mesh. Since the approximation order from S is obviously 6 in this case (the order is realized by local polynomial interpolation), this would provide an example that the two notions of approximation order are, indeed, different. But, Jia became interested in this question and decided to look at it differently using ideas from [8], and has, in the meantime, proved this fact. In fact, in [9] he determines the controlled approximation order from S_{loc} for <u>all</u> choices of k and ρ !

This difference between approximation order and controlled approximation order is sad for those who like clean results. But there is hope. After all, the box splines are not the only locally supported elements in our S. In fact, as [4] (and others) point out, they are usually not even of minimal support, i.e., S contains elements whose support fits strictly inside the support of some box spline in S, and these might even be more interesting than the box splines. At the same time, controlled approximation depends crucially on the choice of Φ . In fact, in the special case of C^1-quintics just cited, it is easy to make up a Φ for which controlled approximation has order 6 . This raises the hope [9] that the approximation order from S can always be realized by some quasi-interpolant, i.e., that approximation order always equals controlled approximation order from some suitably chosen S_{loc} .

Finally, there is a simpler version of this problem of relating approximation power to smoothness, viz. the following question (related by (3) to $(i)_1$):

$$\text{for all } f \in C ,\ \mathrm{dist}(f,S_h) = o(1) \quad \overset{?}{\Longrightarrow} \quad 1 \in S_{loc} \text{ stably}$$

I.e., is the eventual denseness of (S_h) in C sufficient for having a **stable local partition of unity** in S ?

References

1. C. de Boor, Topics in multivariate approximation theory, MRC TSR 2379 (1982); in "Topics in Numerical Analysis", P.R. Turner ed., Springer Lecture Notes in Mathematics **965**, 1982, 39-78.

2. C. de Boor & R. DeVore, Approximation by smooth multivariate splines, Trans.Amer.Math.Soc. **276** (1982) 775-788.

3. C. de Boor & K. Höllig, B-splines from parallelepipeds, MRC TSR 2320 (1982); J.d'Analyse Math., to appear.

4. C. de Boor & K. Höllig, Bivariate box splines and smooth pp functions on a three-direction mesh, J.Comput.Applied Math. **9** (1983) 13-28.

5. C. de Boor & K. Höllig, Approximation order from bivariate C^1-cubics: A counterexample, Proc.Amer. Math.Soc. **85** (1982) 397-400.

6. W. Dahmen & C. A. Micchelli, On the approximation order from certain multivariate spline spaces, preprint, 1983.

7. W. Dahmen & C. A. Micchelli, Recent progress on multivariate splines, to appear in "Approximation Theory IV", C. K. Chui, L. L. Schumaker and J. D. Ward eds., Academic Press, 1983.

8. R.-q. Jia, Approximation by smooth bivariate splines on a three-direction mesh, MRC TSR 2494 (1983); To appear in "Approximation Theory IV", C. K. Chui, L. L. Schumaker and J. D. Ward eds., Academic Press, 1983.

9. R.-q. Jia, On the controlled approximation order from certain spaces of smooth bivariate splines, MRC TSR xxxx (1983).

10. G. Strang & G. Fix, A Fourier analysis of the finite element variational method, C.I.M.E.II, Ciclo Erice, 1971.

THE NUMERICAL SOLUTION OF INTEGRAL EQUATIONS WITH WEAKLY SINGULAR KERNELS

Hermann Brunner

1. INTRODUCTION

Physical modeling processes lead frequently to second-kind integral equations whose kernels contain a weak (integrable) singularity of algebraic or logarithmic type (see, e.g., [16], [21], [26], [6], [13], [31], [22], [3], and the references listed in these papers). In the present paper we shall focus our attention on integral equations whose *singularities* are of *algebraic type*; as for logarithmic singularities, the interested reader is referred to [21], [30], [34], [37], [38], [14]. Of these equations we shall discuss in detail integral equations of *Volterra* type, i.e.,

$$(1.1) \qquad y(t) = g(t) + \int_0^t (t-s)^{-\alpha}\cdot K(t,s)y(s)ds, \quad t \in I := [0,1], \ 0 \leqq \alpha < 1,$$

where it is assumed that the given functions g and K are smooth on their respective domains I and $S_V := \{(t,s) : 0 \leqq s \leqq t \leqq 1\}$. However, in order to put the subsequent analysis and results into proper perspective we shall also, from time to time, consider the analogous *Fredholm* integral equation of the second kind,

$$(1.2) \qquad y(t) = g(t) + \lambda \cdot \int_0^1 |t-s|^{-\alpha}\cdot K(t,s)y(s)ds, \quad t \in I, \ 0 \leqq \alpha < 1;$$

here, we assume that g and K are smooth on I and $S_F := I \times I$, respectively, and that λ^{-1} is not an eigenvalue.

In the following we shall treat various aspects of approximation methods for (1.1) and (1.2) which are based on collocation techniques in certain polynomial, or non-polynomial, spline spaces. Our particular interest lies with the construction of methods which possess a high order of convergence, not only for $\alpha = 0$ but also for $\alpha \in (0,1)$. In order to obtain such methods it will be necessary to know the smoothness properties of the exact solutions of (1.1), (1.2): while, for $\alpha = 0$, smooth g and K imply the existence of a (unique) solution y which is smooth on $[0,1]$, the presence of the weak singularity $(0 < \alpha < 1)$ in the kernel will lead to a (still unique) solution which has, *typically*, unbounded derivatives at $t = 0$ (Volterra equation), or at $t = 0$ and $t = 1$ (Fredholm equation); i.e., if $\alpha \in (0,1)$ then smooth g and K do *not* yield a solution which is smooth on the closed integral I. The practical implication of this observation is that existing numerical methods which exhibit high-order convergence when applied to integral equations with smooth solutions will have to be modified in such a way as to take into account the nature of the singular behavior of the exact solution at the endpoint(s) of the interval of integration. Alternatively, knowledge of the structure of the exact solution near the endpoint(s) may be used to devise new methods especially tailored to integral equations with nonsmooth solutions.

The aim of this paper is to survey recent results on collocation methods for second-kind integral equations with weakly singular kernels and, at the same time, report on some ongoing work, especially for equation (1.1). It is organized as follows. In Section 2 we shall analyze the smoothness properties of the exact solution of (1.1) and compare them with the ones for the Fredholm equation (1.2). Section 3 will contain a review of the basic principles of collocation and iterated collocation methods (and their discretizations) in polynomial spline spaces; here, we also present the basic tools for the subsequent error analysis. This will then be applied, in Section 4, to integral equations with smooth solutions. In Section 5, we shall study two classes of collocation methods for equations with nonsmooth solutions: polynomial spline collocation using a graded mesh, and nonpolynomial spline collocation based on a uniform mesh. In addition, some numerical results will be given.

2. SMOOTHNESS PROPERTIES OF EXACT SOLUTIONS

Suppose first that $\alpha = 0$ in (1.1), and let $R(t,s)$ denote the resolvent kernel associated with the given kernel $K(t,s)$. If $g \in C(I)$, $K \in C(S_V)$, then the (unique) solution $y \in C(I)$ of (1.1) ($\alpha = 0$) is given by

$$(2.1) \qquad y(t) = g(t) + \int_0^t R(t,s)g(s)ds, \qquad t \in I.$$

Moreover, if $g \in C^m(I)$, $K \in C^m(S_V)$, then we have $y \in C^m(I)$: smooth g and K imply a smooth solution y, provided $\alpha = 0$. This follows from the fact that $R(t,s)$ inherits the smoothness of the kernel $K(t,s)$, since we have

$$R(t,s) = \sum_{n=1}^{\infty} K_n(t,s), (t,s) \in S_V$$

(with the Neumann series converging absolutely and uniformly on S_V), where the iterated kernels $\{K_n\}$ are defined by

$$K_n(t,s) := \int_s^t K(t,v)K_{n-1}(v,s)dv, \ n \geqq 2 \quad (K_1(t,s) := K(t,s)).$$

The picture changes rather drastically once we admit weakly singular kernels. Before stating a number of general results we illustrate this by giving two examples. Consider first the integral equation

$$(2.2) \qquad y(t) = 1 + \int_0^t \lambda\cdot(t-s)^{-\alpha}\, y(s)ds, \ t \in I, \ 0 \leqq \alpha < 1.$$

Here, the given functions, $g(t) \equiv 1$ and $K(t,s) \equiv \lambda = \text{const.}$, are obviously smooth. The exact solution of (2.2) is given by

$$(2.3) \qquad y(t) = E_{1-\alpha}(\lambda\cdot\Gamma(1-\alpha)\cdot t^{1-\alpha}), \ t \in I,$$

where

$$E_\beta(z) := \sum_{n=0}^{\infty} z^n/\Gamma(1+n\beta)$$

denotes the MITTAG-LEFFLER function. If $\alpha = 0$ (i.e., $\beta = 1 - \alpha = 1$), then $y(t) = E_1(\lambda t) = \exp(\lambda t)$, which is smooth on I. However, for $0 < \alpha < 1$ we obtain the nonsmooth solution

$$y(t) = 1 + \frac{\lambda}{1-\alpha}\cdot t^{1-\alpha} + \text{"smoother" terms.}$$

Hence, $y'(t) = O(t^{-\alpha})$ as $t \downarrow 0$, while $y(t)$ is smooth in $(0,1]$.

The second example has its origin in the modeling of the temperature $\phi = \phi(x,t)$ of a nonlinearly radiating semi-infinite solid (see [16] and the references mentioned therein; compare also [26] for a different physical problem leading to the same weakly singular Volterra integral equation). It is assumed that energy is radiated away at a rate proportional to $\phi^n(0,t)$, while an external source supplies heat at a rate proportional to a given function $f(t)$. Here, n is a fixed but otherwise arbitrary real number greater than, or equal to, one (the value $n = 4$ corresponds to the Stefan-Boltzmann radiation law). It can be shown that ϕ is governed by the following initial-boundary value problem:

$$\frac{\partial \phi}{\partial t} = \frac{\partial^2 \phi}{\partial x^2}, \; x > 0, \; t > 0;$$

$$\frac{\partial \Phi(0,t)}{\partial x} = c\phi^n(0,t) - f(t), \; t > 0; \; \phi(x,0) = 0, \; x \geqq 0;$$

$$\lim_{x \to \infty} \phi(x,t) = 0, \; t \geqq 0.$$

Here, the dimensional constant $c > 0$ depends only on properties of the given materia The solution of this problem must satisfy the equation

$$(2.4) \qquad \phi(x,t) = \pi^{-1/2} \cdot \int_0^t (t-s)^{-1/2} \cdot \exp\left(\frac{-x^2}{4(t-s)}\right) \{f(s) - c\phi^n(0,s)\} \, ds, \qquad x \geqq 0, \; t \geqq 0.$$

Thus, if $y(t) := \phi(0,t)$ (the temperature on the boundary) is known, $\phi(x,t)$ can be determined from (2.4). By (2.4) ($x=0$), $y(t)$ is the solution of the integral equation

$$(2.5) \qquad y(t) = \pi^{-1/2} \cdot \int_0^t (t-s)^{-1/2} \cdot \{f(s) - c(y(s))^n\} \, ds, \; t \geqq 0.$$

If the given function $f(t)$ (near $t = 0$) has the form $f(t) = d_0 t^{\mu_0 - 1} + d_1 t^{\mu_1 - 1} + \ldots$, with $d_0 > 0$, $0 < \mu_0 < \mu_1 < \ldots$, then the corresponding solution exhibits the following behavior near $t = 0$ ([16]):

$$y(t) = \frac{d_0 \cdot \Gamma(\mu_0)}{\Gamma(\mu_0 + 1/2)} \cdot t^{\mu_0 - 1/2} + \text{"smoother" terms},$$

provided $\mu_0 \geqq 1/2$; the above asymptotic expression is then independent of n. If we have $0 < \mu_0 < 1/2$, then a similar relation holds for all n with $n > (1-\mu_0)/(1/2 - \mu_0)$. Hence, if $f(t)$ is a smooth function (e.g., $\mu_0 = 1$) then the solution of (2.5) has unbounded derivatives at $t = 0$.

In order to deal with the general case, consider the integral equation

$$(2.6) \qquad y(t) = g(t) + \int_0^t (t-s)^{\gamma} \cdot K(t,s) y(s) ds, \; t \in I,$$

where $\gamma > -1$, γ not an integer. Let $R(t,s;\gamma)$ be the resolvent kernel associated with the kernel $(t-s)^{\gamma} \cdot K(t,s)$. As in the case of a regular kernel, $R(t,s;\gamma)$ can be defined in terms of the iterated kernels, i.e. by its Neumann series. The following result then follows readily.

LEMMA 2.1. In (2.6), let $g \in C^m(I)$, $K \in C^m(S_V)$, and assume that $\gamma > -1$, $\gamma \notin \mathbb{N}_0$. Then the resolvent kernel $R(t,s;\gamma)$ has the form

$$(2.7)\qquad R(t,s;\gamma) = (t-s)^{\gamma} \cdot Q(t,s;\gamma),\quad (t,s) \in S_V ,$$

where

$$(2.8)\qquad Q(t,s;\gamma) := \sum_{n=1}^{\infty} (t-s)^{(n-1)(1+\gamma)} \cdot \Phi_n(t,s;\gamma) \in C(S_V),$$

with $\Phi_n \in C^m(S_V)$ $(n \geqq 1)$, and with the series converging absolutely and uniformly on S_V.

According to the classical Volterra theory the (unique) solution of (2.6) is then given by

$$(2.9)\qquad y(t) = g(t) + \int_0^t (t-s)^{\gamma}\, Q(t,s;\gamma)\cdot g(s)ds,\ t \in I.$$

If $\gamma > m-1$ then $y \in C^m(I)$. More generally, we have

THEOREM 2.1. Let the assumptions of Lemma 2.1 be satisfied. Then the solution of (2.6) lies in the space $C^d(I)$, where $d := 1 + [\gamma]$; for $\gamma < m-1$, $y^{(d+1)}(t)$ is unbounded at $t = 0$. In particular, if $\gamma = -\alpha$, $\alpha \in (0,1)$, then $y \in C(I)$ but $y \notin C^1(I)$.

In view of subsequent applications we state a result which sheds more light on the nature of the solution near $t = 0$. We assume that γ is rational, with $1 + \gamma = p/q$, where the positive integers p and q are coprime.

THEOREM 2.2. Let the assumptions of Lemma 2.1 hold, and suppose that $1 + \gamma = p/q$, with $p, q \in \mathbb{N}$ coprime. Then the solution of (2.6) may be written in the form

$$(2.10)\qquad y(t) = \sum_{s=0}^{q-1} v_s(t)\cdot(t^{p/q})^s ,\ t \in I,$$

with $v_s \in C^m(I)$ $(s = 0,\dots,q-1)$.

The proof of this result is based on the representation (2.9), using (2.8). We have, by the substitution $s = vt$,

$$y(t) = g(t) + \sum_{n=1}^{\infty} t^{n(1+\gamma)} \cdot \int_0^1 (1-v)^{n(1+\gamma)-1} \cdot \Phi_n(t,vt;\gamma) g(vt) dv$$

$$=: g(t) + \sum_{n=1}^{\infty} \Psi_n(t;\gamma) \cdot (t^{1+\gamma})^n .$$

Clearly, $\Psi_n \in C^m(I)$ $(n \geqq 1)$ (recall Lemma 2.1). Since $1 + \gamma = p/q$ we have, for $\ell = 1,\dots,q-1$, that $\ell \cdot (1+\gamma)$ is not an integer. Moreover, if $\ell = rq + \ell'$, with $0 < \ell' \leqq q-1$, then $\ell \cdot (1+\gamma) = rp + \ell' \cdot p/q$, where $rp \in \mathbb{N}$, while $\ell' \cdot p/q$ is not an integer. One verifies then easily that the terms in the above expression for $y(t)$ may be rearranged so as to yield

$$v_0(t) = g(t) + \sum_{\ell=1}^{\infty} \Psi_{q\ell}(t;\gamma) \cdot t^{\ell p} ,$$

$$v_s(t) = \sum_{\ell=0}^{\infty} \Psi_{q\ell+s}(t;\gamma) \cdot t^{\ell p} \quad (s=1,\dots,q-1).$$

The representation (2.9) makes it clear that the solution (2.3) of the special integral equation (2.2) reflects the *typical* situation near $t = 0$. Hence, the only hope for obtaining a smooth solution of (1.1) (with smooth $K(t,s)$) is to choose certain special nonsmooth functions $g(t)$. Let

$$g(t) = g_1(t) + t^{\beta} \cdot g_2(t),$$

with $g_1, g_2 \in C^m(I)$, and with $\beta > 0$, β not an integer. Using again (2.9) and (2.8) it is not difficult to show that, in analogy to (2.10), the solution of (2.6) corresponding to this choice of $g(t)$ can be written as

$$(2.11) \quad y(t) = \sum_{s=0}^{q-1} v_s(t) \cdot (t^{p/q})^s + t^{\beta} \cdot \sum_{s=0}^{q-1} w_s(t) \cdot (t^{p/q})^s ,$$

with $1 + \gamma = p/q$, and with $v_s, w_s \in C^m(I)$ $(s = 0,\dots,q-1)$.

As a simple illustration consider again (2.2): replace $g(t) \equiv 1$ by the nonsmooth function $g(t) = g_1 + t^{\beta} \cdot g_2$, where g_1 and g_2 are constants, and choose $\beta = 1 - \alpha$. It is easily verified that a smooth solution exists only of the relation $\lambda \cdot g_1 + (1 - \alpha) \cdot g_2 = 0$ holds; this solution is then $y(t) = g_1$.

Let us, at this point, briefly look at the Fredholm integral equations (1.2), assuming $g \in C^m(I)$, $K \in C^m(S_F)$. If $\alpha = 0$, and if λ^{-1} is not an eigenvalue, the solution may be written in the form

$$y(t) = g(t) + \lambda \cdot \int_0^1 \frac{D(t,s;\lambda)}{D(\lambda)} \cdot g(s)ds,$$

where $D(\lambda)$ is the Fredholm determinant, and where the numerator $D(t,s;\lambda)$ can be shown to inherit the smoothness of $K(t,s)$. Thus, $y \in C^m(I)$. However, if $0 < \alpha < 1$, then the exact solution of (1.2) has, typically, unbounded derivatives at $t = 0$ and at $t = 1$. More precisely, if α is *irrational*, then

(2.12) $$y(t) = v_0(t) \cdot t^{1-\alpha} + v_1(t) \cdot (1-t)^{1-\alpha} + \text{"smoother" terms},$$

with $v_0, v_1 \in C^m(I)$. For *rational* α, $1 - \alpha = p/q$ (with p and q coprime), $y(t)$ contains, in contrast to (2.10), also logarithmic terms; i.e.,

(2.13) $$y(t) = v_0(t) \cdot t^{p/q} + w_0(t) \cdot t^p \ln(t) + $$
$$+ v_1(t) \cdot (1-t)^{p/q} + w_1(t) \cdot (1-t)^p \ln(1-t) + \text{"smoother" terms},$$

with $v_i, w_i \in C^m(I)$ $(i=0,1)$.

Details concerning the smoothness properties of solutions of weakly singular Fredholm equations may be found in [30], [28], [13], [14], [32], [37]; for Volterra equations of the second kind, see [25], [18], [21], [20] (which treats the nonlinear version of (1.1)), and [4].

3. COLLOCATION AND ITERATED COLLOCATION: PRINCIPLES

Let $0 = t_0 < t_1 < \ldots < t_N = 1$ be a partition of the given interval $I = [0,1]$, and set $h_n := t_{n+1} - t_n$, $\sigma_0 := [t_0, t_1]$, $\sigma_n := (t_n, t_{n+1}]$ $(n=1,\ldots,N-1)$. Furthermore, let $Z_N := \{t_n : n=1,\ldots,N-1\}$ denote the interior partition points (which will be the knots of the polynomial splines defined below), and set $\bar{Z}_N := Z_N \cup \{1\}$. The exact solution of (1.1), or of (1.2), will be approximated by an element of the polynomial spline space

(3.1) $$S_{m-1}^{(-1)}(Z_N) := \{u : u|_{\sigma_n} =: u_n \in \pi_{m-1}, \ n=0,\ldots,N-1 \ (m \geq 1)\};$$

the elements of this approximating space may possess finite discontinuities at their knots Z_N. The approximation u will be determined by collocation: for given *collocation points*

$$X(N) := \bigcup_{n=o}^{N-1} X_n \ , \text{ with } \ X_n := \{t_n + c_j h_n \ : \ 0 < c_1 < \ldots < c_m \leq 1\} \subset \sigma_n \ ,$$

we require that u satisfy the given integral equation (1.1) on $X(N)$:

$$(3.2) \qquad u(t) = g(t) + \int_0^t (t-s)^{-\alpha} \cdot K(t,s) u(s) ds, \ t \in X(N).$$

In order to better exhibit the recursive character of the collocation equation (3.2), assume that $t \in X_n$; we may then write (3.2) as

$$u_n(t_n + c_j h_n) = g_{nj} + \int_{t_n}^{t_n + c_j h_n} (t_n + c_j h_n - s)^{-\alpha} \cdot K_{nj}(s) u_n(s) ds + \\ + \sum_{i=o}^{n-1} \int_{t_i}^{t_{i+1}} (t_n + c_j h_n - s)^{-\alpha} \cdot K_{nj}(s) u_i(s) ds$$
$$(j=1,\ldots,m; \ n=0,\ldots,N-1),$$

where $g_{nj} := g(t_n + c_j h_n)$, $K_{nj}(\ . \ ) := K(t_n + c_j h_n \ , \ . \ )$. The substitution $s = t_i + v h_i$ thus leads to

$$(3.3) \qquad u_n(t_n + c_j h_n) = g_{nj} + h_n^{1-\alpha} \cdot \Phi_{nn}^{(j)}[u_n] + \sum_{i=o}^{n-1} h_i^{1-\alpha} \cdot \Phi_{ni}^{(j)}[u_i]$$
$$(j=1,\ldots,m; \ n=0,\ldots,N-1),$$

where we have set

$$(3.4) \qquad \Phi_{ni}^{(j)}[u_i] := \begin{cases} \int_0^1 \left(\frac{t_n - t_i}{h_i} + c_j \frac{h_n}{h_i} - v\right)^{-\alpha} \cdot K_{nj}(t_i + v h_i) u_i(t_i + v h_i) dv, & \text{if } \ i < n, \\ \int_0^{c_j} (c_j - v)^{-\alpha} \cdot K_{nj}(t_n + v h_n) u_n(t_n + v h_n) dv, & \text{if } \ i = n \end{cases}$$
$$(j = 1,\ldots,m).$$

Note that if the partition of I is the uniform one ($h_n = h = N^{-1}$, $t_n = nh$) then the first factor in the integrand of $\Phi_{ni}^{(j)}[u_i]$ $(i < n)$ reduces to $(n-i+c_j-v)^{-\alpha}$.

As a first step towards the full discretization of the collocation equation (3.3) we set (since $u_i \in \pi_{m-1}$)

$$(3.5)\qquad u_i(t_i+vh_i) = \sum_{\ell=1}^{m} L_\ell(v)Y_{i\ell}\ ,\ \text{with } Y_{i\ell} := u_i(t_i+c_\ell h_i),\ L_\ell(v) := \prod_{r\neq\ell}^{m} \frac{v-c_r}{c_\ell-c_r}\ .$$

Hence, the integrals in (3.4) assume the form

$$(3.6)\qquad \Phi_{ni}^{(j)}[u_i] = \begin{cases} \sum_{\ell=1}^{m} Y_{i\ell}\cdot \int_0^1 \left(\frac{t_n-t_i}{h_i} + c_j\frac{h_n}{h_i} - v\right)^{-\alpha}\cdot K_{nj}(t_i+vh_i)L_\ell(v)dv, & \text{if } i<n,\\ \sum_{\ell=1}^{m} Y_{n\ell}\cdot \int_0^{c_j} (c_j-v)^{-\alpha}\cdot K_{nj}(t_n+vh_n)L_\ell(v)dv, & \text{if } i=n. \end{cases}$$

Introduce the integrals

$$(3.7)\qquad \Phi_{ni}^{(j,\ell)} := \begin{cases} \int_0^1 \left(\frac{t_n-t_i}{h_i} + c_j\frac{h_n}{h_i} - v\right)^{-\alpha}\cdot K_{nj}(t_i+vh_i)L_\ell(v)dv, & \text{if } i<n,\\ \int_0^{c_j} (c_j-v)^{-\alpha}\cdot K_{nj}(t_n+vh_n)L_\ell(v)dv, & \text{if } i=n \end{cases} \qquad (j=1,\dots,m)$$

(which do no longer depend on u_i).

With this notation the collocation equation (3.3) becomes

$$(3.8)\qquad Y_{nj} = g_{nj} + h_n^{1-\alpha}\cdot \sum_{\ell=1}^{m} \Phi_{nn}^{(j,\ell)}\cdot Y_{n\ell} + \sum_{i=0}^{n-1} h_i^{1-\alpha}\left(\sum_{\ell=1}^{m} \Phi_{ni}^{(j,\ell)}\cdot Y_{i\ell}\right) \qquad (j=1,\dots,m).$$

Once the quantities $\{Y_{nj}\}$ are known, we may compute the approximation on the subinterval σ_n (recall (3.5)); in particular, we have

$$(3.9)\qquad y_{n+1} := u_n(t_n+h_n) = \sum_{j=1}^{m} L_j(1)Y_{nj} \qquad (n=0,\dots,N-1).$$

However, in order to determine these quantities we need to know the values of the integrals (3.7) (or, equivalently, (3.4)); in most realistic problems these values cannot be found analytically. Suppose then that the integrals (3.4) are approximated by product-type quadrature formulas whose abscissas coincide with the collocation points (if $i < n$), or with their affine images on $[0,c_j]$ (if $i = n$); i.e., we have

$$(3.10)\quad \psi_{ni}^{(j)}[u_i] := \begin{cases} \sum_{\ell=1}^{m} w_{ni}^{(j)}(\alpha)\cdot K_{nj}(t_i+c_\ell h_i)Y_{i\ell}\ , & \text{if } i < n, \\ \sum_{\ell=1}^{m} w_{nn}^{(j)}(\alpha)\cdot K_{nj}(t_n+c_j c_\ell h_n)\cdot \sum_{r=1}^{m} L_r(c_j c_\ell)Y_{nr}\ , & \text{if } i = n \end{cases}$$

$$(j=1,\ldots,m)$$

(note that, in general, the kernel function $K(t,s)$ in (1.1) cannot be smoothly extended to points (t,s) with $s > t$). If we employ these quadrature approximations in the collocation equation (3.8) we generate an element $\tilde{u} \in S_{m-1}^{(-1)}(Z_N)$ which will differ from u. The *fully discretized collocation equation* is then given by

$$(3.10')\quad \tilde{Y}_{nj} = g_{nj} + h_n^{1-\alpha}\cdot\psi_{nn}^{(j)}[\tilde{u}_n] + \sum_{i=0}^{n-1} h_i^{1-\alpha}\psi_{ni}^{(j)}[\tilde{u}_i]$$

$$(j=1,\ldots,m;\quad n=0,\ldots,N-1),$$

where we have set, in analogy to (3.5), $\tilde{Y}_{nj} := \tilde{u}_n(t_n+c_j h_n)$, and where

$$(3.11)\quad \tilde{u}_n(t_n+vh_n) = \sum_{\ell=1}^{m} L_\ell(v)\cdot\tilde{Y}_{n\ell}\ .$$

Hence, the discrete counterpart to (3.8) may be written in the form

$$(3.12)\quad \tilde{Y}_{nj} = g_{nj} + h_n^{1-\alpha}\cdot \sum_{\ell=1}^{m} w_{nn}^{(j)}(\alpha)K_{nj}(t_n+c_j c_r h_n)\cdot \sum_{r=1}^{m} L_\ell(c_j c_r)\tilde{Y}_{n\ell} + $$
$$+ \sum_{i=0}^{n-1} h_i^{1-\alpha}\left(\sum_{\ell=1}^{m}\psi_{ni}^{(j,\ell)}\tilde{Y}_{i\ell}\right)\quad (j=1,\ldots,m),$$

while (3.9) is replaced by

$$(3.13)\quad \tilde{y}_{n+1} := \tilde{u}_n(t_n+h_n) = \sum_{j=1}^{m} L_j(1)\tilde{Y}_{nj}\qquad (n=0,\ldots,N-1).$$

In (3.12) we have set

$$\psi_{ni}^{(j,\ell)} := w_{ni}^{(j)}(\alpha)\ K_{nj}(t_i+c_\ell h_i)\qquad (i < n).$$

The above collocation approximations in $S_{m-1}^{(-1)}(Z_N)$, u and $\tilde{u}$, can be used to compute the corresponding so-called *iterated collocation approximations* u^I and $\tilde{u}^I$. This concept was introduced by Sloan [35] for Fredholm integral equations (compare als [36] and, especially, [11]). For the Volterra integral equation (1.1) these iterates are defined by

$$(3.14) \quad u^I(t) := g(t) + \int_0^t (t-s)^{-\alpha} K(t,s)u(s)ds, \ t \in I,$$

and by

$$(3.15) \quad \tilde{u}^I(t) := g(t) + \int_0^t (t-s)^{-\alpha} K(t,s)\tilde{u}(s)ds, \ t \in I,$$

respectively, with u and $\tilde{u}$ denoting the solutions of (3.3) and (3.10'). In both cases we are again faced with the problem of discretizing the integrals in (3.14) and (3.15): since u is generally not computable (because the integrals in (3.3) are not known exactly) we consider the discretization of (3.15), where we shall be interested in computing the values of $\tilde{u}^I(t)$ for $t \in \bar{Z}_N$. Let the discretization of (3.15) (with $t = t_n \in \bar{Z}_N$) be given by

$$(3.16) \quad \hat{u}^I(t_n) := g(t_n) + \sum_{i=0}^{n-1} h_i^{1-\alpha} \sum_{\ell=1}^{m} w_{n-i,\ell}(\alpha)K(t_n,t_i+c_\ell h_i)\, \tilde{Y}_{i\ell},$$

with $\tilde{Y}_{i\ell}$ defined by the discretized collocation equation (3.12), Here, the quadrature weights are usually chosen as

$$w_{n-i,\ell}(\alpha) := \int_0^1 \left(\frac{t_n - t_i}{h_i} - v\right)^{-\alpha} \cdot L_\ell(v)dv.$$

We add a crucial observation: if, in (3.14), $t \in X(N)$, then comparison with (3.2) shows that

$$(3.17) \quad u^I(t) = u(t) \text{ for all } t \in X(N);$$

i.e. the values of the collocation approximation and the corresponding iterated collocation approximation u^I coincide at the collocation points. The same result holds for $\tilde{u}$ and $\hat{u}^I$ (the discretization of (3.15)), provided the same quadrature formulas (3.10) are used.

The remainder of this section will be devoted to a discussion of how the various collocation errors are related. These results will represent, in Sections 4 and 5, the basic tools in the analysis of the convergence orders (global and local) of the above collocation methods; in particular, they will help to explain the difference in the performance of these methods when applied to integral equations with regular or weakly singular kernels.

We start with a summary of the various error functions; y denotes the exact solution of (1.1).

$e(t) := y(t) - u(t)$, with u defined in (3.3);

$\tilde{e}(t) := y(t) - u(t)$, with $\tilde{u}$ defined in (3.10');

$e^I(t) := y(t) - u^I(t)$, with u^I defined in (3.14);

$\tilde{e}^I(t) := y(t) - \tilde{u}^I(t)$, with $\tilde{u}^I$ defined in (3.15);

$\hat{e}^I(t_n) := y(t_n - \hat{u}^I(t_n)$, with $\hat{u}^I$ defined in (3.16);

$\varepsilon(t) := u(t) - \tilde{u}(t)$;

$\varepsilon^I(t) := u^I(t) - \tilde{u}^I(t)$;

$\hat{\varepsilon}^I(t_n) := u^I(t_n) - \hat{u}^I(t_n)$.

Whenever one of these functions carries a subscript (e.g., ε_n), it denotes the restriction of the function to the subinterval σ_n.

Consider the collocation equation (3.2): it may be written as

(3.18) $\quad u(t) = g(t) + \int_0^t (t-s)^{-\alpha} K(t,s)u(s)ds - \delta(t),\ t \in I$, where the *defect function* δ vanishes on the set $X(N)$ of collocation points. Hence, by (1.1) and (3.18), the collocation error e satisfies the integral equation

$$e(t) = \delta(t) + \int_0^t (t-s)^{-\alpha} K(t,s)e(s)ds,\ t \in I, \tag{3.19}$$

whose solution can be determined according to (2.9) (see (3.21) below). On the other hand, it follows from (1.1) and from (3.14) that the iterated collocation error e^I is related to e by

$$e^I(t) = \int_0^t (t-s)^{-\alpha} \cdot K(t,s)e(s)ds, \quad t \in I. \tag{3.20}$$

THEOREM 3.1. Let $R(t,s;-\alpha)$ denote the resolvent kernel of the kernel $K(t,s;-\alpha):= (t-s)^{-\alpha}\cdot K(t,s)$ of (1.1). Then the collocation error e induced by the solution $u \in S_{m-1}^{(-1)}(Z_N)$ of (3.2) has the form

$$(3.21) \qquad e(t) = \delta(t) + \int_0^t R(t,s;-\alpha)\cdot\delta(s)ds, \quad t \in I,$$

while the corresponding collocation error e^I associated with (3.14) is given by

$$(3.22) \qquad e^I(t) = \int_0^t R(t,s;-\alpha)\cdot\delta(s)ds, \quad t \in I.$$

The proof of (3.22) is a consequence of the first of the so-called *Fredholm identities* for the resolvent kernel of a second-kind Volterra integral equation: it is easily deduced from its Neumann series and reads

$$(3.23) \qquad R(t,v;-\alpha) = K(t,v;-\alpha) + \int_v^t K(t,s;-\alpha)R(s,v;-\alpha)ds, \ (t,v) \in S_V .$$

Using (3.21) in (3.20) and applying Dirichlet's Lemma we find

$$\begin{aligned} e^I(t) &= \int_0^t K(t,s;-\alpha)\cdot\{\delta(s) + \int_0^s R(s,v;-\alpha)\cdot\delta(v)dv\}ds = \\ &= \int_0^t K(t,s;-\alpha)\cdot\delta(s)ds + \int_0^t (\int_v^t K(t,s;-\alpha)R(s,v;-\alpha)ds)\cdot\delta(v)dv, \end{aligned}$$

which, by (3.23), reduces to (3.22).

We recall that, for $0 < \alpha < 1$, the resolvent kernel $Q(t,s;-\alpha)$ is continuous, but not continuously differentiable, on S_V (see Lemma 2.1). Note also that $e^I(t) = e(t)$ for all $t \in X(N)$, since the defect δ vanishes on $X(N)$); this just confirms (3.17).

We now turn our attention to the fully discretized collocation equations (3.10')and (3.16). The error $\tilde{e}$ associated with the solution $\tilde{u} \in S_{m-1}^{(-1)}(Z_N)$ of (3.10') (3.11), can be written in the form $\tilde{e}(t) = e(t) + \varepsilon(t)$, $t \in I$, where $e(t)$ satisfies (3.21). By (3.3) and (3.10')we have (due to the linearity of the functionals $\Phi_{ni}^{(j)}$ and $\Psi_{ni}^{(j)}$)

$$(3.24) \qquad \varepsilon_n(t) = h_n^{1-\alpha}\,\Psi_{nn}^{(j)}[\varepsilon_n] + \sum_{i=0}^{n-1} h_i^{1-\alpha}\,\Psi_{ni}^{(j)}[\varepsilon_i] + \sum_{i=0}^{n} h_i^{1-\alpha}\,E_{ni}^{(j)}[u_i],$$

$t = t_n + c_j h_n \in X_n$ $(n = 0,\dots,N-1)$, where the $E_{ni}^{(j)}[u_i]$ are the quadrature remainder terms,

$$(3.25) \qquad E_{ni}^{(j)}[u_i] := \Phi_{ni}^{(j)}[u_i] - \Psi_{ni}^{(j)}[u_i] \quad (0 \leqq i \leqq n \leqq N-1).$$

Hence, the perturbation $\varepsilon = u - \tilde{u}$ (which is an element of $S_{m-1}^{(-1)}(Z_N)$) will be governed entirely by the choice of the quadrature formulas characterizing the fully discretized collocation equation (3.10'). A similar argument holds for $\hat{e}^I(t_n)$: we have $\hat{e}^I(t_n) = e^I(t_n) + \hat{\varepsilon}^I(t_n)$, with $e^I(t)$ given in (3.22). It follows from (3.14) (with $t = t_n \in \bar{Z}_N$) and from (3.16) that

$$(3.26) \qquad \hat{\varepsilon}^I(t_n) = \sum_{i=0}^{n-1} h_i^{1-\alpha}\, \Psi_{ni}\, [\varepsilon_i] + \sum_{i=0}^{n-1} h_i^{1-\alpha}\, E_{ni}[u_i],\ t \in \bar{Z}_N,$$

where we have set

$$\Psi_{ni}[\varepsilon_i] := \sum_{\ell=1}^{m} w_{n-i,\ell}(\alpha)\cdot K(t_n, t_i + c_\ell h_i)\cdot \varepsilon_i(t_i + c_\ell h_i) \quad (0 \leqq i \leqq n \leqq N-1).$$

Note that we meet again the values $\varepsilon_i(t_i + c_\ell h_i$ $(t_i + c_\ell h_i \in X(N))$ which we have shown to satisfy (3.24).

In summary, we may state the following:

(i) The orders of the "exact" collocation and iterated collocation errors, e and e^I, are determined by the choice of the approximating polynomial spline space (i.e., by the defect function δ in (3.18)) and by the smoothness properties of the resolvent kernel $R(t,s;-\alpha)$ (recall (3.21), (3.22)).

(ii) The orders of the perturbations $\varepsilon(t)$ and $\hat{\varepsilon}^I(t_n)$ (which in turn determine the order of the errors induced by the fully discretized collocation equations) depend on the choice of the quadrature formulas $\Psi_{ni}^{(j)}$ and Ψ_{ni} in (3.24) and (3.26); note that the corresponding integrands contain the (piecewise) smooth factors $K(t_n + c_j h_n, \,.\,)u(\,.\,)$ and $K(t_n, \,.\,)u(\,.\,)$, respectively. Thus, by choosing suitable interpolatory product quadrature formulas we shall be able to obtain error terms whose orders are at least equal to those of the collocation errors e and e^I. We shall return to these observations in the following two sections.

The survey paper [36] (pp. 60-66) gives an illuminating introduction to collocation methods for Fredholm integral equations; in addition, compare also [11], [38] , [2], as well as the relevant references in [36]. Product integration methods (which are related to collocation methods) and Galerkin methods have been treated in, e.g., [9], [10], [13], [11], [34], [15]. As for collocation methods for Volterra integral equations we refer to [3] (also for additional references), [6], [8], [31], [1], [5], while product integration methods may be found in [20], [18], [12], [21], [19]. Finally, the theory of Runge-Kutta methods (to which the fully discretized collocation method (3.12), (3.13) is related) for (1.1) with nonsmooth solution has been given in [22]; product integration is analyzed in [23], [17], [33].

4. INTEGRAL EQUATIONS WITH REGULAR KERNELS

In this and the following section we shall sketch how the results of Section 3 (particularly Theorem 3.1) can be used to establish results on the order of convergence for the various collocation approximations. Assume that the mesh $\{t_n\}$ is quasi-uniform: $\max\limits_{(i)} h_i / \min\limits_{(i)} h_i \leqq \gamma < \infty$ for all $n \in \mathbb{N}$. For given N set $h := \max\limits_{(i)} h_i$; if the mesh is uniform then $\gamma = 1$ and $h = N^{-1}$. Observe that the assumption of quasi-uniformity implies

$$(4.1) \qquad h_i \leqq h \leqq \gamma \cdot \min_{(i)} h_i \ ; \ N \cdot h_i \leqq \gamma N \cdot \min_{(i)} h_i \leqq \gamma \cdot 1 = \gamma .$$

Due to limitations of space we shall deal with global convergence results only in passing; they are generally well known. If, in (1.1), $g \in C^m(I)$, $K \in C^m(S_V)$, then the following global results hold:

(i) $\alpha = 0 : \|y - u\|_\infty = O(h^m)$, with $u \in S_{m-1}^{(-1)}(Z_N)$ given in (3.3);

(ii) $0 < \alpha < 1 : \|y - u\|_\infty = O(h^{1-\alpha})$ (for all $m \geqq 1$), with u as in (i).

The same asymptotic results hold if u is replaced by the $\tilde{u} \in S_{m-1}^{(-1)}(Z_N)$, the solution of the fully discretized collocation equation (3.11). For details, see [8], [11], [3], [1].

Suppose now that $t = t_n \in \overline{Z}_N$: for these values of t, (3.21) assumes the form

$$(4.2) \qquad e(t_n) = \delta(t_n) + \sum_{i=0}^{n-1} h_i^{1-\alpha} \cdot \int_0^1 \left(\frac{t_n - t_i}{h_i} - v\right)^{-\alpha} \cdot Q(t_n, t_i + vh_i, -\alpha)\delta(t_i + vh_i)dv .$$

Replace each of the integrals in (4.2) by m-point interpolatory (product) quadrature (using the collocation points $\{t_i+c_\ell h_i:\ell=1,\ldots,m\}$ as abscissas) and the corresponding remainder term; i.e. by

$$\sum_{\ell=1}^{m} w_{n-i,\ell}(\alpha)\cdot Q(t_n,t_i+c_\ell h_i;-\alpha)\cdot\delta(t_i+c_\ell h_i) + E_{ni}(\alpha).$$

Since $\delta(t_i+c_\ell h_i) = 0$, (4.2) becomes

$$(4.3)\qquad e(t_n) = \delta(t_n) + \sum_{i=0}^{n-1} h_i^{1-\alpha} E_{ni}(\alpha),\quad t_n \in \bar{Z}_N .$$

In an analogous fashion, (3.22) (with $t = t_n$) yields

$$(4.4)\qquad e^I(t_n) = \sum_{i=0}^{n-1} h_i^{1-\alpha} E_{ni}(\alpha),\quad t_n \in \bar{Z}_N .$$

In other words, the order of convergence of the collocation and the iterated collocation approximation on $\bar{Z}_N$ is essentially given by the order of the quadrature error $E_{ni}(\alpha)$: this order will be governed by the available smoothness of the integrand, i.e. by the smoothness of the resolvent kernel $Q(t,s;-\alpha)$ and of the defect function $\delta(t)$.

For $\alpha = 0$, smooth g and K in (1.1) imply that the integrand $R(t_n,\,.\,;0)\cdot\delta(.)$ has the same degree of smoothness on each subinterval σ_i; hence, $E_{ni}(0) = O(h^p)$, with $m \leq p \leq 2m$. This implies, by (4.4), that $e^I(t_n) = O(h^{2m})$ on $\bar{Z}_N$, provided the collocation parameters $\{c_j\}$ are the *Gauss-Legendre* points. In contrast, this choice yields only $e(t_n) = O(h^m)$ on $\bar{Z}_N$, since (4.3) contains the term $\delta(t_n)$ which can be shown to be $O(h^m)$ whenever $t_n \notin X(N)$ (i.e. $c_m < 1$). See [8], [3], [5] for additional details.

For the *Fredholm integral* equation (1.2) with $\alpha = 0$ the expressions corresponding to (3.21) and (3.22) take the form

$$(4.5)\qquad e(t) = \delta(t) + \lambda\cdot\int_0^1 \frac{D(t,s;\lambda)}{D(\lambda)}\cdot\delta(s)ds,\ t \in I,$$

and

$$(4.6)\qquad e^I(t) = \lambda\int_0^1 \frac{D(t,s;\lambda)}{D(\lambda)}\cdot\delta(s)ds,\ t \in I ,$$

where $D(t,s;\lambda)$ is smooth (recall (2.12)). While the Gauss parameters yield again

$e(t_n) = O(h^m)$ on $\bar{Z}_N$, they now imply that the iterated collocation approximation furnishes $e^I(t) = O(h^{2m})$ *globally* for $t \in I$: i.e., $\|e^I\|_\infty = O(h^{2m})$. (Compare also [35], [11]. For (1.1) with $\alpha = 0$ one has, in sharp contrast, only $\|e^I\|_\infty = O(h^m)$ ([5]; a similar result holds for Fredholm equations with Green-type kernels [11].)

According to the remarks at the end of the previous section (recall also (3.24) and (3.26)), the above results on the order of convergence on $\bar{Z}_N$ remain valid for $\tilde{u}$ and $\tilde{u}^I$, since we have

$$|\tilde{e}(t_n)| \leqq |e(t_n)| + |\varepsilon(t_n)| \quad \text{and} \quad |\hat{e}^I(t_n)| \leqq |e^I(t_n)| + |\hat{\varepsilon}^I(t_n)|, \; t_n \in \bar{Z}_N .$$

5. INTEGRAL EQUATIONS WITH WEAKLY SINGULAR KERNELS

Consider now the Volterra equation (1.1) for $0 < \alpha < 1$. For $t = t_n \in \bar{Z}_N$, the errors $e(t_n)$ and $e^I(t_n)$ associated with the collocation approximation $u \in S_{m-1}^{(-1)}(Z_N)$ and its iterate u^I are given by (4.3) and (4.4). In contrast to the case $\alpha = 0$, the factor $Q(t_n, t_{n-1}+vh_{n-1}; -\alpha)\cdot\delta(t_{n-1}+vh_{n-1})$ under the last integral in (4.2) is now no longer smooth, since the derivatives of $Q(t,s;-\alpha)$ become unbounded as s approaches t (recall (2.8)); this fact is independent of the assumed smoothness for $K(t,s)$. As a consequence, the order of the remainder term $E_{n,n-1}(\alpha)$ (which, for $\alpha = 0$, is at least equal to m) is diminished, and it can be shown that, for the uniform mesh, $e(t_n) = O(h^{1-\alpha})$ for all $t_n \in \bar{Z}_N$; this order of convergence holds also globally ([27], [21], [31], [19]; see also [22]). The same asymptotic result is obtained for the Fredholm equation (1.2) $(0 < \alpha < 1)$; compare, e.g., [9], [10], [34], as well as [13], [14] for Galerkin methods. This dramatic drop of the order of convergence is, of course, to be expected since we are essentially approximating the function $f(t) = t^{1-\alpha}$ on the intervall $[0,h]$ by polynomials (recall Theorem 2.2; see also [29]). There are two ways of restoring the order of convergence to the level attained for $\alpha = 0$:

(i) by replacing the uniform mesh by an appropriately *graded mesh* which takes into account the singular behavior of the exact solution near $t = 0$ (or, for Fredholm equations, near $t = 0$ and $t = 1$); or

(ii) by retaining the uniform mesh butby replacing the polynomial spline space $S_{m-1}^{(-1)}(Z_N)$ by a special *non-polynomial spline space* whose elements reflect the structure of the exact solution of the given integral equation near the endpoint(s) of the interval I (recall (2.10); (2.13), (2.14)).

For the numerical solution of Fredholm integral equations (1.2) $(0 < \alpha < 1)$, graded meshes of the form $t_i^{(r)} := \frac{1}{2}\left(\frac{2i}{N}\right)^r$ $(i=0,\dots,[N/2])$, $t_i^{(r)} := 1 - \frac{1}{2}\left(\frac{2(N-i)}{N}\right)^r$ $(i = [N/2]+1,\dots,N)$, with $r := \mu/(1-\alpha)(\mu \leqq m)$ have been used, e.g., in [10], [11], [34], [38]; the basic idea of generating high-order approximations to the function $f(t) = t^{1-\alpha}$ on $[0,1]$ by polynomial splines with variable knots goes back to [29]. For the Volterra integral equation (1.1) the graded mesh is given by

$$(5.1) \qquad t_i^{(r)} := \left(\frac{i}{N}\right)^r \quad (i=0,\dots,N), \text{ with } \; r := \mu/(1-\alpha) \text{ and } \mu \leqq m.$$

(Note that this mesh is no longer quasi-uniform: we have $h_i^{(r)} := t_{i+1}^{(r)} - t_i^{(r)}$, and hence, for $r > 1$, $\max_{(i)} h_i^{(r)} / \min_{(i)} h_i^{(r)} = h_{N-1}^{(r)}/h_o^{(r)} = N^r(1-(1-N^{-1}))^r \to \infty$, as $N \to \infty$.) Convergence results and numerical experiments will be reported in [6]; compare also the examples given below.

Non-polynomial spline spaces have been used successfully, for $\alpha = 1/2$, in [31] (compare also [27] for related ideas). For this case, the basis functions of the approximating space (whose dimension is equal to 2mN, as compared to mN for $S_{m-1}^{(-1)}(Z_N)$) include the functions $\{1,t^{1/2};\, t,t^{3/2};\dots;\, t^{m-1},t^{m-1/2}\}$, and it can be shown that one obtains $\|e\|_\infty = O(h^m)$ on the uniform mesh. The paper [31] presents also special quadrature formulas which are exact for the above fractional powers if t and which are employed in the fully discretized collocation equation corresponding to (3.10'); in addition, it contains a large number of numerical examples. For the general theory of collocation methods in non-polynomial spline spaces based on (2.10) we refer the reader to [4]. As for as Fredholm integral equations with weakly singular kernels are concerned, collocation methods in non-polynomial spline spaces have, to this author's knowledge, not yet been investigated.

We conclude by giving a numerical example, illustrating results obtained by employing graded meshes. The Volterra equation (2.2) was solved, for $\alpha = -1$ and $\alpha \in \{1/10, 1/3, 1/2\}$, by collocation in $S_1^{(-1)}(Z_N)$ (i.e. $m = 2$), using as collocation parameters the Gauss points $c_1 = (3-\sqrt{3})/6$, $c_2 = (3+\sqrt{3})/6$.

Table 1: $\alpha = 1/10$				
N	r = 1 (ungraded)		r = 20/9 (μ = m = 2)	
	e(1)	$e^I(1)$	e(1)	$e^I(1)$
5	1.31-3	2.60-5	5.82-3	1.69-5
10	3.05-4	6.18-6	1.45-3	1.10-5
20	7.36-5	1.82-6	3.60-4	2.30-6

Table 2: $\alpha = 1/3$				
N	r = 1		r = 3 (μ = m = 2)	
	e(1)	$e^I(1)$	e(1)	$e^I(1)$
5	9.16-4	1.47-4	8.09-3	9.52-4
10	1.99-4	4.04-5	1.95-3	1.53-4
20	4.61-5	1.41-5	4.71-4	2.96-5

Table 3: $\alpha = 1/2$				
N	r = 1		r = 4 (μ = m = 2)	
	e(1)	$e^I(1)$	e(1)	$e^I(1)$
5	5.54-4	2.89-4	9.40-3	3.50-3
10	1.07-4	8.74-5	2.33-3	6.52.4
20	2.19-5	3.49-5	3.63-4	1.27-4

It seems worthwile to point out two observations which seem to be valid for other examples, too, and which deserve further analysis:

(i) As one moves away from the left endpoint of the interval [0,1] the collocation approximation on the ungraded mesh is more accurate than the one on the graded mesh; furthermore, it appears to converge with order $p = m = 2$. This fact seems to point to what one might call the "self-correcting" character of nonsmooth polynomial splines $u \in S_{m-1}^{(-1)}(Z_N)$.

(ii) The iterated collocation approximation is more accurate on the ungraded mesh but shows a much better convergence rate on the graded mesh.

REFERENCES

1. Abdalkhani, J.: Collocation and Runge-Kutta-type methods for Volterra integral equations with weakly singular kernels, Ph. D. Thesis, Dalhousie University, Halifax, N.S. (Canada), 1982.

2. Atkinson, K., Graham, I., and Sloan, I.: Piecewise continuous collocation for integral equations, SIAM J. Numer. Anal., 20 (1983), 172-186.

3. Brunner, H.: A survey of recent advances in the numerical treatment of Volterra integral and integro-differential equations, J. Comp. Appl. Math., 8 (1982), 213-229.

4. Brunner, H.: Nonpolynomial spline collocation for Volterra equations with weakly singular kernels, SIAM J. Numer. Anal., 20 (1983).

5. Brunner, H.: Iterated collocation methods and their discretizations for Volterra integral equations, SIAM J. Numer. Anal. (to appear).

6. Brunner, H., and Graham, I.G.: Product integration for weakly singular Volterra integral equation (to appear).

7. Brunner, H., and Evans, M.D.: Piecewise polynomial collocation for Volterra-type integral equations of the second kind, J. Inst. Math. Appl., 20 (1977), 415-423.

8. Brunner, H., and Nørsett, S.P.: Superconvergence of collocation methods for Volterra and Abel integral equations of the second kind, Numer. Math., 36 (1981), 347-358.

9. Chandler, G.A.: Superconvergence of numerical methods to second kind integral equations, Ph. D. Thesis, Australian National University, Canberra, 1979.

10. Chandler, G.A.: Superconvergence for second kind integral equations, in: The Application and Numerical Solution of Integral Equations (R.S. Anderssen, F.R. de Hoog and M.A. Lukas, eds.), Sijthoff & Noordhoff, Alphen/Rijn (Netherlands), 1980. pp. 103-117.

11. Chatelin, F., and Lebbar, R.: The iterated projection solution for the Fredholm integral equation of second kind, J. Austral. Math. Soc. Ser. B, 22 (1981), 439-451.

12. Garey, L.: The numerical solution of Volterra integral equations with singular kernels, BIT, 14 (1974), 33-39.

13. Graham, I.G.: The numerical solution of Fredholm integral equations of the second kind, Ph. D. Thesis, University of New South Wales, Sydney, 1980.

14. Graham, I.G.: Singularity expansions for the solutions of second kind Fredholm integral equations with weakly singular convolution kernels, J. Integral Equations, 4 (1982), 1-30.

15. Graham, I.G.: Galerkin methods for second kind integral equations with singularities, Math. Comp., 39 (1982), 519-533.

16. Handelsman, R.A., and Olmstead, W.E.: Asymptotic solution to a class of nonlinear Volterra integral equations, SIAM J. Appl. Math., 22 (1972), 373-384.

17. de Hoog, F., and Weiss, R.: Asymptotic expansions for product integration, Math. Comp., 27 (1973), 295-306.

18. de Hoog, F.R., and Weiss, R.: High order methods for a class of Volterra integral equations with weakly singular kernels, SIAM J. Numer. Anal., 11 (1974), 1166-1180.

19. Kershaw, D.: Some results for Abel-Volterra integral equations of the second kind, in: Treatment of Integral Equations by Numerical Methods (C.T.H. Baker and G.F. Miller, eds.), Academic Press, London, 1982, pp. 273-282.

20. Linz, P.: Numerical methods for Volterra integral equations with singular kernels, SIAM J. Numer. Anal., 6 (1969), 365-374.

21. Logan, J.E.: The approximate solution of Volterra integral equations of the second kind, Ph. D. Thesis, University of Iowa, Iowa City, 1976.

22. Lubich, C.: Runge-Kutta theory for Volterra and Abel integral equations of the second kind, Preprint Nr. 154, Sonderforschungsbereich 123, University of Heidelberg, 1982.

23. Lyness, J.N., and Ninham, B.W.: Numerical quadrature and asymptotic expansions, Math. Comp., 21 (1967), 162-178.

24. McKee, S.: Generalized discrete Gronwall lemmas, Z. Angew. Math. Mech., 62 (1982), 429-434.

25. Miller, R.K., and Feldstein, A.: Smoothness of solutions of Volterra integral equations with weakly singular kernels, SIAM J. Math. Anal., 2 (1971), 242-258.

26. Olmstead, W.E.: A nonlinear integral equation associated with gas absorption in a liquid, Z. Angew. Math. Phys., 28 (1977), 513-523.

27. Oulès, H.: Résolution numérique d'une équation intégrale singulière, Rev. Française Trait. Inform. (Chiffres), 7 (1964), 117-124.

28. Pitkäranta, J.: On the differential properties of solutions to Fredholm equations with weakly singular kernels, J. Inst. Math. Appl., 24 (1979), 109-119.

29. Rice, J.R.: On the degree of convergence of nonlinear spline approximation, in: Approximation with Special Emphasis on Spline Functions (I.J. Schoenberg, ed.), Academic Press, New York, 1969, pp. 349-365.

30. Richter, G.R.: On weakly singular Fredholm integral equations with displacement kernels, J. Math. Anal. Appl., 55 (1976), 32-42.

31. te Riele, H.J.J.: Collocation methods for weakly singular second-kind Volterra integral equations with non-smooth solution, IMA J. Numer. Anal., 2 (1982), 437-449.

32. Schneider, C.: Regularity of the solution to a class of weakly singular Fredholm integral equations of the second kind, Integral Equations Operator Theory, 2 (1979), 62-68.

33. Schneider, C.: Produktintegration mit nicht-äquidistanten Stützstellen, Numer. Math., 35 (1980), 35-43.

34. Schneider, C.: Product integration for weakly singular integral equations, Math. Comp., 36 (1981), 207-213.

35. Sloan, I.H.: Improvement by interation for compact operator equations, Math. Comp., 30 (1976), 758-764.

36. Sloan, I.H.: A review of numerical methods for Fredholm equations of the second kind, in: The Application and Numerical Solution of Integral Equations (R.S. Anderssen, F.R. de Hoog and M.A. Lukas, eds.), Sijthoff & Noordhoff, Alphen/Rijn (Netherlands), 1980, pp. 51-74.

37. Vainikko, G., and Pedas, A.: The properties of solutions of weakly singular integral equations, J. Austral. Math. Soc. Ser. B, 22 (1981), 419-430.

38. Vainikko, G., and Uba, P.: A piecewise polynomial approximation to the solution of an integral equation with weakly singular kernel, J. Austral. Math. Soc. Ser. B, 22 (1981), 431-438.

ACKNOWLEDGEMENTS

I am grateful to Dr. Ivan Graham (University of Melbourne) with whom I enjoyed many stimulating conversations on Fredholm integral equations with weakly singula kernels. I would also like to thank Mrs. L. Wolf for the careful typing of the manuscript.

A Trust-Region Approach to Linearly Constrained Optimization

David M. Gay

ABSTRACT

This paper suggests a class of trust-region algorithms for solving linearly constrained optimization problems. The algorithms use a "local" active-set strategy to select the steps they try. This strategy is such that degeneracy and zero Lagrange multipliers do not slow convergence (to a first-order stationary point) and that no anti-zigzagging precautions are necessary. (Unfortunately, when there are zero Lagrange multipliers, convergence to a point failing to satisfy second-order necessary conditions remains possible.) We discuss specialization of the algorithms to the case of simple bounds on the variables and report preliminary computational experience.

1. Introduction.

There has been considerable interest recently in "trust-region" algorithms for unconstrained optimization. This paper describes a way to extend some such algorithms to handle problems where there are linear constraints on the variables. For simplicity of notation we only consider inequality constraints.

Suppose $f:\mathbb{R}^p \to \mathbb{R}$ is continuously differentiable and that we are given $C \in \mathbb{R}^{m\times p}$ and $b \in \mathbb{R}^m$ (a real $m\times p$ matrix and real m-vector respectively). The problem of interest is that of minimizing $f(x)$ subject to $Cx \geq b$, i.e., of approximating a point

$$x^* := \arg\min\{f(x)\colon Cx \geq b\} \tag{1.1}$$

(when such a point exists). We assume p to be small enough that it is reasonable to use dense-matrix techniques and to work explicitly with a matrix $H \in \mathbb{R}^{p\times p}$ that "approximates" the Hessian matrix $\nabla^2 f$ (matrix of second partial derivatives) of f.

The way most commonly recommended to solve a problem of the form (1.1) is to use an "active-set" strategy — see, for instance, [Fle81] and [GilMS81]. That is, one decides to regard some of the inequality constraints (those in the current "active set") as equalities for a while (since the task of minimizing a function $x \in \mathbb{R}^p$ subject to x satisfying k linear equations is easily converted to the problem of minimizing an unconstrained function of $p-k$ variables). This is generally a "global" active-set strategy, in that the same active set is usually used for several iterations. For trust-region algorithms, this paper contends that it is reasonable and in some ways desirable to use a "local" active-set strategy, one where the choice of active set is made afresh every iteration.

The rest of this section introduces some notation. For later reference, §2 describes trust-region algorithms for unconstrained minimization. §3 discusses some modifications for linear constraints, and §4 suggests some simple ways to approximate some of the calculations needed in §3. Some convergence theory supporting §§2−4 appears in §5. §6 discusses various details of linear algebra and suggests a further approximation that is sometimes useful when new constraints are encountered. To resolve degeneracy (i.e., linear dependence of the normals to the tight constraints), §3 suggests a generalization of Powell's dogleg strategy [Pow70a,b]; §7 suggests a way to carry out the necessary calculation and relates the generalized dogleg step to Powell's original one. §8 briefly discusses the special but common case of simple bounds and presents some encouraging computational results for it. Some concluding remarks appear in §9, and more details of the testing described in §8 are given in the appendix.

It will be convenient to denote the ith component of b by b_i and the ith row of C by C_i, so that the ith constraint is $C_i x \geq b_i$. We denote the gradient of f at $x \in \mathbb{R}^p$ by $\nabla f(x) \in \mathbb{R}^p$, and superscript T stands for "transpose". $\mathbf{0}$ stands for a matrix or vector of zeros and I stands for an identity matrix, both of dimension dictated by context. The ith standard unit vector, i.e., the ith column of I, will be denoted e_i. If S is a discrete set, then $|S|$ means the number of elements in S.

At any point x, we shall use $A_0(x)$ to denote the index set of the constraints $Cx \geq b$ that are tight at x, i.e., satisfied as equalities. In other words,

$$A_0(x) := \{i\colon C_i x = b_i,\ 1 \leq i \leq m\}. \tag{1.2}$$

For later reference, recall that the first-order necessary conditions for x^* to solve (1.1) are that

$$Cx^* \geq b \tag{1.3a}$$

i.e., that x^* is feasible, and that there exists $\lambda^* \in \mathbb{R}^m$ with

$$\nabla f(x^*) = C^T \lambda^*, \tag{1.3b}$$

$$\lambda^* \geq 0, \tag{1.3c}$$

and $\lambda_i^* = 0$ if $C_i x^* > b_i$, i.e.,

$$(b - Cx^*)^T \lambda^* = 0. \tag{1.3d}$$

We shall be interested in studying convergence to points x^* that are strong local minimizers, i.e., that satisfy not only (1.3), but also the second-order sufficiency condition

(1.4) $\inf\{d^T \nabla^2 f(x^*) d \colon \|d\| = 1,\ C_i d = 0\ \forall\ i \in A_0(x^*) \text{ with } \lambda_i^* > 0\} > 0,$

where $\|\cdot\|$ is an arbitrary norm on $\mathbb{R}^p$. We shall also have occasion to consider the second-order necessary condition

(1.5) $\inf\{d^T \nabla^2 f(x^*) d \colon \|d\| = 1,\ C_i d = 0\ \forall\ i \in A_0(x^*) \text{ with } \lambda_i^* > 0\} \geq 0.$

Optimality conditions (1.3–5) are specializations to the linearly constrained case of criteria described in [McC76]; see [McC76] and [Kuh76] for references on and history of optimality conditions. Sufficient condition (1.4) is a slightly stronger assumption than the corresponding one in [McC76] in that it places no restriction on $C_i d$ for $i \in A_0(x^*)$ having $\lambda_i^* = 0$; it is the appropriate assumption where it is used in §5.

2. Basic unconstrained trust-region algorithm.

It will be useful to have a description of basic unconstrained trust-region algorithms. Such algorithms are iterative and are generally descent algorithms; given current iterate x^c, they seek a new iterate x^+ that gives a better function value, i.e., $f(x^+) < f(x^c)$. To find x^+, these algorithms use some *model*, say $q(s)$, for the function change, $f(x^c+s) - f(x^c)$, that a step s from x^c may achieve. Usually $q(s)$ will only be reliable when $\|s\|$ is sufficiently small, so trust-region algorithms maintain a bound δ on the values of $\|s\|$ for which they deem $q(s)$ reliable. (This bound is often called the *step-bound* or *trust-radius*.) They choose trial iterates x^t of the form $x^t = x^c + s^t$, where s^t is chosen to approximately minimize $q(s)$ subject to the constraint $\|s\| \leq \delta$. Many variations are possible, depending on how one chooses $q(s)$, measures $\|s\|$ (i.e., chooses the norm $\|\cdot\|$), calculates s^t, updates δ, and decides when $f(x^t)$ is sufficiently smaller than $f(x^c)$. This paper often assumes that $\|\cdot\|$ is the Euclidean norm $\|\cdot\|_2$ and that $q(s)$ has the form

(2.1) $$q(s) = g^T s + \tfrac{1}{2} s^T H s,$$

where g is usually the current gradient $\nabla f(x^c)$ and $H = H^c \in \mathbb{R}^{p \times p}$ is either $\nabla^2 f(x^c)$, i.e., the current Hessian, or some approximation to it (perhaps as computed by a secant update, such as the well-known DFP or BFGS updates — see [DenM77] for details and references). Whatever the choice of $\|\cdot\|$ and $q(\cdot)$, q and δ determine a set $S(q,\delta)$ of candidate steps, such as

(2.2) $$S(q,\delta) = \{s \in B(\delta),\ q(s) \leq \eta \cdot \min q(B(\delta))\},$$

where

(2.3) $$B(\delta) = \{s \in \mathbb{R}^p \colon \|s\| \leq \delta\},$$

$0 < \eta < 1$, and $\min q(B(\delta))$ means $\min\{q(s) \colon s \in B(\delta)\}$. The process of finding a suitable next iterate x^t then takes the form

Procedure NEXT_IT:

Repeat {

Compute s^t satisfying

(2.4) $$s^t \in S(q,\delta).$$

Test the inequality

(2.5) $$f(x^c + s^t) - f(x^c) \leq \alpha \cdot q(s^t)$$

and choose δ^+ satisfying

(2.6) $$\begin{cases} \delta \leq \delta^+ \leq \rho_2 \cdot \delta & \text{if (2.5) holds} \\ \rho_0 \|s^t\| \leq \delta^+ \leq \rho_1 \|s^t\| & \text{otherwise.} \end{cases}$$

Set

$$\delta := \delta^+.$$

}

Until (2.5) holds. Accept $x^+ := x^c + s^t$ as the new iterate.

We assume that the constants in (2.6) satisfy

$$0 < \alpha < 1,$$

and

$$0 < \rho_0 < \rho_1 < 1 < \rho_2.$$

See [Gay81] and [MorS81] for discussions of computing s^t satisfying (2.4) with q given by (2.1), S given by (2.2), (2.3), and $\|\cdot\| = \|\cdot\|_2$.

3. Modification for linear inequalities.

When all constraints are linear, as in (1.1), it seems reasonable to insist on preserving feasibility, i.e., to consider only trial iterates x^t that satisfy $Cx^t \geq b$. This paper will not discuss finding an initial feasible x^c (beyond noting that it may, for example, be found by linear programming); we shall simply assume one to be given. An obvious way to modify the trust-region algorithms described in §2 to handle linear inequality constraints is by changing (2.3) into

(3.1) $$B(\delta, x^c) = \{s \in \mathbb{R}^p : \|s\| \leq \delta,\ Cs \geq b - Cx^c\}.$$

If we choose $\|\cdot\| := \|\cdot\|_\infty$, then solving (i.e., finding s^t satisfying) (2.4) when B is given by (3.1) is a quadratic programming problem. This otherwise appealing approach has the drawback that if H is indefinite, then we may have to find a global solution to an indefinite quadratic programming problem to ensure solving

(2.4). This can be an expensive task, but if we do not satisfy (2.4), then we might get limit points x^* that fail to satisfy the first-order necessary conditions (1.3). To preclude such limit points, it suffices to use an extension of Powell's dogleg scheme [Pow70a,b] and to solve versions of (2.4) in which B has the simpler form (2.3), as we shall see below. Doing this when H is indefinite and $\|\cdot\| = \|\cdot\|_\infty$ may require estimating an eigenvector corresponding to the smallest eigenvalue of H. An alternative of primary interest in this paper is to choose $\|\cdot\| := \|\cdot\|_2$; it is still necessary to obtain information about the smallest eigenvalue of H, but this can be done quickly and implicitly, as in [MorS81]. The following is intended mainly for $\|\cdot\| = \|\cdot\|_2$, but we need not assume this choice until §6.

One can use an active-set strategy in an attempt to solve (2.4) with B given by (3.1). The following is one such procedure: starting from $s^{t,0} = \mathbf{0}$ and $k := 1$, repeatedly

(3.2a) choose a subset $A^k \subset A_0(x^c + s^{t,k-1})$ of the tight constraints to be considered active, i.e., to be treated as equality constraints at $x^c + s^{t,k-1}$;

(3.2b) after setting $g := \nabla f(x^c) + Hs^{t,k-1}$ in (2.1), make a linear change of variables so that when the inactive constraints (those whose indices are not in A^k) are ignored, $B(\delta, x^c + s^{t,k-1})$ has the form (2.3);

(3.2c) solve the resulting unconstrained problem (2.4);

(3.2d) translate the solution of (3.2c) back into the original variables, obtaining $s^{t,k}$;

(3.2e) if $s^{t,k}$ is feasible, then use $s^t := s^{t,k}$; otherwise retract $s^{t,k}$ onto $B(\delta, x^c)$, set $k := k+1$, and return to (3.2a).

One way to retract $s^{t,k}$ in (3.2e) is to set

$$s^{t,k} := \zeta s^{t,k} + (1-\zeta)s^{t,k-1}, \tag{3.3}$$

with $\zeta \in [0,1]$ chosen to make $s^{t,k}$ feasible.

Several other parts of procedure (3.2) deserve discussion. First, choosing the set A^k of "active" constraints in (3.2a) may be a nontrivial task, particularly when the tight constraints are degenerate, i.e., when $\{C_i^T: i \in A_0(x^c + s^{t,k-1})\}$ is linearly dependent. In general, to prevent convergence to a nonstationary point, it is necessary to allow dropping constraints, i.e., to allow A^{k+1} to omit an index from A^k. In theory, at least, it is also necessary to guarantee that cycling does not occur when the tight constraints are degenerate. We can avoid much of this complication and arrange to use only nested choices of A^k in (3.2), i.e., to have $A^k \subset A^{k+1}$, by using an extension of Powell's dogleg scheme [Pow70a,b]. The idea is first to compute the "Cauchy step" s^0 defined by

$$s^0 := \arg\min\{\nabla f(x^c)^T s\colon \|s\|_2 \le 1,\ C_i s \ge 0\ \forall\ i \in A_0(x^c)\}, \tag{3.4a}$$

then to run procedure (3.2) with $s^{t,0} := \zeta s^0$, where

(3.4b) $\zeta = \min\{\theta_0\delta, \max\{\tau > 0: \tau Cs^0 \geq b - Cx^c\}, -\theta_1(s^0)^T\nabla f(x^c)/[(s^0)^T Hs^0]\}$,

$\theta_0, \theta_1 \in (0,1]$, and $-(s^0)^T\nabla f(x^c)/[(s^0)^T Hs^0]$ is regarded as $+\infty$ if $(s^0)^T Hs^0 \leq 0$. (We discuss solving (3.4a) and the connection between Powell's dogleg scheme and (3.4) in §7.) As shown in §5, the following scheme allows us to use nested choices of A^k without impairing convergence properties:

(3.5) Let P_0 denote projection onto the orthogonal complement of $\{C_i^T: i \in A_0(x^c)\}$. If $\|P_0\nabla f(x^c)\| > \theta_2\|\nabla f(x^c)\|$, then choose $s^{t,0} := \mathbf{0}$ in (3.2a). Otherwise choose $s^{t,0} := \zeta s^0$ with s^0 and ζ from (3.4). In either case choose $A^1 \subset A_0(x^c)$ arbitrarily, such that $\{C_i^T: i \in A^1\}$ is linearly independent.

This way we must only worry about degeneracy when solving (3.4a). (θ_2 is an arbitrary positive constant. So far as theory is concerned, we could change $\|\cdot\|_2$ to $\|\cdot\|_\infty$ or $\|\cdot\|_1$ in (3.4a), thus converting (3.4a) into a linear programming problem. But this introduces more constraints than does (3.4a) as it stands. Moreover, s^0 is a continuous function of x^c wherever $\nabla f(x^c) \neq \mathbf{0}$. This intuitively appealing property would not hold if $\|\cdot\|_2$ were changed to $\|\cdot\|_\infty$ or $\|\cdot\|_1$.)

When the tight constraints are nondegenerate (see §6), a scheme simpler than (3.4−5) suffices. Let $C_0 = C_0(x^c)$ denote the matrix obtained by dropping from C rows corresponding to the constraints that are slack at x^c, and let C_0^- be a generalized inverse of C_0. Using C_0^-, we obtain a Lagrange multiplier estimate λ for the λ^* of (1.3) by setting

$$\text{(3.6a)} \qquad \lambda_i := \begin{cases} \left(C_0^{-T}\nabla f(x^c)\right)_j & \text{if } C_i \text{ is row } j \text{ of } C_0 \\ 0 & \text{if } i \notin A_0(x^c). \end{cases}$$

We then set

$$\text{(3.6b)} \qquad A^1 := \{i: \lambda_i > 0\}.$$

(3.6a) is the general first-order Lagrange multiplier estimate described in §4 of [GilM77]. The most intuitively appealing choice for λ is probably the least-squares estimate

$$\text{(3.7)} \quad \lambda = \lambda(x^c) := \arg\min\{\|\nabla f(x^c) - \sum_i \lambda_i C_i^T\|_2: \lambda_i = 0 \text{ unless } i \in A_0(x^c)\},$$

but the convergence theory of §5 allows any fixed choice of C_0^- for each possible $A_0(x^c)$.

There are various ways to handle the linear algebra called for in (3.2b) and (3.2d). This detail is discussed briefly in §§5 and 6.

A point glossed over in (3.2c) is what step bound should be used in the "unconstrained" step computation. One possibility would be to modify this computation to enforce a bound of the form $||s+s^{t,k-1}|| \le \delta$ instead of $||s|| \le \delta$. Though this is straightforward, it is simpler either to require $||s|| \le \delta - ||s^{t,k-1}||$ in (3.2c) or to use (3.3) to enforce both feasibility and the step bound.

The choice of $S(q,\delta)$ can have a profound effect on (3.2). A choice of the form (2.2), (3.1) may sometimes require several iterations of (3.2), even with the simplifications proposed in §4. Because of this, some authors in effect suggest stopping (3.2) after one iteration, i.e., taking $s^t := s^{t,1}$. A primary point of this paper is that appropriately taking more iterations of (3.2) has several advantages. First, if one stops after just one iteration, then it may be necessary to introduce anti-zigzagging rules, and rate-of-convergence results generally require assuming the constraints to be nondegenerate at x^*. The scheme sketched in §4 avoids these complications. Second, stopping (3.2) after one iteration may lead to performing more function evaluations — to evaluating f and ∇f at more points. In cases where such evaluations are expensive, it seems better to spend more time on choosing s^t in trade for fewer evaluations of f and ∇f. The scheme proposed in §4 attempts this trade.

4. Approximation.

Each of the schemes described next attempts to approximate

$$s^* := \arg\min\{q(s) \colon s \in B(\delta, x^c)\}$$

in a simple way. There are four parts to the schemes:

(4.1a) If the binding constraints are degenerate (see §6), then choose A^1 by (3.5). Otherwise use either (3.5) or (3.6).

(4.1b) When $k > 1$ in (3.2a), obtain A^k by adjoining to A^{k-1} one of the hitherto inactive constraints encountered in (3.2e). This ensures that $A^{k-1} \subset A^k$.

(4.1c) Either use step bound $\delta - ||s^{t,k-1}||$ in (3.2c) or modify (3.2e) so that it also enforces the step bound $||s^{t,k}|| \le \delta$.

(4.1d) Use (3.3) in (3.2e).

When (3.5) is used in (4.1a), it seems reasonable to let σ and γ denote the Lagrange multipliers associated with (3.4), so that $\nabla f(x^c) = C^T\gamma + \sigma s^0$, and to choose $A^1 := \{i \colon \gamma_i > 0\}$. Otherwise, if the binding constraint normals $\{C_i^T \colon i \in A_0(x^c + s^{t,0})\}$ are linearly independent (i.e., the binding constraints are nondegenerate), it seems reasonable to compute a least-squares Lagrange multiplier estimate λ and to choose $A^1 := \{i \colon \lambda_i > 0\}$. Many other choices are possible, such as the general (3.6a) and the second-order Lagrange multiplier estimates in §4 of [GilM77], but for the convergence results of §5 to hold, it is only necessary that $A^1 \subset A_0(x^c + s^{t,0})$. Even choosing $A^1 := \emptyset$ is allowed, but a more informed choice may lead to fewer iterations of (3.2).

5. Convergence results.

The results in [Gay82] extend readily to the linearly constrained problem. Specifically, we have

LEMMA 1. *Suppose q is given by (2.1) and let (3.2), (3.3), and (4.1) be used in (2.4). If* $f\colon \mathbb{R}^p \to \mathbb{R}$ *is continuously differentiable, if* $x^c \in \mathbb{R}^p$ *is feasible, i.e.,* $Cx^c \geq b$, *but does not satisfy the first-order necessary conditions (1.3), then for any initial* $\delta > 0$, *(2.5) is satisfied after finitely many repetitions of the Repeat loop of Procedure NEXT_IT.*

Proof sketch. First note that (3.2) eventually produces a feasible, nonzero $s^{t,k} =: s^t$. This is clear if $P_0 \nabla f(x^c) \neq \mathbf{0}$, i.e., if (1.3b) cannot hold (for $x^* := x^c$), or if (3.5) is used. If $P_0 \nabla f(x^c) = \mathbf{0}$ and (3.6) is used, on the other hand, then (3.6a) uniquely determines λ and (1.3b) holds. By construction (1.3a) and (1.3d) also hold (for $x^* := x^c$ and $\lambda^* := \lambda$), so we must have $\lambda_i < 0$ for at least one i. Either (3.2) produces a nonzero $s^{t,k}$, or it continues long enough that exactly one such i is not in A^k at (3.2a). Let P_k denote projection onto the orthogonal complement of $\{C_i^T : i \in A^k\}$. At this point, $\lambda_i < 0$ implies that $P_k \nabla f(x^c)$ is a negative multiple of $P_k C_i^T$. Hence, the $s^{t,k}$ obtained in (3.2d) will have $C_i s^{t,k} \geq 0$, so i will not be adjoined to A^k. It is still possible that $C_j s^{t,k} < 0$ for some $j \in A_0(x^c)$ with $j \neq i$, so (3.2) may continue for a few more cycles (at most $p-1$), but eventually a feasible, nonzero $s^{t,k}$ will be produced. Either this $s^t := s^{t,k}$ satisfies (2.5) or Procedure NEXT_IT reduces δ until a feasible, nonzero $s^{t,k}$ is produced with $A^k \subset A_0(x^c)$. The proof of lemma 1 of [Gay82] may now be applied for each possible $A^k \subset A_0(x^c)$ that can deliver s^t in (3.2e). □

THEOREM 2. *Suppose* $f\colon \mathbb{R}^p \to \mathbb{R}$ *is continuously differentiable, and that the sequence* $x^1, x^2, \cdots$ *is generated as in lemma 1. (a) If*

$$\inf\{f(x^k)\} > -\infty$$

then either

$$\lim_{k \to \infty} \|x^{k+1} - x^k\| = 0$$

or else for all positive ϵ_S *and* δ°,

$$\max\{q(s)\colon s \in B(\delta^\circ, x^k)\} \leq \epsilon_S$$

is satisfied infinitely often. Moreover, (b) if $\|H\|$ *in (2.1) is locally bounded, i.e., if each* $x \in \mathbb{R}^p$ *has an open neighborhood* $N \subset \mathbb{R}^p$ *and a bound* $\beta = \beta(x)$ *such that* $\|H\| \leq \beta(x)$ *whenever* $x^c \in N$, *then any limit point* x^* *of* $x^1, x^2, \cdots$ *is a critical point of f, i.e., (1.3) holds.*

Proof sketch. *(a)* is proved by a straightforward generalization of the corresponding proof in [Gay82].

(b) Suppose H is bounded and let x^* be a feasible point that does *not* satisfy (1.3). If x^c is close enough to x^* and δ is small enough, then any A^k determined in (4.1) satisfies $A^k \subset A_0(x^*)$. Extending the corresponding proof in [Gay82], we find first that there exist $\zeta_1 > 0$ and $\xi_1 > 0$ such that any trial step s^t computed from x^c with $\|x^c - x^*\| < \zeta_1$ and $\delta < \zeta_1$ satisfies (2.5) and gives

$$f(x^c) - f(x^+) \geq \xi_1 \delta,$$

where $x^+ = x^c + s^t$. We then find that there exists $\xi_2 > 0$ such that if step s^t is computed from x^c and δ satisfying $\|x^c - x^*\| < \zeta_1$ and $\delta \geq \zeta_1$, and if s^t satisfies (2.5), then $x^+ = x^c + s^t$ satisfies

$$f(x^c) - f(x^+) \geq \xi_2.$$

The same reasoning as in [Gay82] then shows that x^* has a neighborhood that can contain at most finitely many iterates x^k, whence x^* is not a limit point of x^1, x^2, $\ldots$. □

In view of theorem 2, the discussion in [Gay82] of stopping tests in the context of trust-region algorithms for (low dimensional) unconstrained optimization algorithms extends immediately to linearly constrained optimization as in lemma 1. I regard this as an advantage, since the resulting stopping tests should be both easy to interpret and informative. They should, for example, wave a red flag (at least sometimes) when the problem giving rise to f is poorly specified.

Another advantage of the approach in §§3−4 is that, under reasonable conditions, it does not impair fast local convergence. Theorem 3 below illustrates this for the case $H := \nabla^2 f$. It is helpful at this point to expand a bit on (3.2b−d). Let $n := |A^k|$. The change of variables called for in (3.2b) may be done as in [GilM74]: find a matrix $Z \in \mathbb{R}^{p \times n}$ whose columns span the orthogonal complement of $\{C_i^T : i \in A^k\}$; let the unconstrained problem of (3.2c) have step bound δ and gradient $\tilde{g}$, model Hessian $\tilde{H}$, and norm $\|\cdot\|_{\tilde{}}$ given by

(5.1a) $$\tilde{g} := Z^T \nabla f(x^c),$$

(5.1b) $$\tilde{H} := Z^T H Z,$$

(5.1c) $$\|y\|_{\tilde{}} := \|Zy\|.$$

If $\tilde{s}$ is the step computed for (3.2c), then

(5.1d) $$s^{t,k} := Z\tilde{s}.$$

In the following theorem, we require that $\tilde{s} := -\tilde{H}^{-1}\tilde{g}$ whenever $\tilde{H}$ is positive definite and $\|\tilde{H}^{-1}\tilde{g}\|_{\tilde{}} \leq \delta$.

THEOREM 3. *Suppose x^* is a strong local minimizer, i.e., satisfies (1.3) and*

(1.4), and that $\nabla^2 f$ exists everywhere and is Lipschitz continuous. There exists $\zeta > 0$ such that if $\|x^0 - x^*\| < \zeta$ and the sequence $x^1, x^2, \cdots$ is generated from x^0 as in lemma 1 and with (3.c–e) done as described in the preceding paragraph, using $H := \nabla^2 f(x^c)$, then this sequence converges Q-quadratically to x^*.

Proof sketch. Let λ^* be such that (1.3) holds. Without losing generality, we may assume that $A^* := \{i: \lambda_i^* > 0\}$ is such that $\{C_i^T: i \in A^*\}$ is a linearly independent set. Moreover, there exists $\zeta_1 > 0$ such that whenever $x^c \in B(\zeta_1, x^*)$ and $\delta \geq 2\|x^c - x^*\|$, the s^t given by (3.2) satisfies $A^* \subset A_0(x^c + s^t)$. After finitely many iterations, δ is not further decreased and all further sets A^k considered in (3.2) could satisfy $A^* \subset A^k$. Because of degeneracy and because there may be one or more $i \in A_0(x^*)$ with $\lambda_i^* = 0$, these A^k may not be unique, but a simple extension of the proof of local quadratic convergence for Newton's method applied to $\tilde{f}(\hat{s}) := f(x^c + s^{t,k-1} + Z\hat{s})$ for the final k considered in (3.2c) shows (for small enough ζ) that

$$\|x^+ - x^*\| \leq \xi \cdot \|x^c - x^*\|^2$$

for some $\xi \in \mathbb{R}$ independent of x^c. (The simple extension is that instead of $\bar{s} = -[\nabla^2 \tilde{f}(\mathbf{0})]^{-1} \nabla \tilde{f}(\mathbf{0})$, we have

$$\hat{s} = -(Z^T H Z)^{-1} Z^T \left[\nabla f(x^c) + H s^{t,k-1}\right],$$

where $H = \nabla^2 f(x^c)$. But

$$\|\nabla \tilde{f}(\mathbf{0}) - Z^T \left[\nabla f(x^c) + H s^{t,k-1}\right]\| = O(\|x^c - x^*\|^2)$$

and

$$\max\{\|\nabla^2 \tilde{f}(\mathbf{0}) - Z^T \nabla^2 \tilde{f}(x^*) Z\|, \|Z^T H Z - Z^T \nabla^2 \tilde{f}(x^*) Z\|\} = O(\|x^c - x^*\|).$$

Because of (1.4), both $\nabla^2 \tilde{f}(\mathbf{0})$ and $Z^T H Z$ are positive definite and $\|[\nabla^2 \tilde{f}(\mathbf{0})]^{-1} \nabla \tilde{f}(\mathbf{0}) + \hat{s}\| = O(\|x^c - x^*\|^2)$ for x^c near x^*.) □

As with general nonlinear equations, many refinements are possible. For example, if $\nabla^2 f$ is Hölder continuous of order α, i.e., $\|\nabla^2 f(x) - \nabla^2 f(y)\| \leq \|x - y\|^\alpha$, then we get local convergence of Q order $1 + \alpha$, and if $\nabla^2 f(x)$ is just continuous at x^*, then we get local superlinear convergence. There are also analogs of the Kantorovich theorem.

A weakness of Theorem 2 — and of the approach described in §§2–4 — is that when there are zero Lagrange multipliers, they do not exclude convergence to a stationary point at which the second-order necessary condition (1.5) fails. The trouble is that there could be one or more indices i for which $\lambda_i(x^c)$ is always positive in (3.7) but $\lambda_i^* = 0$ in (1.3). Because of this, the "inf" in (1.5) may be over a larger set than the algorithm is ever allowed to see. In other words, (3.2) with (4.1) may eventually fail their goal of satisfying (2.4) when convergence to a critical point having zero Lagrange multipliers occurs. This defect could be overcome by modifying (3.2) and (4.1) so that they globally minimize the model function q in the constrained trust region (3.1). Such modification could considerably complicate — and sometimes considerably slow down — the step-computing process. In practice I expect it is best not to worry about zero (and near-zero) Lagrange multipliers until a convergence test is satisfied. Then one could do the work necessary to globally minimize the current model q on a region $B(\delta^\circ, x^c)$ of "reasonable" prescribed radius δ°; if this delivers a trial step s^t for which $q(s^t)$ is sufficiently less than zero, then one could resume procedure NEXT_IT with $\delta := \delta^\circ$ and with (3.2) modified to compute globally optimal trust-region steps. After realizing a sufficient decrease in f, one could continue iterating with the global trust-region step optimization turned back off. In this way one can construct an algorithm such that if f is twice continuously differentiable, $H = \nabla^2 f(x^c)$, and τ is an arbitrary positive tolerance, then any limit point x^* of the iterates satisfies not only the first-order necessary condition (1.3), but also comes close to satisfying the second-order necessary condition (1.5) in the sense that

(5.2) $\inf\{d^T \nabla^2 f(x^*) d \colon \|d\| = 1,\ C_i d = 0 \text{ for all } i \in A_0(x^*) \text{ with } \lambda_i^* > 0\} \geq -\tau.$

6. Linear algebra.

Assume now that $\|\cdot\| = \|\cdot\|_2$. It is most convenient to use a Z in (5.1) that has orthonormal columns, so that all instances of (3.2) use the same norm, $\|\cdot\| = \|\cdot\|_2$. This is also the best conditioned choice of Z — see [GilM74, §2.8] and [GilMW81, §5.1.3].

For computing least-squares Lagrange multiplier estimates (for choosing A^1), it is convenient to use a QR factorization of the active constraints, as in [GilM77]. Thus we maintain an upper triangular matrix $R \in \mathbb{R}^{n \times n}$ (where $n = |A^k|$ is the number of currently active constraints) and orthogonal matrices $Q \in \mathbb{R}^{p \times n}$ and (as above) $Z \in \mathbb{R}^{p \times (p-n)}$ such that $\{C_i^T \colon i \in A^k\}$ is the set of columns in QR and $Z^T Q = \mathbf{0}$. It is straightforward to update this information using orthogonal transformations, either elementary reflectors (Householder transformations) or plane (Givens) rotations, as in §5 of [GilGMS74]; for illustration, we will use the latter. Briefly, when going from A^k to $A^{k+1} = A^k \cup \{i\}$, we find a sequence of plane rotations $G_1, \ldots, G_\mu$, where $\mu = p-n-1$, such that $G_\mu \cdots G_1 Z^T C_i^T$ is ρ times the first standard unit vector (i.e., has ρ in its first component and zeros in its remaining components). We adjoin to R the column whose first n components are $Q^T C_i^T$ and whose $n+1$st component is ρ; we adjoin the first column of

$$Z G_1^T \cdots G_\mu^T \tag{6.1}$$

to Q and take the remaining columns of (6.1) as the new Z.

If we solve (2.4) with B from (2.3) as in [Gay81] and [MorS81], then we obtain $\alpha \geq 0$ and a Cholesky factor (lower triangular matrix) L such that

$\tilde{H} + \alpha I = LL^T$ is positive definite and

$$(\tilde{H} + \alpha I)\tilde{s} = -\tilde{g}, \tag{6.2}$$

where $\tilde{g}$, $\tilde{H}$, and $\tilde{s}$ are given by (5.1a,b,d). It seems reasonable to augment (4.1) to speed up the linear algebra when (3.2e) returns to (3.2a). The idea is to use the same value of α in (3.2c) for as many successive values of k as possible. When (3.2e) encounters a constraint (i.e., $s^{t,k}$ is infeasible), we update Q, R, and Z as above and determine plane rotations $\tilde{G}_1, \ldots, \tilde{G}_\mu$ such that

$$G_\mu \cdots G_1 L \tilde{G}_1 \cdots \tilde{G}_\mu \tag{6.3}$$

is lower triangular. The leading principal $\mu \times \mu$ submatrix of (6.3) is a Cholesky factor of $\tilde{H} + \alpha I$ for the new $\mu \times \mu$ $\tilde{H}$. Occasionally this will result in a new candidate step $s^{t,k+1}$ that is too long, in which case we can either choose a new α or simply set

$$s^{t,k+1} := \zeta s^{t,k+1} + (1-\zeta)s^{t,k} \tag{6.4}$$

for the appropriate $\zeta \in (0,1)$. Unless a new α is chosen, this device reduces from $O(\mu^3)$ to $O(\mu^2)$ the number of arithmetic operations expended when (3.2) increments k. Since $\alpha = 0$ for Newton steps, this device has no effect on Theorem 3. (The other results in §5 are unaffected as well.)

(If we used an elementary reflector rather than plane rotations in (6.1), then (6.3) would become $G_1 L \tilde{G}_1 \cdots \tilde{G}_{2\mu-1}$, where G_1 is the elementary reflector and the $\tilde{G}_k$ are still plane rotations. We could also use elementary reflectors throughout and proceed as in method 2 of [Gol76].)

When H is a secant approximation to $\nabla^2 f$, such as the BFGS approximation, I like to use the double dogleg scheme of Dennis and Mei [DenM79] (a modification of Powell's dogleg scheme), rather than a computation of the form (6.2). Not only does the dogleg scheme require less linear algebra per step than (6.2), but it also led to slightly fewer function and gradient evaluations both in a comparison I made several years ago and in the comparison reported in §7 of [SchKW82]. For the (double) dogleg it is convenient to maintain not H itself, but rather a Cholesky factor L of H (updated, say, as in [Gol76]). In the present context we would maintain a Cholesky factorization of $[Z{:}Q]^T H [Z{:}Q]$ rather than of H. This can be updated as in (6.3) when A^k changes. Many authors recommend maintaining just (a factorization of) $Z^T HZ$, arguing that when a new column is added to Z, one can replace $Z^T HZ$ with $\begin{bmatrix} Z^T HZ & 0 \\ 0 & \eta \end{bmatrix}$, with $\eta = 1$ unless better information is available, then gain correct information in the new last row and column from subsequent updates. Powell [Pow72] presents some theoretical support of the claim that this costs at most one extra iteration per column added to Z. Contrary to this popular view and aside from equality constraints (i.e., constraints guaranteed never to become inactive), I prefer to update the entire Hessian approximation

$[Z{:}Q]^T H [Z{:}Q]$, so that good Hessian information is immediately available. On problems where f is expensive to evaluate, the extra expense of updating the entire Hessian approximation may be more than repaid by saving even a single function evaluation. Another reason for this preference is that in practice I use a more elaborate rule than (2.6) to update the trust radius, one that sometimes asks how well the predicted change in gradients agrees with the actual one — see [DenGW81] for details. If an approximation to the entire Hessian is available, then this radius update can proceed as though there were no constraints.

In §8.6 of [Fle81], Fletcher points out that when there is degeneracy, roundoff error can lead one to choose a "singular" active set, an A^k for which $\{C_i^T: i \in A^k\}$ is (or is nearly) linearly dependent. We can generally avoid this danger by using fast condition estimators, such as those described in [CliMSW79] and [CliCV82]. When contemplating the choice $A^{k+1} := A^k \cup \{j\}$, we check whether this would result in R having too high a condition number; if so, then we deem C_j^T to be linearly dependent on $\{C_i^T: i \in A^k\}$.

When f arises from a nonlinear least-squares problem, it is sometimes desirable to compute Levenberg-Marquardt steps from the "Gauss-Newton" model, as in [Mor78]. In effect, we are not directly given $\nabla f(x^c)$ and H, but rather a Cholesky factor L of $H = LL^T$ and the quantity $L^{-1}\nabla f(x^c)$. Instead of explicitly computing $g := \nabla f(x^c) + Hs^{t,k-1}$ in (3.2b), we would therefore add $L^{-1}Hs^{t,k-1} = L^T s^{t,k-1}$ to the given $L^{-1}\nabla f(x^c)$ and feed this sum to the Levenberg-Marquardt procedure. (This is done in the subroutine NL2SLB mentioned in §8.)

7. Dogleg steps.

We first discuss solving (3.4a), then describe the connection between (3.4) and Powell's dogleg scheme.

Instead of solving (3.4a) directly, we could solve a related problem,

$$\tilde{s}^0 := \arg\min\{\|s\|_2: \nabla f(x^c)^T s = -1,\ C_i s \geq 0\ \forall\ i \in A_0(x^c)\}. \tag{7.1}$$

If (7.1) has no solution, i.e., if the constraints in (7.1) are inconsistent, then $s^0 = \mathbf{0}$ solves (3.4a); otherwise $s^0 = \tilde{s}^0/\|\tilde{s}^0\|_2$ solves (3.4a). By changing $\|s\|_2$ to $\|s\|_2^2$ in (7.1), we could convert (3.4a) into a quadratic programming problem. Though we could then apply a general quadratic programming algorithm, it may be more efficient to a use special-purpose algorithm, such as Algorithm LDP in [LawH74, ch. 23].

It may be still more efficient to attack (3.4a) directly by the following algorithm (though this remains to be seen). The algorithm is just Rosen's gradient projection algorithm [Ros60], modified to handle the additional constraint $\|s\|_2 \leq 1$. It employs a current active set A^k and uses orthogonal factorizations of the "active" portion of C. As before, we require $\{C_i^T: i \in A^k\}$ to be linearly independent, and we assume that Q, R, and Z correspond to A^k as in §6. In this section, g denotes $\nabla f(x^c)$.

Given A^k and a feasible approximate solution s^k, the algorithm determines a better s^{k+1} and corresponding A^{k+1}. It starts with an arbitrary A^1 and with a feasible s^1 having $C_i s^1 = 0$ for all $i \in A^1$. (We could choose $A^1 := \emptyset$, but it is probably more efficient to make A^1 as large as possible. The most convenient choice for s^1 is $\mathbf{0}$.) Given A^k, it checks whether the trial step

$$(7.2) \qquad \bar{s}^k := \begin{cases} \mathbf{0} \text{ if } Z \text{ is null or } Z^T g = \mathbf{0} \\ -\dfrac{ZZ^T g}{\|ZZ^T g\|_2} \text{ otherwise} \end{cases}$$

is feasible, i.e., whether $C_i \bar{s}^k \geq 0$ for all $i \in A_0(x^c)$. (Z is null if $|A^k| = p$.) If not, then we set $s^{k+1} := \zeta \bar{s}^k + (1-\zeta)s^k$, with $\zeta \in [0,1)$ as large as possible such that s^{k+1} is feasible; after setting $A^{k+1} := A^k \cup \{i\}$, where i is such that the constraint $C_i s \geq 0$ determines ζ, we step k and return to (7.2). If $\bar{s}^k$ is feasible, then we set $s^{k+1} := \bar{s}^k$ and check whether this step is optimal, i.e., whether

$$(7.3) \qquad R^{-1}Q^T g \geq \mathbf{0}.$$

(The Lagrange multiplier corresponding to the constraint $\|s\|_2 \leq 1$ is just $\|Z^T g\|_2 > 0$ when this constraint is active, so it suffices to check (7.3).) If (7.3) holds, then $s^0 := \bar{s}^k$ solves (3.4a); otherwise we must alter A^k. At this point, it may be necessary to resolve degeneracy as in linear programming — we would probably only resort to (3.4) if the constraints were degenerate. To do so, we either determine optimality or find a feasible direction by proceeding as though solving the linear programming problem

$$(7.4) \qquad \begin{array}{c} \text{minimize } g^T s \text{ subject to } (\bar{s}^k)^T s = 0 \text{ and} \\ C_i s \geq 0 \ \forall\, i \in A_0(x^c) \text{ with } C_i \bar{s}^k = 0. \end{array}$$

That is, we select some j with

$$(7.5) \qquad \left(R^{-1}Q^T g\right)_j < 0,$$

note the index i such that C_i^T is column j of QR, and check whether there is an index $h \in A_0(x^c)$ such that $C_h \bar{s}^k = 0$ and $d := QR^{-T}e_j$ has $C_h d < 0$. If so, then it is necessary to exchange i and h in A^k and return to (7.5). Otherwise we can set $A^{k+1} := A^k - \{i\}$ (i.e., drop i from A^k), step k, and return to (7.2). (After changing A^k, there may be no j for which (7.5) holds. In this case $s = \mathbf{0}$ solves (7.4) and $s^0 := \bar{s}^k$ solves (3.4a) as before.) To prevent cycling — returning to (7.5) infinitely often (without stepping k) — it is simplest to use a Bland rule [Bla77] to select j and h (and not much more difficult to use Wolfe's simplified lexicographical rules [Wol63] to select h after an arbitrary choice of j). At any rate, the conventional wisdom from linear programming suggests that cycling would rarely be a problem in practice.

The s^k produced by the algorithm are such that, once $s^k \neq \mathbf{0}$,

$$\frac{g^T s^k}{\|s^k\|_2} \geq \frac{g^T s^{k+1}}{\|s^{k+1}\|_2}, \tag{7.6}$$

and they are always feasible, i.e., always satisfy

$$C_i s^k \geq 0 \;\forall\; i \in A_0(x^c).$$

Moreover, whenever the algorithm returns to (7.2) after checking (7.3), we have strict inequality in (7.6).

A reasonable A^1 can be obtained from an attempt (foiled by linear dependence of $\{C_i^T : i \in A(x^c)\}$) at solving (3.6). The final A^k from this section makes a reasonable choice for the A^1 of (3.5).

For a positive-definite H, Powell's dogleg scheme works as follows. Let s^C denote the steepest-descent step:

$$s^C := -\frac{g^T g}{g^T H g} \cdot g.$$

Consider the piecewise-linear path from x^c to the "Cauchy point" $x^c + s^C$ and thence to the "Newton point" $x_c + s^N = x^c - H^{-1}g$. If the Newton point lies within the trust region — the sphere of radius δ centered at x^c — then the Newton point is the new trial iterate. Otherwise the point where the trust region intersects the path is the new trial iterate. Scheme (3.5) is similar to Powell's dogleg scheme in that it considers a path that starts with a line segment from x^c to a relaxed "Cauchy point", i.e., $x^c + \zeta s^0$, and goes from there along a curvilinear route to a (constrained) Newton point. The rules are the same: if the Newton point lies within the trust region, then it is the next trial iterate; otherwise the intersection of the path and the trust region is the next trial iterate. Figure 7.1 illustrates the similarity between Powell's dogleg scheme and (3.5) when the $s^{t,1}$ of (3.2d) is feasible. (In Figure 7.1, $x^c = \mathbf{0}$ for simplicity, and the θ_0 and θ_1 of (3.4b) are both 1.)

8. Computational experience with simple bounds.

Perhaps the most common constraints are simple bounds on the variables (often just nonnegativity constraints). Thus it seems worthwhile to consider the special case where b and C have the form

$$b = \begin{bmatrix} -\bar{b} \\ \underline{b} \end{bmatrix} \text{ and } C = \begin{bmatrix} -I \\ I \end{bmatrix},$$

so that the constraints reduce to $\underline{b} \leq x \leq \bar{b}$. This case is considerably simpler

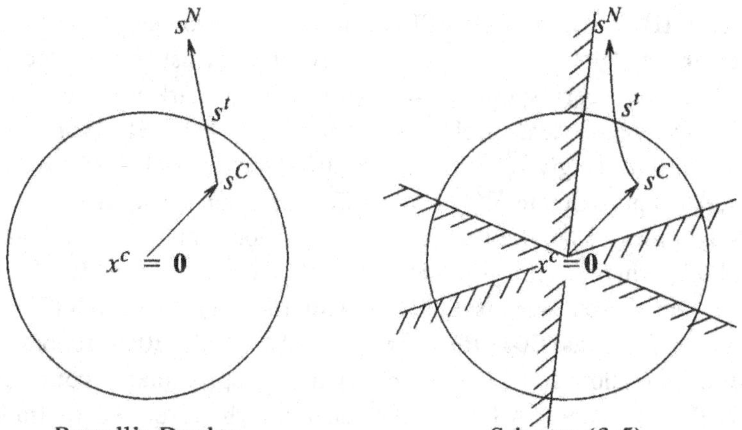

Powell's Dogleg
(unconstrained problem)

Scheme (3.5)
(degenerate constraints)

Figure 7.1

than the general one, in that degeneracy is not an issue and the orthogonal matrices Z and Q of §6 reduce to permutation matrices. Indeed, Gill and Murray [GilM76] claim this case is so simple that separate algorithms for the unconstrained case are unnecessary.

I have so far experimented with the local active-set strategy of §§3 and 4 in three contexts: (*i*) H given explicitly; (*ii*) H approximated by the BFGS secant update; and (*iii*) nonlinear least squares, in which there are two choices of H. In context (*iii*), where f has the form $f(x) := \frac{1}{2} r(x)^T r(x)$, I use the two choices of H described in [DenGW81]: either $H = r'(x)^T r'(x)$ or $H = r'(x)^T r'(x) + S$, where $r'(x)$ is the Jacobian matrix corresponding to the residual vector $r(x)$ and S is a secant-updated approximation to the second-order part of the least-squares Hessian. In all these contexts I use (4.1) with A^1 chosen by (3.6) and use step bound $\delta - \|s^{t,k-1}\|$ in (3.2c). In context (*ii*) I use the double-dogleg scheme [DenM79] to compute s^t, and for (*i*) and (*iii*) I use (6.2) as described in §6. (This includes reusing α after encountering a constraint. When this yields too long a candidate step $s^{t,k+1}$, it appears marginally better, as measured by function and gradient evaluations, to recompute α than to use (6.4); α is thus recomputed in the results reported below.)

The computer codes for contexts (*i*) and (*ii*) are modifications, called HUMSLB and SUMSLB, of the subroutines HUMSL and SUMSL described in [Gay83]. The code for context (*iii*), NL2SLB, is a modified form of NL2SOL [DenGW81]. To test these codes, I have run them on the test problems described in [MorGH81], with arbitrarily chosen simple bounds (generally chosen to have at least one bound active at any solution). Since such details might be useful to other investigators, the appendix gives the bounds I used and a summary of the results I obtained.

To see how HUMSLB and SUMSLB compare with conventional codes that use line searches and a "global" active set strategy (in contrast to the "local" strategy of §§3−6), I ran two corresponding subroutines from mark 9 of the NAG library [NAG82] on the same test problems. These are E04LBF (corresponding to HUMSLB in that it is given $\nabla^2 f$) and E04KBF (which, like SUMSLB, uses a secant update to approximate $\nabla^2 f$). With E04KBF one must choose between two line search routines, E04JBQ, which uses function values only, and E04LBS, which uses both function and gradient values. In what follows, E04KBF-Q ("Q" for quadratic line search) means E04KBF with E04JBQ and E04KBF-C ("C" for cubic line search) means E04KBF with E04LBS. With E04KBF one must also decide whether to allow a "local search" when (approximate) convergence to a questionable point occurs. Allowing the local search never led to finding a different "solution" but sometimes cost more function and gradient evaluations; for the comparisons reported below, the local search was turned off.

Table 8.1 gives some statistics comparing HUMSLB or SUMSLB with the NAG routines on those test problems where both found the same solution (as judged by either agreement in $f(x^{\text{final}})$ values to six decimal places or by both $f(x^{\text{final}})$ values being less than 10^{-14}; note that all test problems in [MorGH81] are sums of squares, so $f(x) \geq 0$ for all these problems). Table 8.1 compares — separately — the numbers of function and gradient evaluations required by routines "A" and "B" to solve the number of test problems specified in the column labeled "No. of Runs". Statistics are given both for the unconstrained problems of [MorGH81] and for the same problems with the simple bounds specified in the appendix: the column labeled "Bounds" has "No" for no constraints and "Yes" for simple bounds. The statistics given are for

$$(N_A - N_B)/\min\{N_A, N_B\}; \tag{8.1}$$

the column labeled "N" tells whether N_A and N_B are the number of function or gradient evaluations for routine "A" and "B" respectively: "F" means function evaluations, "G" means gradient evaluations. A positive value of (8.1) means routine "B" did better than routine "A" as measured by either function or gradient evaluations.

Computing for Table 8.1 was done in single precision on a Cray 1 computer. Stopping tests for HUMSLB and SUMSLB were as described in [DenGW81] and [Gay83], using the default tolerances. Stopping tests for E04KBF and E04LBF were for the default value provided when parameter XTOL is set to zero; the line search tolerance ETA for these routines was set to the recommended value of 0.9, and the STEPMX parameter was 10^5 — see the NAG documentation [NAG82] for details on these parameters.

(Table 8.1 includes a comparison of E04KBF-Q and E04KBF-C. The unconstrained version of this comparison supports a claim in the NAG documentation that E04KBF-C should be preferred to E04KBF-Q unless a gradient evaluation

Table 8.1

A	B	No. of Runs	Bounds	N	$(N_A - N_B)/\min\{N_A, N_B\}$: Mean	St.Dev.	Min.	Max.	Median
E04LBF	HUMSLB	50	Yes	F	−0.06	0.62	−2.67	1.29	0.00
				G	0.43	0.79	−1.59	2.60	0.26
		44	No	F	0.17	1.09	−1.30	4.58	−0.09
				G	0.59	1.39	−1.00	5.70	0.13
E04KBF−Q	SUMSLB	43	Yes	F	1.41	1.11	−0.14	5.85	1.32
				G	0.42	0.65	−0.72	3.00	0.32
		35	No	F	1.00	0.92	−1.00	3.85	0.86
				G	0.11	0.50	−1.33	1.42	0.00
E04KBF−C	SUMSLB	42	Yes	F	0.25	0.65	−1.16	2.63	0.08
				G	0.70	0.83	−0.63	3.08	0.43
		35	No	F	−0.06	0.53	−1.40	1.03	−0.07
				G	0.27	0.56	−0.95	1.60	0.14
E04KBF−Q	E04KBF−C	44	Yes	F	0.94	0.31	0.18	1.73	0.93
				G	−0.18	0.23	−1.00	0.22	−0.12
		37	No	F	0.89	0.35	−0.10	1.52	0.95
				G	−0.22	0.35	−1.87	0.18	−0.13

costs more than about four times a function evaluation.)

The NAG stopping tests and those used in HUMSLB and SUMSLB often seemed to perform comparably, but since they are different, one must take the statistics in Table 8.1 with something of a grain of salt. For what it is worth, Table 8.1 suggests that a trust-region approach combined with a local active set strategy compares favorably — in terms of function and gradient evaluations — with the conventionally recommended approach to low dimensional optimization, with or without simple bounds on the variables.

9. Concluding remarks.

This paper suggest a class of trust-region algorithms for solving low dimensional linearly constrained optimization problems that have a continuously differentiable objective function. The algorithms use a "local" active-set strategy that needs no anti-zigzagging precautions and that allows using the same diagnostic stopping tests (advocated in [Gay82]) as for unconstrained optimization. Computational results in the special but common case of simple bounds on the variables are encouraging.

It is not unreasonable to combine a conventional "global" active-set strategy with trust-region algorithms (in the way Danny Sorensen described in talks at the

1982 SIAM National Meeting and the XI. International Symposium on Mathematical Programming). It will be interesting to see how this compares computationally with the approach of this paper — both on the simply bounded problems considered here and on problems with general linear constraints.

10. References.

[Bla77] Bland, R. G. New finite pivoting rules for the simplex method. *Math. Oper. Res.* **2** (1977), 103–107.

[CliCV82] Cline, A. K., Conn, A. R. and Van Loan, C. F. Generalizing the LINPACK condition estimator. In *Numerical Analysis*, edited by J. P. Hennart, Lecture Notes in Mathematics 909, Springer-Verlag, Berlin, Heidelberg and New York, 1982.

[CliMSW79] Cline, A. K., Moler, C. B., Stewart, G. W. and Wilkinson, J. H. An estimate for the condition number of a matrix. *SIAM J. Numer. Anal.* **16** (1979), 368–375.

[DenGW81] Dennis, J. E., Gay, D. M. and Welsch, R. E. An adaptive nonlinear least-squares algorithm. *ACM Trans. Math. Software* **7** (1981), 348–368.

[DenM79] Dennis, J. E. and Mei, H. H-W. Two new unconstrained optimization algorithms which use function and gradient values. *J. Optimization Theory Appl.* **28** (1979), 453–482.

[DenM77] Dennis, J. E. and Moré, J. J. Quasi-Newton methods, motivation and theory. *SIAM Rev.* **19** (1977), 46–89.

[Fle81] Fletcher, R. *Practical Methods of Optimization Vol. 2, Constrained Optimization*, Wiley, Chichester and New York, 1981.

[Gay81] Gay, D. M. Computing optimal locally constrained steps. *SIAM J. Sci. Statist. Comput.* **2** (1981), 186–197.

[Gay82] Gay, D. M. On Convergence Testing in Model/Trust-Region Algorithms for Unconstrained Optimization. Computing Science Technical Report No. 104, Bell Laboratories.

[Gay83] Gay, D. M. Subroutines for unconstrained minimization using a model/trust-region approach. To appear in *ACM Trans. Math. Software*.

[GilGMS74] Gill, P. E., Golub, G. H., Murray, W. and Saunders, M. A. Methods for modifying matrix factorizations. *Math. Comput.* **28** (1974), 505–535.

[GilM74] Gill, P. E. and Murray, W. Newton-type methods for linearly constrained optimization. In *Numerical Methods for Constrained Optimization*, edited by P. E. Gill and W. Murray, Academic Press, London and New York, 1974.

[GilM76] Gill, P. E. and Murray, W. Minimization subject to bounds on the variables. NPL Report NAC 72, National Physical Laboratory, England, 1976.

[GilM77] Gill, P. E. and Murray, W. The computation of Lagrange-multiplier estimates for constrained optimization. NPL Report NAC 77, National Physical Laboratory, England, 1977.

[GilMW81] Gill, P. E., Murray, W. and Wright, M. H. *Practical Optimization*. Academic Press, London and New York, 1981.

[Gol76] Goldfarb, D. Factorized variable metric methods for unconstrained optimization. *Math. Comput.* **30** (1976), 796–811.

[Kuh76] Kuhn, H. W. Nonlinear programming: a historical view. In *Nonlinear Programming (SIAM–AMS Proceedings, vol. IX)*, edited by R. W. Cottle and C. E. Lemke, American Mathematical Society, Providence, R.I., 1976.

[LawH74] Lawson, C. L. and Hanson, R. J. *Solving Least Squares Problems*. Prentice-Hall, Englewood Cliffs, N.J., 1974.

[McC76] McCormick, G. P. Optimality criteria in nonlinear programming. In *Nonlinear Programming (SIAM–AMS Proceedings, vol. IX)*, edited by R. W. Cottle and C. E. Lemke, American Mathematical Society, Providence, R.I., 1976.

[Mor78] Moré, J. J. The Levenberg-Marquardt algorithm: implementation and theory. In *Numerical Analysis*, edited by G. A. Watson, Lecture Notes in Mathematics 630, Springer-Verlag, Berlin, Heidelberg and New York, 1978.

[MorGH81] Moré, J. J., Garbow, B. S., and Hillstrom, K. E. Testing unconstrained optimization software. *ACM Trans. Math. Software* **7** (1981), 17–41.

[MorS81] Moré, J. J., and Sorensen, D. C. Computing a Trust Region Step. Technical Report ANL–81–83, Applied Mathematics Division, Argonne National Lab.

[NAG82] *NAG FORTRAN Library Manual Mark 9*, vol. 3, Numerical Algorithms Group, Oxford, 1982.

[Pow70a] Powell, M. J. D. A new algorithm for unconstrained optimization. In *Nonlinear Programming*, edited by J. B. Rosen, O. L. Mangasarian, and K. Ritter; Academic Press, New York, 1970.

[Pow70b] Powell, M. J. D. A FORTRAN subroutine for unconstrained minimization, requiring first derivatives of the objective function. Report AERE-R.6469, A.E.R.E. Harwell, Oxon., England, 1970.

[Pow72] Powell, M. J. D. Quadratic termination properties of minimization algorithms I. Statement and discussion of results. *J. Inst. Math. Applic.* **10** (1972), 333–342.

[Ros60] Rosen, J. B. The gradient projection method for non-linear programming, Part I: linear constraints. *J. SIAM* **8** (1960), 181–217.

[SchKW82] Schnabel, R. B., Koontz, J. E., and Weiss, B. E. A modular system of algorithms for unconstrained minimization. Technical report CU-CS-240-82, Department of Computer Science, University of Colorado at Boulder.

[Wol63] Wolfe, P. A technique for resolving degeneracy in linear programming. *J. SIAM* **11** (1963), 205–211.

11. Appendix: testing details.

Table A.1 gives the simple bounds used in testing HUMSLB and SUMSLB. The problems are identified by the run number that appears in Tables A.2 and A.3. See [MorGH81] for a description of the problems themselves — the column labeled "MGH#" gives the number of the problem description in §3 of [MorGH81]. Tables A.2 and A.3 give a summary of the performance of HUMSLB and SUMSLB on these problems. The column labels have the following meanings: "P" is the number of variables p; "LS" is the common logarithm of the factor by which the standard starting guess listed in [MorGH81] was multiplied: if $\hat{x}^0$ denotes this standard start, then the starting guess x^0 used was $\hat{x}^0$, $10\hat{x}^0$, or $100\hat{x}^0$ for LS = 0, 1, or 2; "NF" and "NG" are the numbers of function and gradient evaluations expended respectively; "F" is the objective function value $f(x^{final})$ achieved; "G" is $\|\nabla f(x^{final})\|_2$; "PRELDF" is the predicted relative function reduction, $|q(s^t)/f(x^c)|$, for the last step tried; and "RELDX" is the relative stepsize,

$$\text{RELDX} := \max\{|s_i^t|: 1 \le i \le p\}/\max\{|x_j^c| + |(x^c+s^t)_j|: 1 \le j \le p\},$$

again for the last step tried; finally, "C" tells why the code stopped: R means relative function convergence: the PRELDF value for a full Newton step was at most $\text{MACHEPS}^{2/3} \doteq 3.7\times10^{-10}$, where MACHEPS, the unit roundoff, is about 7.1×10^{-15} on the Cray 1 used for this testing; X means X convergence: RELDX for a full Newton step was at most $\text{MACHEPS}^{1/2} \doteq 8.4\times10^{-8}$; and B means both

R and X. (The stopping tolerances just described are the default tolerances described in [DenGW81] and [Gay83].) Tables A.4–6 summarize the results of running E04LBF, E04KBF-Q, and E04KBF-C on the same test problems used for Tables A.2 and A.3; the column labeled "IFAIL" give a return code that has, roughly, the following meanings: 0 means the convergence tests (whatever they are) were satisfied; 2 means function evaluation limit exceeded; 3 means (quoting the NAG documentation) "the conditions for a minimum have not all been met, but a lower point could not be found", perhaps because the stopping tolerance was too tight; 5 means "a local search has failed to find a feasible point which gives a significant change of function value", possibly because of extreme ill conditioning or because "steps in the subspace of free variables happened to meet a bound before they changed the function value". [One advantage of the approach in this paper is that it cannot experience this last difficulty.]

Figure A.7 extracts a graphical comparison of HUMSLB and E04LBF from Tables A.2 and A.4: it separately compares the relative numbers of function and gradient evaluations these routines expended on the simply bounded test problems for which both found the same solution. (Sameness of solutions is judged as for Table 8.1.) Similar comparisons (extracted from Tables A.3, A.5, and A.6) of SUMSLB with E04KBF-Q and E04KBF-C appear in Figures A.8 and A.9.

Table A.10 is analogous to Table A.1; it gives the simple bounds used for the least-squares test problems on which NL2SLB ran. Again the problems are identified by run number; this time it is the run number from table A.11, which summarizes the performance of NL2SLB on the nonlinear least-squares test problems of [MorGH81]. The results for Table A.11 were computed using double precision arithmetic on a VAX computer; the default stopping tolerances are slightly different from those listed above for the Cray 1: the relative function convergence tolerance (on PRELDF) is 10^{-10} (the default is actually $\max\{10^{-10}, \mathrm{MACHEP}^{2/3}\}$) and the X convergence tolerance is about 3.7×10^{-9}. The column labeled "N" in Table A.11 tells the number of residual components (the sum of whose squares is $2f(x)$) for each problem.

There is an additional entry in the column labeled "C" of Table A.11: "S" means singular convergence, i.e., more favorable tests not satisfied and PRELDF no more than 10^{-10} for a trust radius of $\delta = 1$. This is the correct diagnostic on runs 3–6; on runs 40, 41, and 47 it occurs because of a scaling difficulty. The norm $\|\cdot\|$ used in NL2SLB has the form $\|x\| := \|Dx\|_2$, where D is a positive-definite diagonal scaling matrix. (HUMSLB and SUMSLB also use such a norm, but in the testing considered here, D was always the identity matrix.) By default, NL2SLB, like NL2SOL, estimates D from the diagonal of the Jacobian matrix (see [DenGW81] for details). The technique used results in so large a D that the singular convergence test is satisfied on runs 40, 41, and 47. With D changed to I (a drastic change from the default), NL2SLB finds a local minimizer and gives a favorable return code on the problems in question.

Here is a list of the run numbers omitted from Table 8.1, followed by some comments on the omitted problems.

E04LBF versus HUMSLB: *runs skipped* = *4, 47.*

E04KBF-Q versus SUMSLB: *runs skipped* = *6, 13, 14, 20–24, 46.*

E04KBF-C versus SUMSLB: *runs skipped* = *6, 14, 20–25, 46, 48.*

Comments: BIGGS EXP6 (run 4) and WOOD (runs 46–48) have at least two local minimizers, one or the other of which all runs found. On POWELL BADLY SCALED (run 6), both variants of E04KBF stopped because they exceeded the function evaluation limit of 1000. WATSON 12 (runs 13 and 14) appears to be quite flat near the minimizer having $f(x^*) \doteq 0.0716428$; at least on run 13 E04KBF-Q gives the favorable return IFAIL = 0 (at a point from which HUMSLB can reduce f from 0.0716432 to 0.0716428), and if SUMSLB on run 14 is given a sufficiently small relative function convergence tolerance, i.e., 10^{-14}, then it, too, continues on to achieve this function value. The PENALTY II functions with p = 4 and 9 (runs 20–25) similarly appear quite flat near the minimizers found.

Table A.1. Simple Bounds Given to HUMSLB and SUMSLB

Runs	Problem	P	MGH#	Feasible Region
1−3	HELICAL VALLEY	3	7	$[-100,.8]\times[-1,1]\times[-1,1]$
4	BIGGS EXP6	6	18	$[0,2]\times[0,8]\times[0,1]\times[1,7]\times[0,5]^2$
5	GAUSSIAN	3	9	$[.398,4.2]\times[1,2]\times[-.5,.1]$
6	POWELL BADLY SCALED	2	3	$[0,1]\times[1,9]$
7	BOX 3-D	3	12	$[0,2]\times[5,9.5]\times[0,20]$
8−10	VARIABLY DIMENSIONED	10	25	$[0,10]\times[0,20]\times[0,30]\times[0,40]\times$ $\times[0,50]\times[0,60]\times[0,70]\times[0,80]\times$ $\times[0,90]\times[0,.5]$
11−12	WATSON 9	9	20	$[-10^{-5},10^{-5}]\times[0,.9]\times[0,.1]\times[0,1]\times$ $\times[0,1]\times[-3,0]\times[0,4]\times[-3,0]\times[0,2]$
13−14	WATSON 12	12	20	$[-1,0]\times[0,.9]\times[-1,0]\times[-1,.3]\times$ $\times[-1,0]\times[0,1]\times[-3,0]\times[0,10]\times$ $\times[-10,0]\times[0,10]\times[-5,0]\times[0,1]$
15−17	PENALTY I	10	23	$[0,100]\times[1,100]\times[0,100]\times[0,100]\times$ $\times[0,100]\times[1,100]\times[0,100]\times[0,100]\times$ $\times[0,100]\times[1,100]$
18−19	PENALTY II 1	1	24	$[-1,1]$
20−22	PENALTY II 4	4	24	$[-10,50]\times[.3,50]\times[0,50]\times[-1,.5]$
23−25	PENALTY II 10	10	24	$[-10,50]\times[.1,50]\times[0,50]\times[.05,50]\times$ $\times[0,50]\times[-10,50]\times[0,50]\times[.2,50]\times$ $\times[0,50]\times[0,.5]$
26−28	BROWN BADLY SCALED	2	4	$[0,10^6]\times[3\times10^{-5},100]$
29−31	BROWN AND DENNIS	4	16	$[-10,100]\times[0,15]\times[-100,0]\times[-20,.2]$
32−33	GULF R AND D	3	11	$[0,10]\times[0,10]\times[0,10]$
34−36	TRIGONOMETRIC	10	26	$[0,10]\times[10,20]\times[20,30]\times[30,40]\times$ $\times[40,50]\times[50,60]\times[60,70]\times[70,80]\times$ $\times[80,90]\times[90,100]$
37−39	EXTENDED ROSENBROCK	2	21	$[-50,.5]\times[0,100]$
40−42	EXTENDED POWELL SING	4	22	$[.1,100]\times[-20,20]\times[-1,1]\times[-1,50]$
43−44	BEALE	2	5	$[.6,10]\times[.5,100]$
46−48	WOOD	4	14	$[-100,0]\times[-100,10]\times[-100,100]^2$
49	CHEBYQUAD 7	7	35	$[0,.05]\times[0,.23]\times[0,.333]\times[0,1]\times$ $\times[0,1]^3$
50	CHEBYQUAD 8	8	35	$[0,.04]\times[0,.2]\times[.1,.3]\times[0,1]\times$ $\times[0,1]^4$
51	CHEBYQUAD 9	9	35	$[0,1]\times[0,.2]\times[.1,.23]\times[0,.4]\times$ $\times[0,1]^5$
52	CHEBYQUAD 10	10	35	$[0,1]\times[.1,.2]\times[.2,.3]\times[0,.4]^2\times[.5,1]^5$

Table A.2. HUMSLB Test Runs

Run	PROBLEM	P	LS	NF	NG	C	F	G	PRELDF	RELDX
1	HELICAL VALLEY	3	0	11	7	B	.990422+00	.36-08	.18-18	.55-10
2	HELICAL VALLEY	3	1	13	10	R	.990422+00	.14-08	.34-09	.24-05
3	HELICAL VALLEY	3	2	15	11	B	.990422+00	.38-08	.21-18	.60-10
4	BIGGS EXP6	6	0	75	55	R	.242808+00	.56-07	.45-10	.40-05
5	GAUSSIAN	3	0	4	3	B	.112793-07	.97-10	.55-12	.14-09
6	POWELL BADLY SCALED	2	0	159	101	X	.151259-09	.12-08	.13-03	.87-13
7	BOX 3-D	3	0	14	11	B	.114348-03	.86-14	.27-24	.37-15
8	VARIABLY DIMENSIONED	10	0	16	15	B	.337413+00	.19-07	.96-18	.78-11
9	VARIABLY DIMENSIONED	10	1	19	18	B	.337413+00	.60-11	.15-12	.30-08
10	VARIABLY DIMENSIONED	10	2	28	24	B	.337413+00	.60-11	.92-25	.00+00
11	WATSON 9	9	0	19	15	R	.374014-01	.12-11	.13-11	.88-07
12	WATSON 9	9	1	21	15	B	.374014-01	.79-06	.29-13	.13-07
13	WATSON 12	12	0	31	21	B	.716428-01	.41-08	.36-17	.45-08
14	WATSON 12	12	1	29	23	B	.716428-01	.41-08	.36-17	.45-08
15	PENALTY I	10	0	13	11	B	.756257+01	.45-06	.12-14	.12-07
16	PENALTY I	10	1	20	17	R	.756257+01	.81-16	.96-12	.33-06
17	PENALTY I	10	2	22	19	B	.756257+01	.00+00	.00+00	.00+00
18	PENALTY II 1	1	0	8	7	B	.489395+00	.28-13	.15-27	.22-14
19	PENALTY II 1	1	1	7	6	B	.489395+00	.28-13	.15-27	.22-14
20	PENALTY II 4	4	0	89	65	R	.943420-05	.47-09	.36-10	.82-05
21	PENALTY II 4	4	1	167	138	R	.943420-05	.49-08	.17-09	.27-04
22	PENALTY II 4	4	2	188	150	R	.943420-05	.16-10	.49-12	.15-05
23	PENALTY II 10	10	0	90	80	R	.294344-03	.80-08	.94-10	.26-04
24	PENALTY II 10	10	1	108	91	R	.294344-03	.14-07	.51-10	.46-04
25	PENALTY II 10	10	2	109	96	R	.294344-03	.46-10	.11-10	.27-05
26	BROWN BADLY SCALED	2	0	16	4	B	.784000+03	.13-07	.56-19	.19-14
27	BROWN BADLY SCALED	2	1	14	5	X	.784000+03	.58-08	.90-09	.42-09
28	BROWN BADLY SCALED	2	2	22	5	B	.784000+03	.28-07	.25-18	.56-14
29	BROWN AND DENNIS	4	0	12	10	B	.888605+05	.23-09	.34-29	.00+00
30	BROWN AND DENNIS	4	1	16	14	R	.888605+05	.16-07	.13-11	.26-06
31	BROWN AND DENNIS	4	2	16	14	R	.888605+05	.59-08	.16-12	.91-07
32	GULF R AND D	3	0	11	8	B	.512169+01	.48-09	.38-10	.59-07
33	GULF R AND D	3	1	6	6	R	.512169+01	.14-08	.11-09	.99-07
34	TRIGONOMETRIC	10	0	12	8	B	.609515+03	.10-09	.44-25	.25-17
35	TRIGONOMETRIC	10	1	11	7	B	.609515+03	.82-06	.27-17	.22-10
36	TRIGONOMETRIC	10	2	11	7	B	.915741+03	.95-09	.11-10	.41-07
37	EXTENDED ROSENBROCK	2	0	15	13	B	.250000+00	.36-12	.13-26	.18-14
38	EXTENDED ROSENBROCK	2	1	49	36	B	.250000+00	.00+00	.00+00	.00+00
39	EXTENDED ROSENBROCK	2	2	82	58	B	.250000+00	.00+00	.00+00	.00+00
40	EXTENDED POWELL SING	4	0	13	12	B	.187820-03	.43-09	.49-15	.24-08
41	EXTENDED POWELL SING	4	1	21	19	B	.187820-03	.26-14	.22-27	.00+00
42	EXTENDED POWELL SING	4	2	24	21	B	.187820-03	.11-13	.88-13	.30-07
43	BEALE	2	0	11	6	B	.000000+00	.00+00	.10+01	.00+00
44	BEALE	2	1	30	26	B	.000000+00	.00+00	.10+01	.00+00
45	BEALE	2	2	29	26	B	.000000+00	.00+00	.10+01	.00+00
46	WOOD	4	0	55	36	B	.155670+01	.18-08	.37-20	.57-11
47	WOOD	4	1	26	22	B	.709384+01	.68-11	.16-12	.13-07
48	WOOD	4	2	23	20	B	.709384+01	.68-11	.29-13	.27-07
49	CHEBYQUAD 7	7	0	9	7	B	.602610-03	.85-13	.29-24	.95-15
50	CHEBYQUAD 8	8	0	15	11	B	.359224-02	.23-08	.48-16	.79-10
51	CHEBYQUAD 9	9	0	18	16	R	.188847-03	.89-10	.33-10	.22-06
52	CHEBYQUAD 10	10	0	14	10	R	.477271-02	.38-09	.13-09	.53-06

Table A.3. SUMSLB Test Runs

Run	PROBLEM	P	LS	NF	NG	C	F	G	PRELDF	RELDX
1	HELICAL VALLEY	3	0	15	12	R	.990422+00	.69-08	.28-11	.21-06
2	HELICAL VALLEY	3	1	15	13	B	.990422+00	.40-07	.10-12	.42-07
3	HELICAL VALLEY	3	2	18	13	B	.990422+00	.22-06	.18-11	.29-07
4	BIGGS EXP6	6	0	110	86	B	.532099-03	.17-07	.72-11	.18-08
5	GAUSSIAN	3	0	7	6	X	.112793-07	.18-08	.95-07	.89-08
6	POWELL BADLY SCALED	2	0	183	142	X	.151259-09	.16-03	.49-05	.17-13
7	BOX 3-D	3	0	15	12	X	.114348-03	.31-06	.21-07	.52-07
8	VARIABLY DIMENSIONED	10	0	22	22	B	.337413+00	.42-07	.55-13	.50-08
9	VARIABLY DIMENSIONED	10	1	27	26	R	.337413+00	.32-08	.19-09	.26-06
10	VARIABLY DIMENSIONED	10	2	45	40	R	.337413+00	.84-05	.24-09	.61-06
11	WATSON 9	9	0	77	61	R	.374014-01	.51-08	.39-12	.16-06
12	WATSON 9	9	1	61	52	R	.374014-01	.73-08	.97-11	.29-06
13	WATSON 12	12	0	186	156	R	.716428-01	.11-03	.22-09	.98-05
14	WATSON 12	12	1	182	147	R	.717550-01	.50-02	.15-10	.17-05
15	PENALTY I	10	0	17	14	B	.756257+01	.54-07	.25-16	.73-08
16	PENALTY I	10	1	18	13	R	.756257+01	.36-04	.92-11	.12-05
17	PENALTY I	10	2	31	24	R	.756257+01	.33-11	.18-09	.79-05
18	PENALTY II 1	1	0	8	8	B	.489395+00	.58-11	.56-14	.20-07
19	PENALTY II 1	1	1	9	8	R	.489395+00	.25-08	.99-11	.84-06
20	PENALTY II 4	4	0	283	219	R	.943420-05	.16-07	.13-09	.23-04
21	PENALTY II 4	4	1	471	355	R	.943436-05	.16-05	.79-10	.14-04
22	PENALTY II 4	4	2	347	270	R	.943420-05	.15-07	.21-09	.26-04
23	PENALTY II 10	10	0	266	212	R	.294426-03	.25-05	.18-09	.21-04
24	PENALTY II 10	10	1	511	407	R	.294361-03	.37-05	.93-10	.16-04
25	PENALTY II 10	10	2	395	316	R	.294350-03	.13-05	.69-10	.32-06
26	BROWN BADLY SCALED	2	0	14	5	X	.784000+03	.58-08	.90-09	.42-09
27	BROWN BADLY SCALED	2	1	13	3	X	.784000+03	.12-05	.90-09	.42-09
28	BROWN BADLY SCALED	2	2	31	10	X	.784000+03	.61-07	.90-09	.42-09
29	BROWN AND DENNIS	4	0	41	31	R	.888605+05	.63-02	.88-11	.16-06
30	BROWN AND DENNIS	4	1	28	20	R	.888605+05	.13-01	.17-10	.79-06
31	BROWN AND DENNIS	4	2	30	23	R	.888605+05	.20-02	.88-12	.20-06
32	GULF R AND D	3	0	21	19	B	.512169+01	.48-10	.80-24	.83-14
33	GULF R AND D	3	1	9	8	B	.512169+01	.44-07	.21-10	.44-07
34	TRIGONOMETRIC	10	0	25	14	R	.609515+03	.29-02	.57-10	.15-06
35	TRIGONOMETRIC	10	1	21	11	R	.609515+03	.29-03	.78-11	.75-07
36	TRIGONOMETRIC	10	2	24	14	R	.915741+03	.52-03	.38-11	.74-07
37	EXTENDED ROSENBROCK	2	0	22	17	B	.250000+00	.00+00	.00+00	.00+00
38	EXTENDED ROSENBROCK	2	1	62	50	B	.250000+00	.00+00	.00+00	.00+00
39	EXTENDED ROSENBROCK	2	2	114	86	B	.250000+00	.36-12	.13-26	.18-14
40	EXTENDED POWELL SING	4	0	25	24	R	.187820-03	.11-07	.10-10	.89-07
41	EXTENDED POWELL SING	4	1	37	30	B	.187820-03	.70-08	.57-11	.29-07
42	EXTENDED POWELL SING	4	2	42	35	R	.187820-03	.95-07	.24-09	.15-05
43	BEALE	2	0	5	4	X	.454384-27	.20-12	.15+01	.12-14
44	BEALE	2	1	53	44	X	.599276-22	.24-10	.10+01	.32-07
45	BEALE	2	2	39	36	X	.444286-26	.64-12	.24+01	.47-14
46	WOOD	4	0	48	38	R	.709384+01	.11-04	.32-10	.11-05
47	WOOD	4	1	49	38	B	.155670+01	.77-06	.29-12	.15-07
48	WOOD	4	2	46	35	B	.709384+01	.49-07	.53-14	.20-08
49	CHEBYQUAD 7	7	0	27	19	B	.602610-03	.19-06	.94-10	.47-07
50	CHEBYQUAD 8	8	0	41	28	B	.359224-02	.36-06	.74-10	.61-07
51	CHEBYQUAD 9	9	0	43	34	B	.188847-03	.28-07	.39-11	.44-07
52	CHEBYQUAD 10	10	0	49	37	R	.477271-02	.17-05	.34-09	.83-06

Table A.4. E04LBF Test Runs

Run	PROBLEM	P	LS	NF	NG	C	F	G
1	HELICAL VALLEY	3	0	11	11	0	.990422+00	.52−08
2	HELICAL VALLEY	3	1	12	12	0	.990422+00	.52−08
3	HELICAL VALLEY	3	2	12	12	0	.990422+00	.52−08
4	BIGGS EXP6	6	0	33	33	0	.532099−03	.31−11
5	GAUSSIAN	3	0	3	3	0	.112793−07	.19−09
6	POWELL BADLY SCALED	2	0	257	257	5	.151259−09	.12−08
7	BOX 3-D	3	0	9	9	0	.114348−03	.37−10
8	VARIABLY DIMENSIONED	10	0	16	16	0	.337413+00	.82−09
9	VARIABLY DIMENSIONED	10	1	19	19	0	.337413+00	.60−11
10	VARIABLY DIMENSIONED	10	2	37	37	3	.337413+00	.57−06
11	WATSON 9	9	0	19	19	0	.374014−01	.74−12
12	WATSON 9	9	1	21	21	0	.374014−01	.65−09
13	WATSON 12	12	0	22	22	0	.716428−01	.29−09
14	WATSON 12	12	1	45	45	3	.716428−01	.41−08
15	PENALTY I	10	0	13	13	0	.756257+01	.93−18
16	PENALTY I	10	1	18	18	0	.756257+01	.34−10
17	PENALTY I	10	2	23	23	0	.756257+01	.13−12
18	PENALTY II 1	1	0	5	5	0	.489395+00	.27−09
19	PENALTY II 1	1	1	6	6	0	.489395+00	.23−08
20	PENALTY II 4	4	0	78	78	0	.943420−05	.36−09
21	PENALTY II 4	4	1	81	81	0	.943420−05	.36−09
22	PENALTY II 4	4	2	58	58	0	.943420−05	.24−09
23	PENALTY II 10	10	0	120	120	0	.294344−03	.46−09
24	PENALTY II 10	10	1	126	126	0	.294344−03	.46−09
25	PENALTY II 10	10	2	121	121	0	.294344−03	.45−09
26	BROWN BADLY SCALED	2	0	14	14	0	.784000+03	.58−08
27	BROWN BADLY SCALED	2	1	16	16	0	.784000+03	.58−08
28	BROWN BADLY SCALED	2	2	18	18	0	.784000+03	.58−08
29	BROWN AND DENNIS	4	0	10	10	0	.888605+05	.24−09
30	BROWN AND DENNIS	4	1	15	15	0	.888605+05	.23−09
31	BROWN AND DENNIS	4	2	19	19	0	.888605+05	.24−09
32	GULF R AND D	3	0	10	10	0	.512169+01	.48−11
33	GULF R AND D	3	1	9	9	0	.512169+01	.31−05
34	TRIGONOMETRIC	10	0	15	15	0	.609515+03	.19−07
35	TRIGONOMETRIC	10	1	16	16	0	.609515+03	.19−07
36	TRIGONOMETRIC	10	2	16	16	0	.915741+03	.25−09
37	EXTENDED ROSENBROCK	2	0	19	19	0	.250000+00	.36−12
38	EXTENDED ROSENBROCK	2	1	71	71	0	.250000+00	.00+00
39	EXTENDED ROSENBROCK	2	2	74	74	0	.250000+00	.36−12
40	EXTENDED POWELL SING	4	0	12	12	0	.187820−03	.55−09
41	EXTENDED POWELL SING	4	1	25	25	0	.187820−03	.43−11
42	EXTENDED POWELL SING	4	2	24	24	0	.187820−03	.49−10
43	BEALE	2	0	3	3	5	.454384−27	.53−13
44	BEALE	2	1	31	31	5	.000000+00	.00+00
45	BEALE	2	2	24	24	5	.000000+00	.00+00
46	WOOD	4	0	39	39	3	.155670+01	.52−06
47	WOOD	4	1	40	40	3	.155670+01	.17−06
48	WOOD	4	2	22	22	0	.709384+01	.44−09
49	CHEBYQUAD 7	7	0	12	12	3	.602610−03	.49−08
50	CHEBYQUAD 8	8	0	23	23	0	.359224−02	.60−10
51	CHEBYQUAD 9	9	0	21	21	0	.188847−03	.22−10
52	CHEBYQUAD 10	10	0	32	32	3	.477271−02	.20−08

Table A.5. E04KBF-Q Test Runs

Run	PROBLEM	P	LS	NF	NG	C	F	G
1	HELICAL VALLEY	3	0	41	17	3	.990422+00	.10−05
2	HELICAL VALLEY	3	1	37	17	0	.990422+00	.61−06
3	HELICAL VALLEY	3	2	42	18	3	.990422+00	.10−05
4	BIGGS EXP6	6	0	158	66	0	.532099−03	.46−07
5	GAUSSIAN	3	0	13	7	0	.112793−07	.32−07
6	POWELL BADLY SCALED	2	0	1002	361	2	.806743−06	.89+00
7	BOX 3-D	3	0	26	13	0	.114348−03	.51−09
8	VARIABLY DIMENSIONED	10	0	72	35	0	.337413+00	.31−07
9	VARIABLY DIMENSIONED	10	1	185	76	0	.337413+00	.18−06
10	VARIABLY DIMENSIONED	10	2	270	107	0	.337413+00	.34−07
11	WATSON 9	9	0	186	81	0	.374014−01	.10−06
12	WATSON 9	9	1	208	88	0	.374014−01	.19−09
13	WATSON 12	12	0	369	164	0	.716432−01	.16−04
14	WATSON 12	12	1	478	199	0	.716428−01	.15−07
15	PENALTY I	10	0	57	25	0	.756257+01	.77−09
16	PENALTY I	10	1	73	32	0	.756257+01	.81−09
17	PENALTY I	10	2	72	32	0	.756257+01	.81−09
18	PENALTY II 1	1	0	23	12	5	.489395+00	.68−07
19	PENALTY II 1	1	1	21	11	5	.489395+00	.68−07
20	PENALTY II 4	4	0	22	11	0	.943741−05	.41−06
21	PENALTY II 4	4	1	330	131	0	.943423−05	.11−06
22	PENALTY II 4	4	2	1002	398	2	.967858−05	.19−03
23	PENALTY II 10	10	0	1001	404	2	.294379−03	.33−04
24	PENALTY II 10	10	1	1002	414	2	.294722−03	.16−04
25	PENALTY II 10	10	2	1003	416	2	.294350−03	.14−04
26	BROWN BADLY SCALED	2	0	30	12	0	.784000+03	.00+00
27	BROWN BADLY SCALED	2	1	31	12	0	.784000+03	.00+00
28	BROWN BADLY SCALED	2	2	35	12	0	.784000+03	.00+00
29	BROWN AND DENNIS	4	0	36	18	0	.888605+05	.33−03
30	BROWN AND DENNIS	4	1	67	30	3	.888605+05	.80−04
31	BROWN AND DENNIS	4	2	62	28	3	.888605+05	.83−04
32	GULF R AND D	3	0	37	18	5	.512169+01	.10−07
33	GULF R AND D	3	1	21	8	5	.512169+01	.18−06
34	TRIGONOMETRIC	10	0	51	19	0	.609515+03	.12−04
35	TRIGONOMETRIC	10	1	50	19	0	.609515+03	.50−04
36	TRIGONOMETRIC	10	2	46	18	0	.915741+03	.46−04
37	EXTENDED ROSENBROCK	2	0	55	23	0	.250000+00	.00+00
38	EXTENDED ROSENBROCK	2	1	206	80	0	.250000+00	.36−12
39	EXTENDED ROSENBROCK	2	2	240	94	0	.250000+00	.00+00
40	EXTENDED POWELL SING	4	0	44	22	0	.187820−03	.83−08
41	EXTENDED POWELL SING	4	1	71	34	0	.187820−03	.12−05
42	EXTENDED POWELL SING	4	2	98	47	0	.187820−03	.64−06
43	BEALE	2	0	6	3	0	.121169−26	.85−13
44	BEALE	2	1	111	49	5	.155951−14	.37−06
45	BEALE	2	2	62	31	0	.454384−27	.53−13
46	WOOD	4	0	133	58	0	.155670+01	.37−06
47	WOOD	4	1	141	64	0	.155670+01	.19−06
48	WOOD	4	2	74	33	0	.709384+01	.57−07
49	CHEBYQUAD 7	7	0	36	17	0	.602610−03	.14−07
50	CHEBYQUAD 8	8	0	65	27	0	.359224−02	.39−06
51	CHEBYQUAD 9	9	0	77	36	0	.188847−03	.25−06
52	CHEBYQUAD 10	10	0	110	49	0	.477271−02	.13−06

Table A.6. E04KBF-C Test Runs

Run	PROBLEM	P	LS	NF	NG	C	F	G
1	HELICAL VALLEY	3	0	15	15	0	.990422+00	.20−06
2	HELICAL VALLEY	3	1	18	18	0	.990422+00	.20−06
3	HELICAL VALLEY	3	2	16	16	0	.990422+00	.20−06
4	BIGGS EXP6	6	0	86	86	0	.532099−03	.25−08
5	GAUSSIAN	3	0	7	7	0	.112793−07	.32−07
6	POWELL BADLY SCALED	2	0	1001	1001	2	.895930−06	.37+01
7	BOX 3-D	3	0	14	14	0	.114348−03	.16−08
8	VARIABLY DIMENSIONED	10	0	37	37	3	.337413+00	.40−07
9	VARIABLY DIMENSIONED	10	1	98	98	0	.337413+00	.29−07
10	VARIABLY DIMENSIONED	10	2	126	126	0	.337413+00	.31−08
11	WATSON 9	9	0	94	94	0	.374014−01	.16−07
12	WATSON 9	9	1	100	98	0	.374014−01	.58−06
13	WATSON 12	12	0	253	253	0	.716428−01	.24−05
14	WATSON 12	12	1	225	225	0	.716428−01	.17−07
15	PENALTY I	10	0	31	31	0	.756257+01	.24−08
16	PENALTY I	10	1	53	53	0	.756257+01	.60−07
17	PENALTY I	10	2	61	54	0	.756257+01	.19−08
18	PENALTY II 1	1	0	12	12	5	.489395+00	.55−07
19	PENALTY II 1	1	1	9	9	0	.489395+00	.12−07
20	PENALTY II 4	4	0	12	12	0	.943741−05	.41−06
21	PENALTY II 4	4	1	30	30	0	.948879−05	.11−05
22	PENALTY II 4	4	2	625	625	0	.944097−05	.90−06
23	PENALTY II 10	10	0	530	530	0	.294344−03	.21−08
24	PENALTY II 10	10	1	501	501	0	.294345−03	.26−06
25	PENALTY II 10	10	2	777	776	0	.294344−03	.41−08
26	BROWN BADLY SCALED	2	0	14	12	0	.784000+03	.00+00
27	BROWN BADLY SCALED	2	1	14	12	0	.784000+03	.00+00
28	BROWN BADLY SCALED	2	2	21	19	0	.784000+03	.00+00
29	BROWN AND DENNIS	4	0	19	19	0	.888605+05	.41−04
30	BROWN AND DENNIS	4	1	40	40	3	.888605+05	.75−03
31	BROWN AND DENNIS	4	2	31	31	0	.888605+05	.55−02
32	GULF R AND D	3	0	22	22	0	.512169+01	.14−09
33	GULF R AND D	3	1	10	10	0	.512169+01	.22−06
34	TRIGONOMETRIC	10	0	26	26	0	.609515+03	.11−04
35	TRIGONOMETRIC	10	1	26	26	0	.609515+03	.53−04
36	TRIGONOMETRIC	10	2	36	36	3	.915741+03	.37−05
37	EXTENDED ROSENBROCK	2	0	24	24	0	.250000+00	.00+00
38	EXTENDED ROSENBROCK	2	1	97	97	0	.250000+00	.00+00
39	EXTENDED ROSENBROCK	2	2	104	104	0	.250000+00	.00+00
40	EXTENDED POWELL SING	4	0	28	28	0	.187820−03	.54−06
41	EXTENDED POWELL SING	4	1	31	31	0	.187820−03	.83−08
42	EXTENDED POWELL SING	4	2	50	50	0	.187820−03	.69−07
43	BEALE	2	0	4	3	0	.121169−26	.85−13
44	BEALE	2	1	49	48	0	.000000+00	.00+00
45	BEALE	2	2	32	31	0	.454384−27	.53−13
46	WOOD	4	0	70	70	0	.155670+01	.12−07
47	WOOD	4	1	77	77	3	.155670+01	.34−06
48	WOOD	4	2	51	51	0	.155670+01	.56−06
49	CHEBYQUAD 7	7	0	20	20	0	.602610−03	.11−06
50	CHEBYQUAD 8	8	0	40	40	0	.359224−02	.48−06
51	CHEBYQUAD 9	9	0	40	40	0	.188847−03	.10−05
52	CHEBYQUAD 10	10	0	57	57	0	.477271−02	.29−06

Run	PROBLEM	LS
1	HELICAL VALLEY	0
2	HELICAL VALLEY	1
3	HELICAL VALLEY	2
5	GAUSSIAN	0
6	POWELL BADLY SCALED	0
7	BOX 3-D	0
8	VARIABLY DIMENSIONED	0
9	VARIABLY DIMENSIONED	1
10	VARIABLY DIMENSIONED	2
11	WATSON 9	0
12	WATSON 9	1
13	WATSON 12	0
14	WATSON 12	1
15	PENALTY I	0
16	PENALTY I	1
17	PENALTY I	2
18	PENALTY II 1	0
19	PENALTY II 1	1
20	PENALTY II 4	0
21	PENALTY II 4	1
22	PENALTY II 4	2
23	PENALTY II 10	0
24	PENALTY II 10	1
25	PENALTY II 10	2
26	BROWN BADLY SCALED	0
27	BROWN BADLY SCALED	1
28	BROWN BADLY SCALED	2
29	BROWN AND DENNIS	0
30	BROWN AND DENNIS	1
31	BROWN AND DENNIS	2
32	GULF R AND D	0
33	GULF R AND D	1
34	TRIGONOMETRIC	0
35	TRIGONOMETRIC	1
36	TRIGONOMETRIC	2
37	EXTENDED ROSENBROCK	0
38	EXTENDED ROSENBROCK	1
39	EXTENDED ROSENBROCK	2
40	EXTENDED POWELL SING	0
41	EXTENDED POWELL SING	1
42	EXTENDED POWELL SING	2
43	BEALE	0
44	BEALE	1
45	BEALE	2
46	WOOD	0
48	WOOD	2
49	CHEBYQUAD 7	0
50	CHEBYQUAD 8	0
51	CHEBYQUAD 9	0
52	CHEBYQUAD 10	0

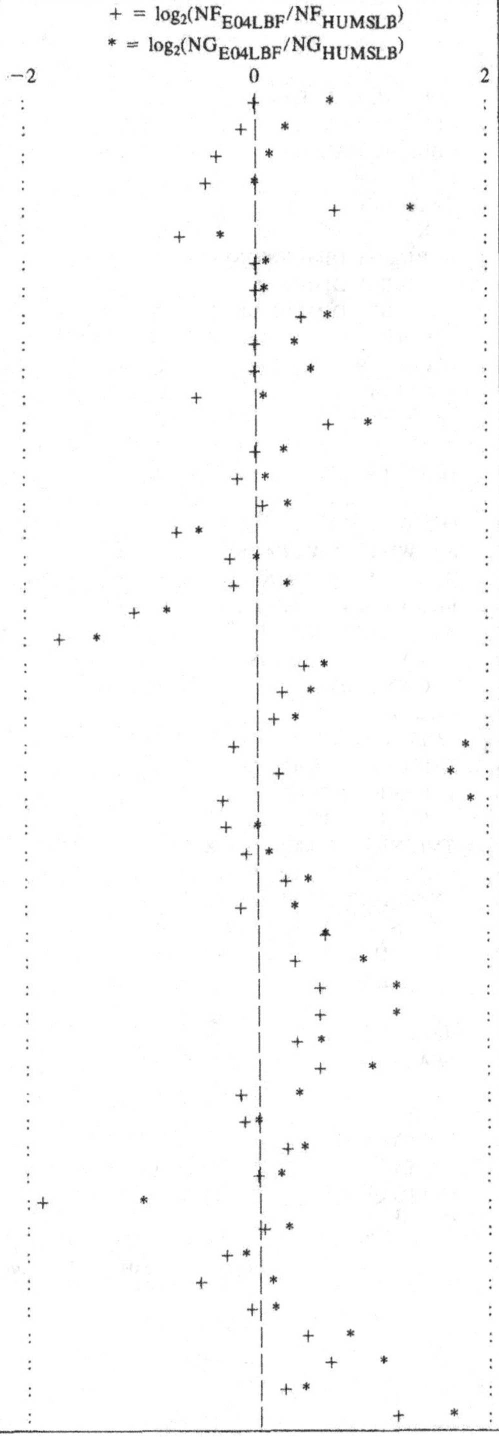

Figure A.7: E04LBF *versus* HUMSLB.

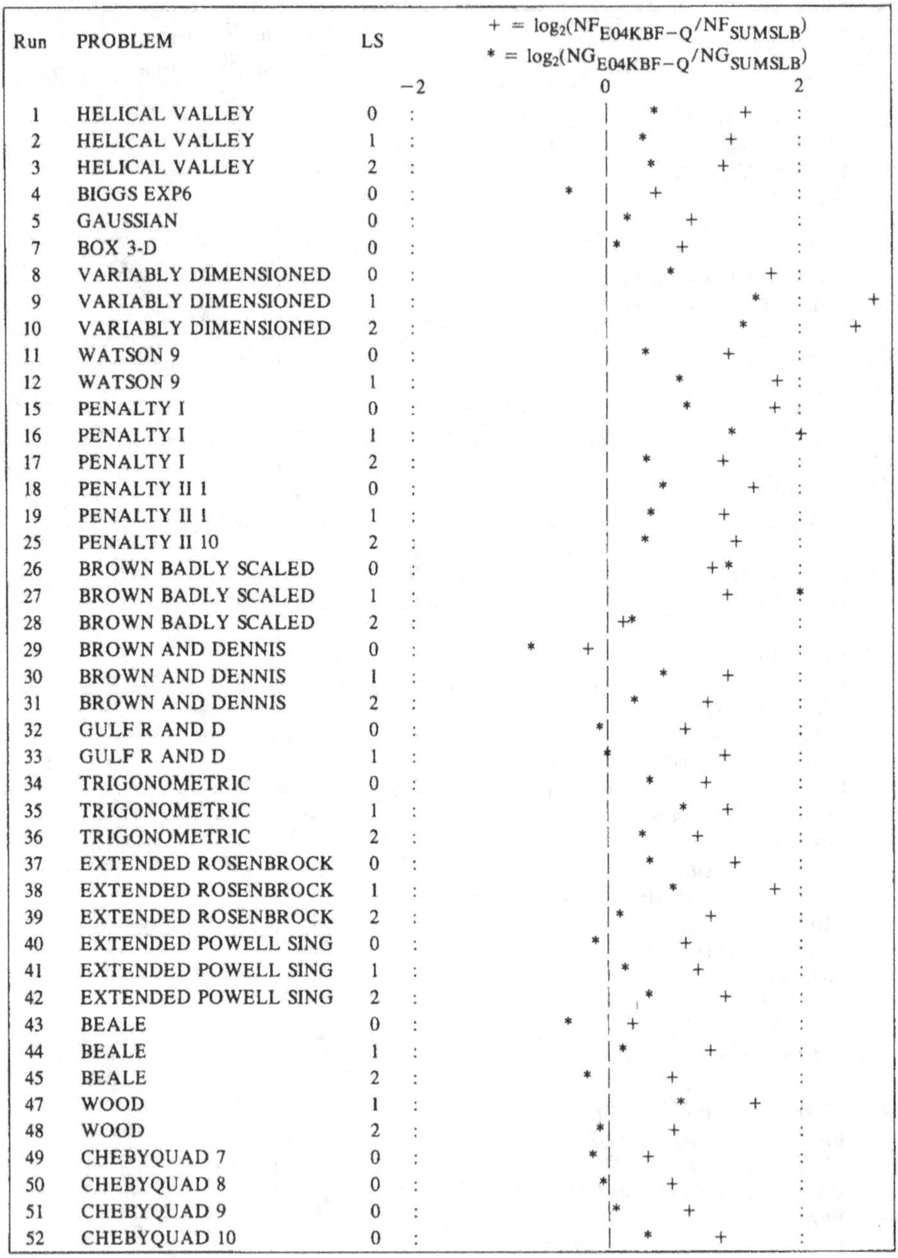

Run	PROBLEM	LS	$+ = \log_2(NF_{E04KBF-Q}/NF_{SUMSLB})$ $* = \log_2(NG_{E04KBF-Q}/NG_{SUMSLB})$

Figure A.8: E04KBF-Q *versus* SUMSLB.

Run	PROBLEM	LS	$+ = \log_2(NF_{E04KBF-C}/NF_{SUMSLB})$ $* = \log_2(NG_{E04KBF-C}/NG_{SUMSLB})$
			-2 $\qquad$ 0 $\qquad$ 2
1	HELICAL VALLEY	0	$:$ $\qquad$ $+$ $*$ $\qquad$ $:$
2	HELICAL VALLEY	1	$:$ $\qquad$ $\vert$ $+$ $*$ $\qquad$ $:$
3	HELICAL VALLEY	2	$:$ $\qquad$ $+$ $\vert$ $*$ $\qquad$ $:$
4	BIGGS EXP6	0	$:$ $\qquad$ $+$ $*\!\vert$ $\qquad$ $:$
5	GAUSSIAN	0	$:$ $\qquad$ $+$ $*$ $\qquad$ $:$
7	BOX 3-D	0	$:$ $\qquad$ $+\!\vert$ $*$ $\qquad$ $:$
8	VARIABLY DIMENSIONED	0	$:$ $\qquad$ $+$ $\qquad$ $:$
9	VARIABLY DIMENSIONED	1	$:$ $\qquad$ $+*:$
10	VARIABLY DIMENSIONED	2	$:$ $\qquad$ $+$ $*$ $:$
11	WATSON 9	0	$:$ $\qquad$ $\vert$ $+$ $*$ $\qquad$ $:$
12	WATSON 9	1	$:$ $\qquad$ $\vert$ $+$ $*$ $\qquad$ $:$
13	WATSON 12	0	$:$ $\qquad$ $\vert$ $+$ $*$ $\qquad$ $:$
15	PENALTY I	0	$:$ $\qquad$ $\vert$ $+$ $*$ $\qquad$ $:$
16	PENALTY I	1	$:$ $\qquad$ $\vert$ $+$ $*$
17	PENALTY I	2	$:$ $\qquad$ $\vert$ $+$ $*$ $\qquad$ $:$
18	PENALTY II 1	0	$:$ $\qquad$ $\vert$ $*$ $\qquad$ $:$
19	PENALTY II 1	1	$:$ $\qquad$ $+$ $*$ $\qquad$ $:$
26	BROWN BADLY SCALED	0	$:$ $\qquad$ $+$ $*$ $\qquad$ $:$
27	BROWN BADLY SCALED	1	$:$ $\qquad$ $\vert+$ $*$
28	BROWN BADLY SCALED	2	$:$ $\qquad$ $+$ $\vert$ $*$ $\qquad$ $:$
29	BROWN AND DENNIS	0	$:$ $\qquad$ $+$ $*$ $\vert$ $\qquad$ $:$
30	BROWN AND DENNIS	1	$:$ $\qquad$ $\vert$ $+$ $*$ $\qquad$ $:$
31	BROWN AND DENNIS	2	$:$ $\qquad$ $\vert\!+$ $*$ $\qquad$ $:$
32	GULF R AND D	0	$:$ $\qquad$ $\vert\!+$ $*$ $\qquad$ $:$
33	GULF R AND D	1	$:$ $\qquad$ $\vert$ $+$ $*$ $\qquad$ $:$
34	TRIGONOMETRIC	0	$:$ $\qquad$ $\vert\!+$ $*$ $\qquad$ $:$
35	TRIGONOMETRIC	1	$:$ $\qquad$ $\vert$ $+$ $*$ $\qquad$ $:$
36	TRIGONOMETRIC	2	$:$ $\qquad$ $\vert$ $+$ $*$ $\qquad$ $:$
37	EXTENDED ROSENBROCK	0	$:$ $\qquad$ $\vert+$ $*$ $\qquad$ $:$
38	EXTENDED ROSENBROCK	1	$:$ $\qquad$ $\vert$ $+$ $*$ $\qquad$ $:$
39	EXTENDED ROSENBROCK	2	$:$ $\qquad$ $+$ $\vert$ $*$ $\qquad$ $:$
40	EXTENDED POWELL SING	0	$:$ $\qquad$ $\vert$ $+*$ $\qquad$ $:$
41	EXTENDED POWELL SING	1	$:$ $\qquad$ $+$ $\vert\!*$ $\qquad$ $:$
42	EXTENDED POWELL SING	2	$:$ $\qquad$ $\vert$ $+$ $*$ $\qquad$ $:$
43	BEALE	0	$:$ $\qquad$ $*+$ $\vert$ $\qquad$ $:$
44	BEALE	1	$:$ $\qquad$ $+$ $\vert$ $*$ $\qquad$ $:$
45	BEALE	2	$:$ $\qquad$ $+*$ $\vert$ $\qquad$ $:$
47	WOOD	1	$:$ $\qquad$ $\vert$ $+$ $*$ $\qquad$ $:$
49	CHEBYQUAD 7	0	$:$ $\qquad$ $+$ $\vert*$ $\qquad$ $:$
50	CHEBYQUAD 8	0	$:$ $\qquad$ $+\!\vert$ $*$ $\qquad$ $:$
51	CHEBYQUAD 9	0	$:$ $\qquad$ $+\!\vert$ $*$ $\qquad$ $:$
52	CHEBYQUAD 10	0	$:$ $\qquad$ $\vert$ $+$ $*$ $\qquad$ $:$

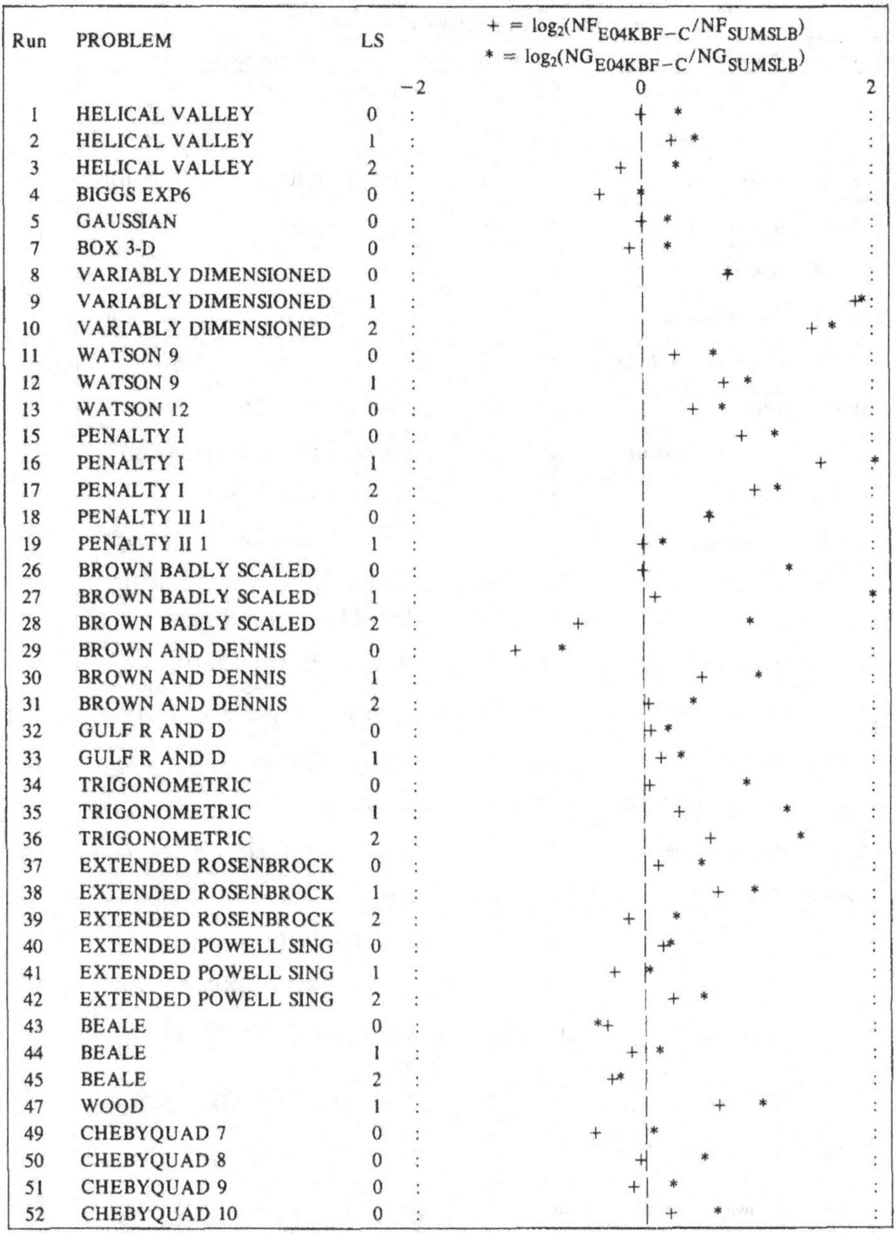

Figure A.9: E04KBF-C *versus* SUMSLB.

Table A.10. Simple Bounds Given to NL2SLB

Runs	Problem	P	MGH#	Feasible Region
1–2	Linear -- full rank	5	32	$[-.5,100]\times[-2,100]^4$
3–4	Linear -- rank 1	5	33	$[-.9,100]\times[-2,100]^4$
5–6	Linear -- rank1, 0's	5	34	$[0,100]\times[-.4,100]\times[.3,100]\times[0,100]^2$
7–9	Rosenbrock	2	1	$[-50,.5]\times[0,100]$
10–12	Helical Valley	3	7	$[-100,.8]\times[-1,1]^2$
13–15	Powell Singular	4	13	$[.1,100]\times[-20,20]\times[-1,1]\times[-1,50]$
16–18	Freudenstrein & Roth	2	2	$[0,20]\times[-30,-.9]$
19–21	Bard	3	8	$[.1,50]\times[0,100]\times[0,50]$
22–24	Kowalik & Osborne	4	15	$[0,10]\times[-1,12]\times[.13,13]\times[.12,12]$
25–26	Meyer	3	10	$[.006,2]\times[0,3\times10^5]\times[0,3\times10^4]$
27–29	Watson 6	6	20	$[-.015,10]\times[-10,100]^4\times[-10,.99]$
30–31	Watson 9	9	20	$[-10^{-5},10^{-5}]\times[0,.9]\times[0,.1]\times[0,1]^2\times$ $\times[-3,0]\times[0,4]\times[-3,0]\times[0,2]$
32–33	Watson 12	12	20	$[-1,0]\times[0,.9]\times[-1,0]\times[-1,.3]\times$ $\times[-1,0]\times[0,1]\times[-3,0]\times[0,10]\times$ $\times[-10,0]\times[0,10]\times[-5,0]\times[0,1]$
34	Box 3D	3	12	$[0,2]\times[5,9.5]\times[0,20]$
35	Jennrich & Sampson	2	6	$[.26,10]\times[0,20]$
36–38	Brown & Dennis	4	16	$[-10,100]\times[0,15]\times[-100,0]\times[-20,.2]$
39–41	Chebyquad 8x1	1	35	$[.5,100]$
42	Chebyquad 8	8	35	$[0,.04]\times[0,.2]\times[.1,.3]\times[0,1]^5$
43	Chebyquad 9	9	35	$[0,1]\times[0,.2]\times[.1,.23]\times[0,.4]\times[0,1]^5$
44	Chebyquad 10	10	35	$[0,1]\times[.1,.2]\times[.2,.3]\times[0,.4]\times$ $\times[0,.4]\times[.5,1]^5$
45–47	Brown almost linear	10	27	$[0,100]\times[1,100]\times[0,.9]\times[0,100]^7$
48	Brown almost linear	30	27	$\left([0,100]\times[1,100]\times[0,.9]\times[0,100]^7\right)^3$
49	Brown almost linear	40	27	$\left([0,100]\times[1,100]\times[0,.9]\times[0,100]^7\right)^4$
50	Osborne1	5	17	$[0,50]\times[0,1.9]\times[-50,-.1]\times[0,10]^2$
51	Osborne2	11	19	$[1,150]\times[.5,100]\times[0,100]\times[.6,100]\times$ $\times[0,100]\times[0,500]^3\times[0,100]\times[0,10]\times$ $\times[0,0]$

Table A.11. NL2SLB Test Runs											
Run	PROBLEM	N	P	LS	NF	NG	C	F	G	PRELDF	RELDX
1	Linear -- full rank	10	5	0	4	2	B	.262500+01	.28−15	.99−32	.83−16
2	Linear -- full rank	50	5	0	6	4	B	.226250+02	.88−16	.16−33	.35−16
3	Linear -- rank 1	10	5	0	8	4	S	.107143+01	.15−14	.17−14	.65+00
4	Linear -- rank 1	50	5	0	10	5	S	.606436+01	.38−12	.27−13	.28+00
5	Linear -- rank1, 0's	10	5	0	7	3	S	.182353+01	.12−13	.12−28	.25−15
6	Linear -- rank1, 0's	50	5	0	7	3	S	.975200+02	.00+00	.00+00	.00+00
7	Rosenbrock	2	2	0	24	15	B	.125000+00	.69−18	.24−35	.00+00
8	Rosenbrock	2	2	1	46	35	B	.125000+00	.69−18	.24−35	.00+00
9	Rosenbrock	2	2	2	47	35	B	.125000+00	.69−18	.24−35	.00+00
10	Helical Valley	3	3	0	13	9	B	.495211+00	.46−10	.12−20	.16−11
11	Helical Valley	3	3	1	13	9	B	.495211+00	.42−10	.26−16	.15−09
12	Helical Valley	3	3	2	16	10	B	.495211+00	.43−10	.93−21	.14−11
13	Powell Singular	4	4	0	14	12	B	.939098−04	.40−12	.11−14	.53−09
14	Powell Singular	4	4	1	19	14	B	.939098−04	.51−13	.74−15	.38−09
15	Powell Singular	4	4	2	24	17	R	.939098−04	.16−11	.12−11	.21−07
16	Freudenstrein & Roth	2	2	0	15	9	B	.244926+02	.31−15	.10−32	.36−17
17	Freudenstrein & Roth	2	2	1	19	11	B	.244926+02	.31−15	.10−32	.88−19
18	Freudenstrein & Roth	2	2	2	20	11	B	.244926+02	.31−15	.10−32	.41−19
19	Bard	15	3	0	8	8	R	.479114−02	.46−12	.31−13	.45−07
20	Bard	15	3	1	21	14	B	.479114−02	.94−11	.71−18	.22−09
21	Bard	15	3	2	32	20	R	.479114−02	.17−13	.10−11	.26−06
22	Kowalik & Osborne	11	4	0	10	9	R	.153915−03	.18−09	.16−11	.16−06
23	Kowalik & Osborne	11	4	1	43	30	B	.153915−03	.14−10	.97−15	.15−08
24	Kowalik & Osborne	11	4	2	44	31	B	.153915−03	.91−10	.59−13	.25−08
25	Meyer	16	3	0	138	127	B	.637259+02	.16−07	.24−15	.31−11
26	Meyer	16	3	1	28	16	B	.637259+02	.10−11	.56−12	.15−09
27	Watson 6	31	6	0	10	9	R	.114480−02	.21−09	.12−11	.27−07
28	Watson 6	31	6	1	17	12	R	.114480−02	.26−10	.29−10	.16−06
29	Watson 6	31	6	2	27	18	R	.114480−02	.21−12	.13−12	.12−07
30	Watson 9	31	9	0	16	13	R	.187007−01	.27−09	.75−10	.33−05
31	Watson 9	31	9	1	27	19	R	.187007−01	.46−10	.10−10	.95−06
32	Watson 12	31	12	0	67	42	R	.358214−01	.74−06	.78−10	.10−04
33	Watson 12	31	12	1	48	35	R	.358214−01	.14−08	.25−10	.95−05
34	Box 3D	10	3	0	10	8	R	.571741−04	.10−13	.37−10	.79−07
35	Jennrich & Sampson	10	2	0	12	10	B	.621916+02	.99−12	.83−15	.34−08
36	Brown & Dennis	20	4	0	23	15	R	.444302+05	.39−03	.11−10	.20−06
37	Brown & Dennis	20	4	1	22	14	R	.444302+05	.53−07	.20−13	.68−08
38	Brown & Dennis	20	4	2	25	16	R	.444302+05	.35−05	.39−10	.42−06
39	Chebyquad 8x1	8	1	0	1	1	B	.177895+01	.00+00	.00+00	.00+00
40	Chebyquad 8x1	8	1	1	42	26	S	.177520+01	.16−13	.65−06	.92−05
41	Chebyquad 8x1	8	1	2	2	1	S	.697247+36	.12+19	.15−17	.00+00
42	Chebyquad 8	8	8	0	28	20	R	.179612−02	.48−07	.24−10	.47−07
43	Chebyquad 9	9	9	0	13	12	B	.944235−04	.20−08	.16−11	.35−08
44	Chebyquad 10	10	10	0	17	15	R	.238636−02	.12−07	.54−11	.38−07
45	Brown almost linear	10	10	0	11	10	R	.304659−02	.91−07	.20−10	.15−06
46	Brown almost linear	10	10	1	30	21	B	.304659−02	.86−15	.15−13	.69−09
47	Brown almost linear	10	10	2	2	2	S	.154495+31	.56+16	.33−14	.85−16
48	Brown almost linear	30	30	0	10	9	R	.144179−01	.75−08	.46−12	.32−07
49	Brown almost linear	40	40	0	11	9	R	.203189−01	.15−07	.12−10	.25−06
50	Osborne1	33	5	0	21	18	B	.273514−04	.57−10	.17−15	.38−10
51	Osborne2	65	11	0	39	30	R	.255115+00	.30−05	.30−10	.60−06

MULTIGRID METHODS FOR PROBLEMS WITH A SMALL PARAMETER IN THE HIGHEST DERIVATIVE

P.W. Hemker

1. INTRODUCTION

Much progress has been made recently in developing multigrid (MG-) methods to solve the systems of equations that arise from discretization of truly elliptic PDEs. Often the emphasis lies upon the search for the most efficient variant. However, for the MG-methods to be generally applied, it is important that the methods are not only efficient, but also that they do not fail or do not need particular adaptation for special cases of the general elliptic equation. Therefore, in this paper, we consider the elliptic equation when it degenerates because a coefficient in the highest derivative tends to zero and we study the behaviour of some MG-methods under these circumstances. Related problems are studied in [2,4,5,12,15,22].

Our main objective is the development of methods for the general linear 2nd order elliptic PDE with variable coefficients

$$Lu \equiv -\nabla(\bar{\bar{\varepsilon}}\nabla u) + \bar{b}\nabla u + cu = f \quad \text{on} \quad \Omega, \tag{1.1}$$

$$u \quad = g \quad \text{on} \quad \Gamma_D, \qquad \Omega \subset \mathbb{R}^2 \text{ bounded},$$

$$\bar{n}\bar{\bar{\varepsilon}}\nabla u = h \quad \text{on} \quad \Gamma_N, \qquad \Gamma_N \cup \Gamma_D = \delta\Omega.$$

Here $\nabla = (\partial/\partial x, \partial/\partial y)$, and $\bar{\bar{\varepsilon}}$ is symmetric positive definite 2×2 matrix. The coefficients $\bar{b}$ and c and the data f,g and h are real functions on Ω or $\partial\Omega$.

In particular our interest goes to cases where general methods easily fail: (i) $\bar{\bar{\varepsilon}}$ has *one* small eigenvalue and (ii) $\bar{\bar{\varepsilon}}$ has *two* small eigenvalues w.r.t. $|\bar{b}h|$, where h is a characteristic length (e.g. the meshsize). To investigate these cases in detail we consider two constant coefficient model problems. The first is the *anisotropic diffusion equation*:

$$L_\varepsilon u \equiv -(\varepsilon c^2+s^2)u_{xx} - 2(\varepsilon-1)scu_{xy} - (\varepsilon s^2+c^2)u_{yy} = f, \tag{1.2}$$

with $c = \cos(\alpha)$ and $s = \sin(\alpha)$. This equation is obtained from $-\varepsilon u_{xx} - u_{yy} = f$ by rotation with an angle α. The eigenvalues of $\bar{\bar{\varepsilon}}$ are 1 and ε. The second problem is the *convection-diffusion equation:*

$$L_\varepsilon u \equiv -\varepsilon(u_{xx}+u_{yy}) + cu_x + su_y = f. \tag{1.3}$$

Here ε is a scalar coefficient and the convection direction is given by α.

We must keep in mind that in applications $\bar{b}$ and c are variable coefficients and the direction of the anisotropy or the convection is a priori unknown. Therefore we keep α as a parameter and we disregard the possibility of alignement of coordinate axes to the special direction in the equation.

Solutions of (1.2) and (1.3) may show layers, i.e. regions in which the solution varies rapidly. For (1.2) these layers may appear along lines in the direction of the strong diffusion. For (1.3) they may appear along the subcharacteristics or at the outflow boundary.

For the discretization of (1.1) we use methods of the finite element (FE) type. We assume that Ω can be covered by a triangularization T_h in a regular rectangular grid

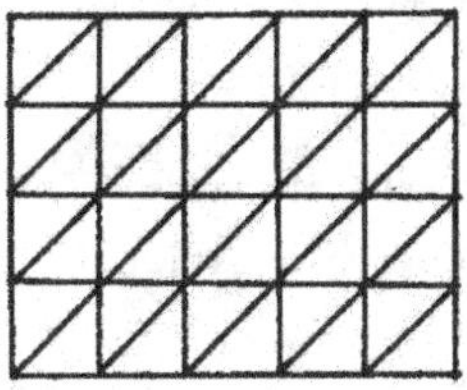

and we use spaces of trialfunctions S^h and testfunctions V^h, such that the support of a basisfunction ϕ_i (or ψ_i) in S^h (or V^h) consists of only the triangles that are connected with the nodal point x_i.

For simple functions ϕ_i and ψ_i these discretizations

$$L_h u_h = f_h \tag{1.4}$$

yield coefficient matrices L_h with a regular 7-diagonal structure. The standard (FEM) method is with both ϕ_i and ψ_i continuous piecewise linear. The 7-point discretizations are the simplest ones by which also a cross-term derivative u_{xy} can be represented.

2. THE MULTIGRID ALGORITHM

The multigrid method considered here is an iterative process for the solution of (1.4). It makes use of a sequence of discretizations on grids coarser than used for $L_h u_h = f_h$. Each next coarser grid has a doubled meshsize and is obtained by leaving out each second meshline.

In the multigrid method (MGM) each iteration cycle consists of:

1.) p (pre-) relaxation sweeps;

2.) a coarse grid correction;

3.) q (post-) relaxation sweeps.

The coarse grid correction (GGC) consists of:

a) the computation of the current residual, $r_h := f_h - L_h u_h$;

b) the restriction of the residual to the next coarser grid, $r_H := \bar{R}_{Hh} r_h$;

c) the computation of $\tilde{c}_H$, the approximate solution of the correction equation on a coarser grid:

$$L_H c_H = r_H, \tag{2.1}$$

by application of s MGM iteration cycles to this equation;

d) an update of the current solution u_h by addition of the prolongated (interpolated) correction

$$u_h := u_h + P_{hH}\tilde{c}_H.$$

By the recursive structure of this algorithm a coarsest grid exists on which the correction equation (2.1) has to be solved by another method (at choice). The coarse grid discrete operators L_H can be contained either by discretization, analogous to L_h, or by the construction of the *Galerkin approximation*

$$L_H = \bar{R}_{Hh} L_h P_{hH}. \tag{2.2}$$

We see that, beside the choice of the operator L_H, for a CGC we have to choose operators for the restriction ($\bar{R}_{Hh}$) and prolongation (P_{hH}). These operators are discussed in section 3.

If (2.1) is solved exactly, no coarser discretizations than L_H are involved, and the algorithm is a two-grid method (TGM). Its CGC is described by

$$u_h := u_h + P_{hH} L_H^{-1} \bar{R}_{Hh} (f_h - L_h u_h). \tag{2.3}$$

It can be shown [8] that under suitable conditions, for s large enough (roughly $s \geq 2$), the convergence behaviour of the MGM is almost the same as of the TGM. In practice, also $s = 1$ is often a good choice.

Most essential for the efficiency of the MGM is the choice of the relaxation method. Methods that are often used in this context are Point Gauss-Seidel relaxations (scanning the points in some order e.g. red-black or various lexicographical orderings) Line Gauss-Seidel relaxations (with different possible line-orderings, e.g. zebra or lexicographical [19]). Other relaxation methods are based on incomplete decompositions of the coefficient matrix, viz. Incomplete LU-decomposition (ILU-) relaxation or Incomplete Line-LU- (ILLU-) relaxation. All these relaxation methods are of the form

$$\tilde{L}_h u_h^{(i+1)} = \tilde{L}_h u_h^{(i)} - L_h u_h^{(i)} + f_h, \tag{2.4}$$

where $\tilde{L}_h$ is an approximation to L_h.

For ILU relaxation, in each sweep a linear system is solved of the form

$$L\, U\, u_h^{(i+1)} = f_h + R\, u_h^{(i)},$$

where $\tilde{L}_h = LU = L_h - R$ is an approximate Crout-decomposition of L_h, with L and U lower and upper triangular matrices with the same sparsity structure as L_h [17,20].

For ILLU relaxation [15,16] in each sweep $u_h^{(i+1)}$ is solved from a system

$$(L+\bar{D})\bar{D}^{-1}(\bar{D}+U)(u_h^{(i+1)} - u_h^{(i)}) = f_h - L_h u_h^{(i)}. \tag{2.5}$$

The matrices L, $\bar{D}$ and U are obtained from the coefficient matrix L_h, written in block-tridiagonal form as

$$L_h = L + D + U = \begin{array}{|c|c|c|c|c|} \hline D_1 & U_1 & & & \\ \hline L_2 & D_2 & U_2 & & \\ \hline & L_3 & D_3 & \ddots & \\ \hline & & \ddots & \ddots & U_{n-1} \\ \hline & & & L_n & D_n \\ \hline \end{array} \, .$$

The block-diagonal matrix $\bar{D}$, with the same sparsity pattern as D, is computed recursively from

$$\begin{cases} \bar{D}_1 = D_1, \\ \bar{D}_j = D_j - \underline{\text{tridiag}} \; (L_j \bar{D}_{j-1}^{-1} U_{j-1}), \quad j = 2,3,\ldots,n, \end{cases} \tag{2.6}$$

where tridiag is the operator which selects the tridiagonal submatrix from a dense matrix.

3. NESTED DISCRETIZATIONS

The relation between the discretizations on the different grids in a MGM can be considered analogous to the relation between the continuous and a discrete problem. For discretization of an equation $Lu = f$, $(L: X \to Y$, X and Y Banach spaces), we relate to it the discrete equation $L_h u_h = f_h$, $(L_h: X_h \to Y_h)$. The relation between the two equations is made by the prolongation $P_h: X_h \to X$ (a linear injection) and the restrictions $R_h: X \to X_h$ and $\bar{R}_h: Y \to Y_h$ (linear surjections).

In the same way coarser discretizations $L_H u_H = f_H$ are related to finer $L_h u_h = f_h$ by a prolongation $P_{hH}: X_H \to X_h$ (linear injection), and restrictions $R_{Hh}: X_h \to X_H$ and $\bar{R}_{Hh}: Y_h \to Y_H$ (linear surjections). The coarse grid Galerkin approximation (2.2) is the analogue of the Galerkin discretization $L_h = \bar{R}_h L P_h$. A sequence of nested discretizations of a continuous equation $Lu = f$ is obtained by selecting prolongations and restrictions such that

$$P_H = P_h P_{hH}, \; R_H = R_{Hh} R_h, \; \bar{R}_H = \bar{R}_{Hh} \bar{R}_h. \tag{3.1}$$

In the standard FE discretization (sect. 1) the prolongation $P_h: X_h \to X$ is defined by linear interpolation over the triangles of T_h; $R_h: X \to X_h$ is defined by injection (i.e. restriction of the function values to nodal points) and $\bar{R}_h$ is defined by weighting by the continuous piecewise linear basis-functions $\phi_i^h \in V^h$:

$$(\bar{R}_h f)_i \equiv (f_h)_i \equiv \int f(x) \phi_i^h(x) d\Omega. \tag{3.2}$$

The FE discretization corresponds to the Galerkin discretization $L_h = \bar{R}_h L P_h$.

To obtain a sequence of nested discretizations related with the FE discretization, we use a corresponding P_{hH} and $\bar{R}_{Hh}$. The prolongation P_{hH} should satisfy (3.1) with linear interpolation for P_h and P_H. Hence,

$$P_H u_H = \sum_j u_j^H \phi_j^H = \sum_{j,i} u_j^H r_{ji} \phi_i^h = \sum_i \left(\sum_j r_{ji} u_j^H\right) \phi_i^h = P_h P_{hH} u_h .$$

Therefore, P_{hH} is given by the prolongation molecule

$$P_{hH}^{*} = (r_{ji}) = \begin{bmatrix} & \frac{1}{2} & \frac{1}{2} \\ \frac{1}{2} & 1 & \frac{1}{2} \\ \frac{1}{2} & \frac{1}{2} & \end{bmatrix} . \tag{3.3}$$

For $\bar{R}_{Hh}$ we have

$$(\bar{R}_{Hh} f_h)_i = (f_H)_i = \int f(x)\phi_i^H(x)d\Omega = \int f \sum_j r_{ij}\phi_j^h d\Omega = \sum_j r_{ij}(f_h)_j ;$$

the restriction molecule $\bar{R}_{Hh}^{*}$ of $\bar{R}_{Hh}$ is also given by (3.3). In this case $\bar{R}_{Hh} = (P_{hH})^T$. These P_{hH} and $\bar{R}_{Hh}$ are the same 7-point prolongation and restriction as introduced in [20,21]. For points near the boundary obvious modifications of the molecules have to be made.

With the P_{hH} and $\bar{R}_{Hh}$ given by (3.3), the FE discretization on the different levels form a nested sequence and L_H can be computed from L_h by (2.2):

$$L_H = \bar{R}_H L P_H = \bar{R}_{Hh} \bar{R}_h L_h P_h P_{hH} = \bar{R}_{Hh} L_h P_{hH} . \tag{3.4}$$

Starting with the discretization L_h on N meshpoints, and using (3.3) for $\bar{R}_{Hh}$ and P_{hH}, it takes less than 29 N additions and 7/3 N multiplications to compute the discrete operators on *all* coarser grids.

Application of FEM to the constant coefficient equation (1.1) on an equidistant regular T_h yields the 7-point difference molecules

$$-h^2\left(\frac{\partial}{\partial x}\right)^2 \sim \begin{bmatrix} & -1 & 0 \\ 0 & 2 & 0 \\ 0 & -1 & \end{bmatrix} = A_{11}^{*}, \quad -h^2\left(\frac{\partial}{\partial y}\right)^2 \sim \begin{bmatrix} & 0 & 0 \\ -1 & 2 & -1 \\ 0 & 0 & \end{bmatrix} = A_{22}^{*},$$

$$-2h^2\left(\frac{\partial}{\partial x}\right)\left(\frac{\partial}{\partial y}\right) \sim \begin{bmatrix} & 1 & -1 \\ 1 & -2 & 1 \\ -1 & 1 & \end{bmatrix} = A_{12}^{*}, \tag{3.5}$$

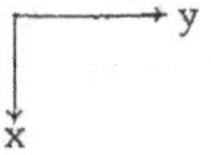

the 2nd order terms; and

$$6h\left(\frac{\partial}{\partial x}\right) \sim \begin{bmatrix} & -2 & -1 \\ -1 & 0 & 1 \\ 1 & 2 & \end{bmatrix} = A_1^{*}, \quad 6h\left(\frac{\partial}{\partial y}\right) \sim \begin{bmatrix} & -1 & 1 \\ -2 & 0 & 2 \\ -1 & 1 & \end{bmatrix} = A_2^{*}, \tag{3.6}$$

the 1st order terms; and

$$12I \sim \begin{bmatrix} & 1 & 1 \\ 1 & 6 & 1 \\ 1 & 1 & \end{bmatrix} = A_0^{*}. \tag{3.7}$$

the 0th order term.

For each p-th order difference molecule A_h^{*} in (3.5)-(3.7) we find (H=2h)

$$\bar{R}_{Hh}^{*} * A_h^{*} * P_{hH}^{*} = 2^{2-p} A_{2h}^{*}, \tag{3.8}$$

where .*.*. denotes the combined application of the prolongation and restriction (i.e.

convolution and contraction of the molecules). This means that the difference molecules (3.5)-(3.7) are all invariant under Galerkin approximation. The factor 2^{2-p} takes into account the difference in meshsize on the different levels. A 7th linearly independent 7-point molecule,

$$A_3^* = \begin{bmatrix} & -1 & 1 \\ 1 & 0 & -1 \\ -1 & 1 & \end{bmatrix} \sim h^3(\frac{\partial}{\partial t})(\frac{\partial}{\partial y})(\frac{\partial}{\partial x} - \frac{\partial}{\partial y}), \tag{3.9}$$

satisfies (3.8) with $p = 3$.

It follows that other than FEM molecules are *not* invariant under Galerkin approximation with $\bar{R}_{Hh}$ and P_{hH} given by (3.3). Examples are:

1) the central difference operator

$$6h(\frac{\partial}{\partial x}) \sim A_1^* + A_3^*; \tag{3.10}$$

2) the upwind difference operator

$$6h(\frac{\partial}{\partial x}) \sim U_1^* = A_1^* + A_3^* + 3A_{11}^*. \tag{3.11}$$

For any of these discretizations on the finest grid, the repeated use of (2.2) with the P_{hH} and $\bar{R}_{Hh}$ given by (3.3), yields discretizations on coarser grids that tend to the FE discretization. E.g., for (3.11) k times application of (3.8) yields

$$(R^**)^k(hU_1^*)(*P^*)^k = 2^k h[A_1^* + 2^{-2k}A_3^* + 3.2^{-k}A_{11}^*], \tag{3.12}$$

which tends to $2^k hA_1^*$ as k increases.

4. THE ANISOTROPIC DIFFUSION EQUATION

The 7-point molecule for (1.2) obtained by FEM reads

$$L_{h,\varepsilon}^* = \begin{bmatrix} & s(c-s) & -sc \\ c(s-c) & 2-2sc & c(s-c) \\ -sc & s(c-s) & \end{bmatrix} + \varepsilon\begin{bmatrix} & -c(s+c) & sc \\ -s(s+c) & 2+2sc & -s(s+c) \\ sc & -c(s+c) & \end{bmatrix}. \tag{4.1}$$

To investigate the relaxation methods for (4.1) we use Local Mode Analysis [19], i.e. we consider the discretization on an infinite domain (or on a finite domain with periodic boundary conditions). For the linear constant coefficient (difference) operator L (or L_h) its symbol $\hat{L}(\omega)$ (or $\hat{L}_h(\omega)$) is introduced by

$$Lu_\omega = \hat{L}(\omega)u_\omega \quad \text{or} \quad L_h u_\omega = \hat{L}_h(\omega)u_\omega,$$

where

$$u_\omega(x,y) = e^{i(\omega_1 x + \omega_2 y)},$$

$$\hat{L}: \mathbb{R}^2 \to \mathbb{C},$$

$$\hat{L}_h: T_h^2 \equiv [-\pi/h, \pi/h]^2 \to \mathbb{C}.$$

T_h^2 is the domain of all frequencies ω that are visible on a grid with meshsize h. For convenience we use also the notation $\phi = h\omega_1$, $\theta = h\omega_2$, and $u_\omega = u_{\phi,\theta}$.

For equation (1.2) we find the symbol

$$h^2\hat{L}_\varepsilon(\phi,\theta) = (s\phi-c\theta)^2 + \varepsilon(c\phi+s\theta)^2. \tag{4.2}$$

For $\varepsilon > 0$, $(\phi,\theta) \neq (0,0)$ we have $\hat{L}_\varepsilon(\omega) > 0$, which shows the ellipticity of L_ε. For the reduced case, $\varepsilon = 0$, we have

$$L_\varepsilon(\phi,\theta) = 0 \text{ iff } s\phi = c\theta. \tag{4.3}$$

This problem is not longer elliptic and it has *unstable modes* $u_{\phi,\theta}$ for (ϕ,θ) satisfying (4.3). For the discretized problem (1.2), we derive from (4.1)

$$\begin{aligned}\hat{L}_{h,\varepsilon}(\phi,\theta) &= [s+(c-s)\cos\phi-c\cos(\phi-\theta)]^2 + [(c-s)\sin\phi-c\sin(\phi-\theta)]^2 \\ &\quad + \varepsilon[c-(c+s)\cos\phi+s\cos(\phi-\theta)]^2 + \varepsilon[(c+s)\sin\phi-s\sin(\phi-\theta)]^2.\end{aligned} \tag{4.4}$$

Again $\hat{L}_{h,\varepsilon}(\phi,\theta) > 0$ for $\varepsilon > 0$, $(\phi,\theta) \neq (0,0)$, but for $\varepsilon = 0$ we see

$$\hat{L}_{h,0}(\phi,\theta) = 0 \text{ iff } \begin{cases} \text{(i)} & \phi = \theta = 0, \\ \text{or (ii)} & \phi = 0 \text{ and } c = 0, \\ \text{or (iii)} & \theta = 0 \text{ and } s = 0, \\ \text{or (iv)} & \phi = \theta \text{ and } s = c. \end{cases} \tag{4.5}$$

Except for $\phi = \theta = 0$, the discrete operator has unstable modes *only* if the direction of the strong diffusion is along one of the (three) gridline directions. For all other α we find $L_{L,0}(\phi,\theta) > 0$ for $(\phi,\theta) \neq (0,0)$. If the strong diffusion is *not* along the gridlines, the discrete scheme is elliptic where the original operator is not. The discretization introduces artificial cross-diffusion.

For $\alpha \neq 0, \pi/4, \pi/2$, this extra stability guarantees the existence of a relaxation method for its solution [6] (viz. a properly damped Jacobi relaxation). How to find an efficient relaxation, however, is not immediately clear. This is particularly so because (4.1) does not yield an L-matrix (non-negative off-diagonal elements) for all ε and α, and hence e.g. ILU-relaxation may diverge. The domain is the α-ε-plane where $L_{h,\varepsilon}$ is an L-matrix is shown in figure 4.1.

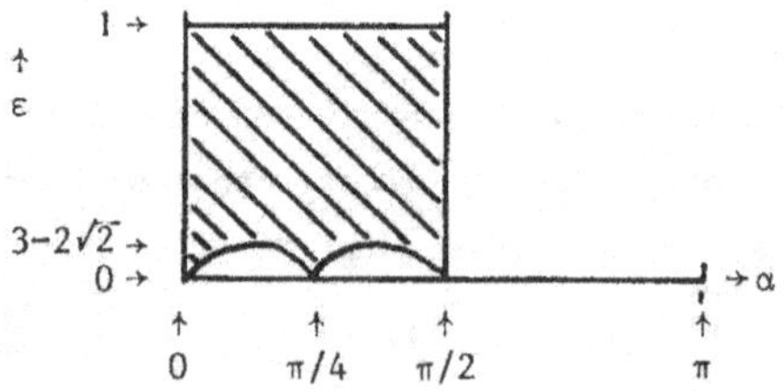

Fig. 4.1. The shaded area denotes the domain in the (α,ε)-plane where (4.1) yields an L-matrix.

We see that $\alpha = 0, \pi/4, \pi/2$ play again a special role. Therefore in numerical experiments more angles α have to be considered to obtain an insight in the general behaviour of the smoothing processes. Here, to examine the relaxation methods, we compute the smoothing factor (cf. [19] sect. 7) for various (α,ε), for lexicographical Gauss-Seidel, zebra, ILU and ILLU relaxation. The smoothing factor gives a first impression of the rate of convergence of a MGM with $p + q = 1$.

	PGS		zebra		ILU		ILLU	
α \ ε	1.0(-2)	1.0(-4)	1.0(-2)	1.0(-4)	1.0(-2)	1.0(-4)	1.0(-2)	1.0(-4)
0°	0.980	1.000	0.125	0.125	0.607	0.946	0.1759	1.97(-1)
$7\frac{1}{2}^{o}$	0.963	0.984	0.472	0.660	2.469 !	6.855 !	0.0607	2.97(-5)
15°	0.928	0.948	0.659	0.727	1.352 !	1.735 !	0.0152	3.05(-6)
22.5	0.902	0.924	0.751	0.803	0.711	0.767	0.0069	1.19(-6)
30	0.885	0.910	0.838	0.884	0.701	0.767	0.0127	2.81(-6)
37.5	0.908	0.957	0.911	0.960	1.138 !	2.283 !	0.0418	2.04(-5)
45	0.925	0.999	0.943	0.999	0.497	0.92	0.1323	1.64(-1)
52.5	0.897	0.949	0.887	0.943	0.044	1.68(-5)	0.0536	4.08(-5)
60	0.871	0.908	0.838	0.874	0.0192	4.02(-6)	0.0196	5.35(-6)
67.5	0.882	0.910	0.867	0.894	0.0168	3.29(-6)	0.0165	3.94(-6)
75	0.921	0.944	0.915	0.940	0.0264	6.56(-6)	0.0239	6.94(-6)
82.5	0.962	0.983	0.961	0.983	0.0728	5.26(-5)	0.0641	5.25(-5)
90	0.980	1.000	0.980	1.000	0.1709	0.1716	0.1490	0.1692
97.5	0.967	0.983	0.966	0.982	0.0420	2.36(-5)	0.0350	1.96(-5)
105	0.927	0.940	0.921	0.935	6.92(-3)	1.27(-6)	5.17(-3)	4.69(-7)
112.5	0.874	0.884	0.845	0.857	1.71(-3)	2.28(-7)	1.16(-3)	1.06(-7)
120	0.819	0.826	0.763	0.773	6.56(-4)	7.78(-8)	3.82(-4)	1.19(-7)
127.5	0.769	0.774	0.661	0.668	3.33(-4)	3.76(-8)	1.69(-4)	6.50(-8)
135	0.724	0.727	0.547	0.553	2.20(-4)	2.41(-8)	1.87(-4)	6.53(-8)
142.5	0.685	0.691	0.427	0.431	3.13(-4)	3.47(-8)	2.76(-4)	1.19(-7)
150	0.777	0.786	0.397	0.411	5.63(-4)	6.44(-8)	5.26(-4)	1.15(-7)
157.5	0.859	0.870	0.432	0.454	1.32(-3)	1.63(-7)	1.38(-3)	2.11(-7)
165	0.923	0.937	0.451	0.501	4.42(-3)	6.50(-7)	5.62(-3)	9.50(-7)
172.5	0.966	0.983	0.394	0.543	0.0275	7.99(-6)	0.0396	1.62(-5)

Table 4.1 Smoothing factors by Local Mode Analysis of lexicographic Point Gauss-Seidel, zebra, ILU and ILLU relaxation for (4.1).

From table 4.1 we see that Gauss-Seidel relaxation is slow for small ε, zebra smoothing is essentially better. For angles $\pi/4 \le \alpha \le \pi$ the ILU is an excellent smoother, but for $0 < \alpha < \pi/4$ it is unreliable and may diverge. This divergence, found by local mode analysis, appears for modes $u_{\phi,\theta}$ with $(\phi,\theta) \approx (0,\pi)$ if $0 < \alpha < \pi/8$ and for $(\phi,\theta) \approx (-\pi,\pi)$ if $\pi/8 < \alpha < \pi/4$. The same modes are also found to diverge in real MGM-iterations if a fine enough grid is used. The ILLU relaxation converges rapidly in all cases.

That ILU is a good smoother for $\pi/4 < \alpha < \pi$ for all ε can be explained by the fact that (4.1) with $\varepsilon = 0$ can be decomposed into a product of two molecules corresponding to the LU decomposition. Two of these decompositions are possible

$$\begin{bmatrix} & s(c-s) & -cs \\ c(s-c) & 2-2cs & c(s-c) \\ -cs & s(c-s) & \end{bmatrix} = \begin{bmatrix} & s(c-s) & 0 \\ c(s-c) & (s-c)^2 & 0 \\ 0 & 0 & \end{bmatrix} * \begin{bmatrix} & 0 & 0 \\ 0 & 1 & \frac{c}{s-c} \\ 0 & \frac{-s}{s-c} & \end{bmatrix} \tag{4.6}$$

$$= \begin{bmatrix} & s(c-s) & -cs \\ 0 & s^2 & 0 \\ 0 & 0 & \end{bmatrix} * \begin{bmatrix} & 0 & 0 \\ 0 & 1 & 0 \\ -c/s & \frac{c-s}{s} & \end{bmatrix} \tag{4.7}$$

Decomposition (4.7) has a bounded inverse for $\pi/4 < \alpha < \pi/2$ and (4.6) for $\pi/2 < \alpha < \pi$ For these ranges of α, the corresponding decompositions are also those found by the LU-decomposition algorithm as described in [20,21]. However, neither (4.6) nor (4.7) is found for $0 < \alpha < \pi/4$. For small ε the decompositions (4.6) and (4.7) are found asymptotically. For $\varepsilon \to 0$ they approximate L and U up to $O(\varepsilon)$. Therefore, for $\pi/4 < \alpha < \pi$, the ILU-decomposition LU is an accurate approximation to $L_{h,\varepsilon}$ for small ε. Hence ILU is a good smoother (only) in these cases.

To explain the small smoothing factor for the ILLU relaxation we consider (2.5) -(2.6) and we introduce the molecules

$$L^* = [0, \quad s(c-s), \ -sc \quad],$$

$$D^* = [c(s-c), \ 2-2cs, \quad c(s-c)],$$

$$U^* = [-sc, \quad s(c-s), \ 0 \quad].$$

We compute $\bar{D}^*$ from $\bar{D}^* * (\bar{D}^* - D^*) + L^* * U^* = 0$ and obtain

$$\bar{D}^* = [c(s-c), \ (c-s)^2 + c^2, \ c(s-c)].$$

This $\bar{D}^*$ corresponds to a tridiagonal matrix and has a bounded inverse for all α. Therefore, it is also a solution of

$$\bar{D}^* = D^* - \underline{\text{tridiag}}\ (L^* * (\bar{D}^*)^{-1} * U^*),$$

and we find that this ILU-decomposition is exact for $\varepsilon = 0$:

$$\tilde{L}^*_{h,0} = (L^* + \bar{D}^*)(\bar{D}^*)^{-1}(\bar{D}^* + U^*) = L^* + D^* + U^* = L^*_{h,0}.$$

For small ε the ILLU-decomposition algorithm generates this decomposition asymptotically (away from the boundary). This explains why ILLU is an excellent smoother for small ε and all α.

5. THE CONVECTION DIFFUSION EQUATION

The direct application of the standard FEM to equation (1.3) yields inadequate discretizations for small ε/h [3,13,18]. The same is true for central differences. Therefore, either direction-dependent (upwind-) differences are used or an artificial diffusion is introduced. For the solution of the discrete systems the MG-method can be applied. However, due to the special character of the equations, typical difficulties may arise. These have been studied for upwind differences in [5,12,16] and for artificial diffusion in [4,5,22].

In this section we make some remarks on the application of artificial diffusion. In section 6 we consider a MG-variant consistent with the Streamline-Upwind FE-method [7,14].

Molecules for the discretization of (1.3) are given by

$$L^*_{\gamma,h} = \gamma(A^*_{11}+A^*_{22}) + \frac{hc}{6}(A^*_1+pA^*_3) + \frac{hs}{6}(A^*_2-pA^*_3). \tag{5.1}$$

For finite elements $p = 0$, for central differences $p = 1$; $\gamma = \gamma(h) = \varepsilon + \beta h$ is the diffusion coefficient, β or $\beta(h)$ is the coefficient of artificial diffusion.

The symbol of $L_{\gamma,h}$ is given by

$$\hat{L}_{\gamma,h}(\phi,\theta) = 4\gamma[\sin^2(\phi/2)+\sin^2(\theta/2)] + ihT(\phi,\theta), \tag{5.2}$$

with

$$\begin{aligned} T(\phi,\theta) &= c\ \sin(\phi/2)[2\ \cos(\phi/2)+(1-p)\cos(\theta-\phi/2)] \\ &\quad + s\ \sin(\theta/2)[2\ \cos(\theta/2)+(1-p)\cos(\phi-\theta/2)]. \end{aligned} \tag{5.3}$$

For the continuous operator (1.3) we have

$$\hat{L}_\varepsilon(\phi,\theta) = \varepsilon(\phi^2+\theta^2) + ih(c\phi+s\theta).$$

Hence, for (1.3) with $\varepsilon = 0$, unstable modes $u_{\phi,\theta}$ exist for (ϕ,θ) with $c\phi + s\theta = 0$. I.e. the reduced continuous operator has one set of unstable modes. (In the solution these components are determined by the inflow boundary data.) For (5.2) with $\gamma = 0$ we find two branches of unstable modes. A branch of spurious unstable modes is found in the high frequency domain (figure 5.1). This implies that relaxation methods of the form (2.4) do not damp these high frequency components.

Fig. 5.1

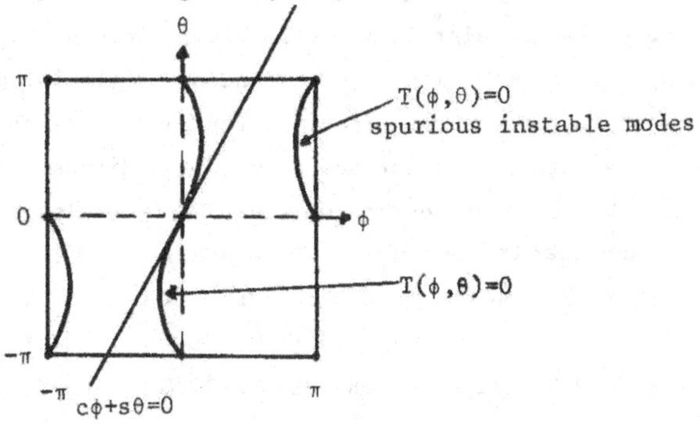

From (5.2) it follows that the discretization (5.1) is stable if $\gamma \geq Ch > 0$ (cf. [22]). For such stable discretizations the existence of a relaxation method which damps the high frequencies in the error is guaranteed (cf. [6]). It is of practical importance to find relaxations that work efficiently. To compare the effect of some relaxations, in table 5.1 we give smoothing factors for zebra, ILU and ILLU relaxation, when applied to (5.1) with $p = 0$; $\gamma(h) = \varepsilon+\beta h$; $\varepsilon = 0$; $\beta = 0.5, 1$.

	zebra		ILU		ILLU	
α	$\beta = 0.5$	$\beta = 1.0$	$\beta = 0.5$	$\beta = 1.0$	$\beta = 0.5$	$\beta = 1.0$
0°	0.711	0.280	0.0777	0.0937	0.0221	0.0257
30°	0.584	0.245	0.0148	0.0837	0.0021	0.0366
60°	0.286	0.243	0.0754	0.100	0.0601	0.0611
90°	0.252	0.250	0.195	0.127	0.1063	0.0746
120°	0.290	0.260	0.237	0.139	0.0674	0.0607
150°	0.573	0.263	0.160	0.123	0.0324	0.0364

Table 5.1 Smoothing rates for FE discretization of (1.3) with $\varepsilon = 0$ and with artificial diffusion $\gamma = \beta h$.

In the MGM not only the smoothing should efficiently damp the high frequencies in the error, but also the CGC should work properly to reduce the low frequencies. For equation (1.3) this CGC needs special attention. From section 3 we know that application of (2.2) to $L_{\gamma(h),h}$, with P_{hH} and $\bar{R}_{Hh}$ given by (3.3), yields on a coarser level

$$\bar{R}_{2h,h} L_{\gamma(h),h} P_{h,2h} = L_{\gamma(h),2h},$$

i.e. the FE discrete operator on the grid 2h with diffusion coefficient $\gamma(h)$. This means that the Galerkin approximation gives an amount of diffusion on the coarse grids that is equal to the amount used at the finer grids. When repeatedly applied, this produces a FE discretization with negligible artificial diffusion on the coarsest grids. Hence, the coarser grid operators become unstable, and diverging corrections will appear in the CGCs.

To avoid the unstable Galerkin approximations, we can discretize the problem on each grid - with meshsize H - with a corresponding artificial diffusion $\gamma(H)$. This is studied in [22], where suggestions are given for the choice of $\gamma(h)$ on the different levels. However, the lack of consistency between the diffusion terms in the discrete operators affects the convergence rate of the CGC. By the same argument as used in [5, p40], it is found that the reduction of some low-frequency components is only by a factor $(\gamma(H)-\gamma(h))/\gamma(H)$ when in a CGC the operators $L_{\gamma(h),h}$ and $L_{\gamma(H),H}$ are used.

6. A STREAMLINE-UPWIND RESTRICTION FOR THE CONVECTION DIFFUSION EQUATION

In this section we introduce a new, asymmetric restriction. This restriction is applied in combination with the Streamline-Upwind Petrov-Galerkin (SU-PG) FE method of discretization [7,14] and, in fact, it is the discrete analogue of the asymmetric weight-function in that method. With this $\bar{R}_{Hh}$ in (2.2) we obtain Galerkin coarse grid operators that are again of the Streamline-Upwind type. Other asymmetric restrictions have been studied for finite difference methods in [1,12,16]. These restrictions satisfy $\bar{R}_{Hh} = P^T_{hH}$ where the interpolation P_{hH} is deduced from the difference equation (matrix-weighted interpolation). The comparison of the different asymmetric methods might be the subject of future study. Here, with the new restriction, we remain consistent with the Petrov-Galerkin approach [7,14]: the prolongation is kept unchanged and only the restriction is adapted to the differential equation.

The SU-PG method is a FE method for the solution of (1.1) with trialfunctions $S^h = \text{span}\{\phi^h_j\}$ and testfunctions $V^h = \text{span}\{\phi^h_j + k\bar{b}\,\nabla\,\phi^h_j\}$. The functions ϕ^h_j are standard FE basis-functions. We apply the method with piecewise linear ϕ^h_j on the triangularization T^h; $k = k(h,\varepsilon)$ is a scalar parameter. It can be shown [11] that a good choice of $k(h,\varepsilon)$ should satisfy

$$k(h) = O(h) \quad \text{if} \quad \varepsilon/h \le C, \; k(h) = O(h^2/\varepsilon) \quad \text{if} \quad \varepsilon/h \ge C.$$

For (1.3) we obtain the discrete equations

$$\sum_j B(\phi_i,\phi_j)u^h_j = \ell(\phi_i), \tag{6.1}$$

where

$$B(\phi_i,\phi_j) = \sum_e \int \nabla\phi_i(\varepsilon I + k\bar{b}^T\bar{b})\nabla\phi_j + \phi_i\bar{b}\nabla\varphi_j d\Omega_e, \tag{6.2}$$

$$\ell(\phi_i) = \sum_e \int (\phi_i + k\bar{b}\nabla\phi_i) f \; d\Omega_e. \tag{6.3}$$

The differences with standard FEM are:

1) an anisotropic (streamline directed) artificial diffusion appears:

$$\bar{\bar{\gamma}}(h) = \varepsilon I + k(h)\bar{b}^T\bar{b}; \tag{6.4}$$

2) the functions in the space Y are weighted by an asymmetric (upwind weighted) weight function.

The weighting of the space Y defines an asymmetric restriction $\bar{R}_h$ and for a piecewise polynomial approximation of functions in Y, restriction molecules can be derived. For instance, if Y is approximated by piecewise linear functions on T_h, this restriction molecule reads

$$\bar{R}^*_h = \frac{h^2}{6}\,\{\tfrac{1}{2}A^*_0 + kb_1A^*_1 + kb_2A^*_2\}. \tag{6.5}$$

This asymmetric molecule suggests an asymmetric restriction $\bar{R}_{Hh}$ with a molecule

$$\bar{R}^*_{Hh} = P^*_{hH} + \mu_1(A^*_1 + pA^*_3) + \mu_2(A^*_2 - pA^*_3) \tag{6.6}$$

(e.g. $p = 0$ or $p = 1$). The difference molecule corresponding with the discrete operator (6.2) reads

$$h^2 L_h^* = \gamma_{11} A_{11}^* + \gamma_{12} A_{12}^* + \gamma_{22} A_{22}^* + \frac{hb_1}{6} A_1^* + \frac{hb_2}{6} A_2^* . \tag{6.7}$$

When (6.6), (6.7) and (3.3) are used for the construction of a Galerkin approximation (2.2) we find the molecule

$$\begin{aligned}(2h)^2 \bar{R}_{Hh}^* * L_h^* * P_{hH}^* = {} & (\gamma_{11} - \frac{3h\mu_1 b_1}{2}) A_{11}^* + (\gamma_{12} - \frac{3h}{2} \frac{\mu_1 b_2 + \mu_2 b_1}{2}) A_{12}^* + \\ & + (\gamma_{22} - \frac{3h}{2} \mu_2 b_2) A_{22}^* + \frac{h}{3}(b_1 A_1^* + b_2 A_2^*) + r(\vec{b},p) A_3^* .\end{aligned} \tag{6.8}$$

This is a discretization on the mesh $H = 2h$ of the same form as (6.7) except for the remainder term $r(\vec{b},p)A_3^*$. We see that (6.8) has the additional amount of artificial diffusion

$$\gamma(2h) - \gamma(h) = -\frac{3h}{2} \binom{b_1}{b_2} (\mu_1, \mu_2) .$$

This is accounted for by the h-dependence of the parameter k. For (6.7) and (6.8) to be consistent with (6.4) the following relation is to be satisfied

$$[k(2h) - k(h)] \bar{b}^T \bar{b} = \frac{-3h}{2} \binom{b_1}{b_2} (\mu_1, \mu_2) .$$

Introducing the notation $\mu_j = -\frac{2}{3} \mu(h) b_j$, $j = 1,2$, this relation reads

$$k(2h) - k(h) = h\mu(h) . \tag{6.9}$$

Thus, our restriction (6.6) is upwind weighted and equation (6.9) shows how the parameters μ_k in the asymmetric restriction are related to the choice of the artificial streamline-diffusion parameter $k(h)$.

With this asymmetric $\bar{R}_{Hh}$ we expect the MGM to improve for the SU-PG discretization of (1.3) because (i) the CGCs use streamline-upwind Galerkin approximations as coarse grid operators and (ii) by the asymmetric restriction downstream residuals have less upstream influence.

An experiment was made to see the effects. The problem (1.3) was solved for $\varepsilon = 10^{-3}$, on the unit square, using a MGM cycle with $(p,q,s) = (1,0,2)$ and with 5 levels of discretization ($h=1/32$ on the finest mesh; $k(h) = 2h/3$). Both the solution and the initial error were smooth. To see the effect of the new CGC a relaxation was used (zebra) with little capacity to reduce the low frequency error in the direction $\alpha = 0^\circ$ or 180°. Another experiment was made with the ILLU-relaxation (table 6.1).

L_H / $\bar{R}_{Hh}$ / α	zebra (2.2) (3.3)	zebra (2.2) (6.6) p=0	zebra (6.2) (3.3)	zebra (6.2) (6.6) p=0	ILLU (2.2) (3.3)	ILLU (2.2) (6.6) p=0	ILLU (6.2) (3.3)	ILLU (6.2) (6.6) p=0
0°	div	2.1	2.1	2.1	88.4	72.8	78.5	75.2
22.5^o	2.0	1.8	1.9	1.8	29.9	14.6	24.5	14.7
45	2.1	2.6	2.5	2.6	23.7	10.7	16.1	10.7
67.5	2.4	5.4	2.9	4.3	19.0	13.8	16.8	14.8
90	4.9	5.0	4.5	4.9	29.7	25.8	26.7	25.9
112.5	2.0	2.9	2.1	2.9	27.9	25.3	23.7	25.0
135	div	3.9	1.5	4.0	124.	111.	112.	111.
157.5	div	2.9	1.7	2.9	116.	104.	104.	104.
180	div	2.4	1.7	2.3	41.4	36.2	35.8	36.4
202.5	1.7	2.4	1.9	2.3	19.3	11.8	14.1	14.4
225	1.7	2.9	1.7	2.9	21.7	14.3	15.0	14.3
247.5	2.1	3.5	2.5	3.2	21.2	17.6	15.8	16.4
270	4.4	4.2	3.8	4.2	29.1	25.6	25.7	25.5
292.5	div	2.2	1.4	2.2	17.1	13.9	15.4	14.1
315	div	3.6	1.8	3.6	87.6	74.7	78.0	74.7
337.5	div	3.4	2.3	3.3	150.	124.	125.	126.

Table 6.1 Residual convergence factors
$\sqrt[3]{\| f_h - L_h u_h^{(2)} \|_2 / \| f_h - L_h u_h^{(5)} \|_2}$.

Table 6.1 shows that the Galerkin approximation (2.2) with the symmetric $\bar{R}_{Hh}$, eq. (3.3), may diverge indeed. When the asymmetric $\bar{R}_{Hh}$, eq. (6.6), is used, little difference is seen between the CGCs with discretizations (2.2) or (6.2), as was expected from (6.8). In the case of zebra-relaxation the use of the asymmetric restriction has a positive effect. (Similar results were obtained for problems with boundary layers.) If we use the more powerful ILLU-relaxation, we see that the new CGC becomes of little importance. It even has an adverse effect. Now Galerkin approximation with the symmetric $\bar{R}_{Hh}$ shows the best convergence rate. (This effect may disappear if more levels of discretization are used, cf. [22].) Apparently the ILLU-relaxation reduces the total error very efficiently. It also takes care of the low frequency components that are produced by the less stable (and more accurate) CGC obtained by the symmetric Galerkin approximation. This effect is seen for all flow directions α.

REFERENCES

[1] ALCOUFFE, R.E., A. BRANDT, J.E. DENDY Jr. & J.W. PAINTER, *The multigrid method for the diffusion equation with strongly discontinuous coefficients,* SIAM J.S.S.C. 2 (1981) 430-454.

[2] ASSELT, E.J. van, *The multigrid method and artificial viscosity,* In [9], pp. 313-326.

[3] AXELSSON, O., L.S. FRANK & A. van der SLUIS, *Analytical and Numerical Approaches to Asymptotic Problems in Analysis,* North-Holland Publ. Comp., Amsterdam-New York, 1981.

[4] BÖRGERS, C., *Mehrgitterverfahren für eine Mehrstellendiskretisierung der Poisson gleichung und für eine zweidimensionale singulär gestörte Aufgabe,* Diplom-arbeit, Institut für Angewandte Mathematik, Universität Bonn, 1981.

[5] BRANDT, A., *Multigrid solvers for non-elliptic and singular perturbation steady-state problems,* Research Report, Dept. of Applied Mathematics, Weizmann Institute of Science, Rehovot, Israel, 1981.

[6] BRANDT, A., *Numerical stability and fast solutions to boundary value problems,* In [18], pp. 29-49.

[7] BROOKS, A.N. & T.J.R. HUGHES, *Streamline-Upwind Petrov-Galerkin Formulations for Convection Dominated Flows with Particular Emphasis on the Incompressible Navier-Stokes Equations,* Comp. Meth. Appl. Mech. Engng. 32 (1982) pp. 199-259.

[8] HACKBUSCH, W., *On the convergence of a multigrid iteration applied to finite element equations,* Report 77-8, Inst. Angew. Math., Univ. Köln, 1977.

[9] HACKBUSCH, W. & U. TROTTENBERG (eds), *Multigrid Methods,* Lecture Notes in Mathematics 960, Springer-Verlag, Berlin, Heidelberg, New York, 1982.

[10] HEMKER, P.W., *On the comparison of line-Gauss-Seidel and ILU relaxation in multigrid algorithms,* In: J.J.H. Miller (ed.), Computational and asymptotic methods for boundary and interior layers, pp. 269-277. Boole Press, Dublin, 1982.

[11] HEMKER, P.W., *Numerical aspects of singular perturbation problems,* In: Asymptotic Analysis II (F. Verhulst ed.) Springer LNM 985, pp. 267-287, 1983.

[12] HEMKER, P.W., R. KETTLER, P. WESSELING & P.M. de ZEEUW, *Multigrid methods: development of fast solvers,* To appear in: Applied Mathematics and Computation.

[13] HEMKER, P.W. & J.J.H. MILLER (eds), *Numerical Analysis of Singular Perturbation problems,* Academic Press, London, 1979.

[14] JOHNSON, C. & U. NÄVERT, *Analysis of some finite element methods for advection-diffusion problems*, In [3], pp. 99-116.

[15] KETTLER, R., *Analysis and comparison of relaxation schemes in robust multigrid and preconditioned conjugate gradient methods*, In [9], pp. 502-534.

[16] KETTLER, R. & J.A. MEIJERINK, *A multigrid method and a combined multigrid conjugate gradient method for elliptic problems with strongly discontinuous coefficients in general domains*, KSEPL Publication 604, Kon. Shell Expl. and Prod. Lab., Rijswijk, The Netherlands, 1981.

[17] MEIJERINK, J.A. & H.A. van der VORST, *An iterative solution method for linear systems of which the coefficient matrix is a symmetric M-matrix*. Math. Comp 31, 148-162, 1977.

[18] MILLER, J.J.H. (ed.), *Boundary and Interior Layers-Computational and Asymptotic Methods*, Boole Press, Dublin, 1980.

[19] STÜBEN, K. & U. TROTTENBERG, *Multigrid Methods: Fundamental Algorithms, Model Problem Analysis and Applications*. In [9], pp. 1-176.

[20] WESSELING, P., *Theoretical and practical aspects of a multigrid method*. SIAM J.S.S.C. 3 (1982) 387-407.

[21] WESSELING, P., *A robust and efficient multigrid method*. In [9], pp. 614-630.

[22] DE ZEEUW, P.M. & E.J. van ASSELT, *The convergence rate of multi-level algorithms applied to the convection-diffusion equations*. Report NW 142/82, Mathematical Center, Amsterdam, 1982.

NONCONVEX MINIMIZATION CALCULATIONS AND THE CONJUGATE GRADIENT METHOD

M.J.D. Powell

Abstract We consider the global convergence of conjugate gradient methods without restarts, assuming exact arithmetic and exact line searches, when the objective function is twice continuously differentiable and has bounded level sets. Most of our attention is given to the Polak-Ribière algorithm, and unfortunately we find examples that show that the calculated gradients can remain bounded away from zero. The examples that have only two variables show also that some variable metric algorithms for unconstrained optimization need not converge. However, a global convergence theorem is proved for the Fletcher-Reeves version of the conjugate gradient method.

1. Introduction

The conjugate gradient method is highly useful for calculating local unconstrained minima of differentiable functions of many variables because it does not require the storage of any matrices. We consider a Polak-Ribière version of the method with exact line searches, which is as follows. We let $F(.)$ from $\mathbb{R}^n$ to $\mathbb{R}$ be the objective function, we let $\underline{x}\in\mathbb{R}^n$ be the vector of variables, we let $\{\underline{x}_k;\ k=1,2,3,\ldots\}$ be the sequence of points that is generated by the algorithm, and $\underline{g}_k$ denotes the gradient $\underline{\nabla}F(\underline{x}_k)$.

Step 0 $\underline{x}_1\in\mathbb{R}^n$ is chosen by the user and $k=1$ is set.

Step 1 Calculate $\underline{g}_k=\underline{\nabla}F(\underline{x}_k)$. If $\underline{g}_k=0$ then end.

Step 2 If $k=1$ then set $\underline{d}_k=-\underline{g}_k$.

Step 3 If $k>1$ then set $\underline{d}_k=-\underline{g}_k+\beta_k\underline{d}_{k-1}$ where

$$\beta_k=\underline{g}_k^T(\underline{g}_k-\underline{g}_{k-1})/\|\underline{g}_{k-1}\|^2 . \tag{1.1}$$

Step 4 Find the least positive number α_k such that the function of one variable

$$\phi_k(\alpha)=F(\underline{x}_k+\alpha\underline{d}_k),\qquad \alpha\in\mathbb{R} , \tag{1.2}$$

has a local minimum at $\alpha=\alpha_k$, or make an error return if the search for α_k is unsuccessful.

Step 5 Set $\underline{x}_{k+1}=\underline{x}_k+\alpha_k\underline{d}_k$, increase k by one, and branch back to Step 1.

Due to the choice of $\underline{d}_k$, this algorithm finds the exact minimum of a convex quadratic function in at most n steps (see Fletcher, 1980), and it is generally far more efficient than the method of steepest descents. Of course in practice one relaxes the condition on the step-length α_k, and also a restart procedure is often used (Powell, 1977, for example). Further, the definition of $\underline{d}_{k+1}$ may be modified when α_k is not a local minimum of the function (1.2) (Buckley, 1982). Thus one can develop some very useful procedures, but the purpose of this paper is to answer a fundamental question about the convergence of the given basic method.

In order that the required value of α_k is well-defined for all k , we assume that the level set

$$L=\{\underline{x} : F(\underline{x}) \leq F(\underline{x}_1)\}\subset\mathbb{R}^n \tag{1.3}$$

is bounded. We assume also that $F(.)$ is twice continuously differentiable. We ask whether these conditions are sufficient for the given method to provide the limit

$$\liminf_{k\to\infty} \|\underline{g}_k\| = 0 . \tag{1.4}$$

It is well-known that this limit is obtained if the directional derivatives

$$\rho_k=\underline{d}_k^T\underline{g}_k/(\|\underline{d}_k\| \; \|\underline{g}_k\|) < 0 , \qquad k=1,2,3,..., \tag{1.5}$$

are bounded away from zero. Specifically, letting all vector norms be Euclidean, and letting Ω be an upper bound on the induced matrix norms $\{\|\nabla^2F(\underline{x})\| ; \underline{x}\epsilon L\}$, we have the relation

$$F(\underline{x})\leq F(\underline{x}_k)+(\underline{x}-\underline{x}_k)^T\underline{g}_k+\tfrac{1}{2}\Omega\|\underline{x}-\underline{x}_k\|^2, \qquad \underline{x}\epsilon L , \tag{1.6}$$

which, after some consideration of rates of change of first derivatives, gives the inequality

$$\begin{aligned} F(\underline{x}_{k+1})&\leq \min_{\alpha}[F(\underline{x}_k)+\alpha\underline{d}_k^T\underline{g}_k+\tfrac{1}{2}\Omega\alpha^2\|\underline{d}_k\|^2] \\ &= F(\underline{x}_k)-\rho_k^2\|\underline{g}_k\|^2/(2\Omega) . \end{aligned} \tag{1.7}$$

Thus, because $F(.)$ is bounded below, $\Sigma\rho_k^2\|\underline{g}_k\|^2$ is convergent. Hence condition (1.4) fails only if $\Sigma\rho_k^2$ is finite (Zoutendijk, 1970).

Therefore, noting that the definitions of α_{k-1} and $\underline{d}_k$ imply the value $\rho_k=-\|\underline{g}_k\|/\|\underline{d}_k\|$, the gradient norms $\{\|\underline{g}_k\| ; k=1,2,3,...\}$ are bounded away from zero only if the sequence $\{\|\underline{d}_k\| ; k=1,2,3,...\}$ is divergent. In this case, due to Step 3 and the existence of an upper bound on $\{\|\underline{\nabla}F(\underline{x})\| ; \underline{x}\epsilon L\}$, the direction $\underline{d}_k$ tends to be parallel to $\underline{d}_{k-1}$. Therefore, if one is seeking counter-examples to the limit (1.4), it is suitable to consider cases when the points $\{\underline{x}_k;\ k=1,2,3,\ldots\}$ tend to lie on a straight line, which we take as the first co-ordinate direction in $\mathbb{R}^n$. There are no counter-examples in which the sequence $\{\underline{x}_k;\ k=1,2,3,\ldots\}$ converges, because in this case, due to first derivative continuity, the numbers $\{\beta_k;\ k=1,2,3,\ldots\}$ would tend to zero, which would keep $\{\|\underline{d}_k\|;\ k=1,2,3,\ldots\}$ uniformly bounded.

For many finite sequences of distinct points $\{\underline{x}_k;\ k=1,2,\ldots,\ell\}$ one can find gradients $\{\underline{g}_k;\ k=1,2,\ldots,\ell-1\}$ such that the points can be generated by the conjugate gradient method. Specifically, we let $\underline{g}_1 = \underline{x}_1 - \underline{x}_2$, and, for $k \geq 2$, because of the definition of α_{k-1} and $\underline{d}_k$, $\underline{g}_k$ has to be a multiple of the vector in the space spanned by $(\underline{x}_{k+1} - \underline{x}_k)$ and $(\underline{x}_k - \underline{x}_{k-1})$ that is orthogonal to $(\underline{x}_k - \underline{x}_{k-1})$. In order to determine the sign and length of $\underline{g}_k$, we note that the value (1.1) implies the conjugacy condition

$$(\underline{x}_{k+1} - \underline{x}_k)^T (\underline{g}_k - \underline{g}_{k-1}) = 0, \qquad k \geq 2. \tag{1.8}$$

Thus, starting with $\underline{g}_1 = \underline{x}_1 - \underline{x}_2$, the gradients $\{\underline{g}_k;\ k=1,2,3,\ldots\}$ can usually be found recursively, but the descent conditions $\{\underline{g}_k^T(\underline{x}_{k+1} - \underline{x}_k) < 0;\ k=2,3,\ldots,\ell-1\}$ may not hold. Thus not all sequences $\{\underline{x}_k;\ k=1,2,\ldots,\ell\}$ are admissible. Further restrictions on the sequences occur if one lets $\ell \to \infty$, in particular from the aim of keeping the gradient norms $\{\|\underline{g}_k\|;\ k=1,2,3,\ldots\}$ bounded and bounded away from zero.

In spite of these difficulties, we seek sequences that provide counter-examples to the limit (1.4). To simplify the analysis we impose the conditions

$$\left.\begin{aligned} (\underline{x}_{k+m})_1 &= (\underline{x}_k)_1 \\ \hat{\underline{x}}_{k+m} &= \theta \hat{\underline{x}}_k \end{aligned}\right\}, \qquad k = 1,2,3,\ldots, \tag{1.9}$$

where m is a small positive integer, where $(\underline{x})_1$ denotes the first component of $\underline{x}$, where $\hat{\underline{x}}$ is the vector in $\mathbb{R}^{n-1}$ whose components are the last $(n-1)$ components of $\underline{x}$, and where θ is a constant from $(0,1)$. Thus the distance from $\underline{x}_k$ to the first co-ordinate direction tends to zero as $k \to \infty$. Having chosen m and n, there are only a finite number of parameters in the sequence $\{\underline{x}_k;\ k=1,2,3,\ldots\}$. One can express the conditions for consistency with the conjugate gradient method as inequality constraints on the parameters, and one can investigate whether the inequalities have a solution. Some interesting cases for $n=2$ and $n=3$ are reported in Sections 2 and 3 respectively.

Two examples are given for $n=2$. For $m=3$ we find that gradients can stay bounded away from zero if one gives up the second derivative continuity of the objective function. Secondly, by letting $m=8$, it is shown that one can preserve the second derivative continuity if one modifies Step 4 of the algorithm by allowing α_k

to be any local minimum of the function (1.2) that satisfies $\phi_k(\alpha_k) < \phi_k(0)$. This is an important case because in practice one can usually accept any local minimum that reduces the objective function. However, the constraints on α_k in Step 4 are present because in this example the choice of α_k is so contrived that it is unlikely to occur.

For $n=3$ and $m=4$ we give an example that has properties that are similar to the $n=2$ and $m=8$ case of Section 2, and, by letting $n=3$ and $m=6$, we answer the main question of this paper. We find the gradients $\{\underline{g}_k;\ k=1,2,3,\ldots\}$ can remain bounded away from zero when all the conditions that have been stated are satisfied.

Section 4 includes a brief discussion of the examples and their implications. Further, we ask whether it is helpful to replace the multiplier (1.1) by the value

$$\beta_k = \|\underline{g}_k\|^2 / \|\underline{g}_{k-1}\|^2 , \tag{1.10}$$

which is suggested by Fletcher and Reeves (1964). It is proved that this version of the conjugate gradient algorithm has the strong advantage over the Polak-Ribière version that it always provides the limit (1.4), assuming that $F(.)$ is twice continuously differentiable and that the level set (1.3) is bounded.

2. Two variable examples

Let $n=2$, and let the sequence $\{\underline{x}_k;\ k=1,2,3,\ldots\}$ satisfy expression (1.9) for $m=3$, where θ is a constant from $(0,1)$. Then there exist real parameters a_i and b_i $(i=1,2,3)$ such that, for each non-negative integer j, we have

$$\underline{\delta}_{3j+1} = a_1\begin{pmatrix} 1 \\ b_1\theta^j \end{pmatrix}, \quad \underline{\delta}_{3j+2} = a_2\begin{pmatrix} 1 \\ b_2\theta^j \end{pmatrix}, \quad \underline{\delta}_{3j+3} = a_3\begin{pmatrix} 1 \\ b_3\theta^j \end{pmatrix},$$

where $\{\underline{\delta}_k = \underline{x}_{k+1} - \underline{x}_k;\ k=1,2,3,\ldots\}$ and where $a_1 + a_2 + a_3 = 0$. Thus, due to the line searches, the gradients have the form

$$\underline{g}_{3j+1} = c_{3j+1}\begin{pmatrix} b_3\theta^{j-1} \\ -1 \end{pmatrix}, \quad \underline{g}_{3j+2} = c_{3j+2}\begin{pmatrix} b_1\theta^j \\ -1 \end{pmatrix}, \quad \underline{g}_{3j+3} = c_{3j+3}\begin{pmatrix} b_2\theta^j \\ -1 \end{pmatrix},$$

where $\{c_k;\ k=1,2,3,\ldots\}$ are real multipliers. Therefore the conjugacy condition (1.8) implies the equations

$$\left.\begin{aligned}
c_{3j+1}(b_3\theta^{j-1} - b_1\theta^j) &= c_{3j}(b_2\theta^{j-1} - b_1\theta^j) \\
c_{3j+2}(b_1\theta^j - b_2\theta^j) &= c_{3j+1}(b_3\theta^{j-1} - b_2\theta^j) \\
c_{3j+3}(b_2\theta^j - b_3\theta^j) &= c_{3j+2}(b_1\theta^j - b_3\theta^j)
\end{aligned}\right\} , \tag{2.1}$$

which gives the ratio

$$\frac{c_{3j+3}}{c_{3j}} = \frac{b_2 - b_1\theta}{b_3 - b_1\theta} \; \frac{b_3 - b_2\theta}{\theta(b_1 - b_2)} \; \frac{b_1 - b_3}{b_2 - b_3} = 1 , \tag{2.2}$$

where the right hand side is set to one because otherwise the gradients $\{\underline{g}_k;\ k=1,2,3,\ldots\}$ either diverge or tend to zero.

We assume without loss of generality that $a_1 > 0$, $a_2 > 0$ and $a_3 = -a_1 - a_2$. Therefore all the search directions are downhill if and only if the inequalities

$$\left.\begin{aligned}
c_{3j+1}(b_3 - b_1\theta) &< 0 \\
c_{3j+2}(b_1 - b_2) &< 0 \\
c_{3j+3}(b_2 - b_3) &> 0
\end{aligned}\right\} \tag{2.3}$$

hold. Thus, remembering $c_{3j+3} = c_{3j}$, the equations (2.1) imply the conditions

$$\left.\begin{aligned}
(b_2 - b_3)(b_2 - b_1\theta) &< 0 \\
(b_3 - b_1\theta)(b_3 - b_2\theta) &> 0 \\
(b_1 - b_2)(b_1 - b_3) &< 0
\end{aligned}\right\} . \tag{2.4}$$

We consider values of b_1, b_2 and b_3 that satisfy expressions (2.2) and (2.4).

In particular the values $b_1 = 3$, $b_2 = 2$, $b_3 = 2 + \sqrt{2}$ and $\theta = 1/3$ are suitable. In this case, from the relation

$$\underline{x}_{3j+4} - \underline{x}_{3j+1} = \begin{pmatrix} 0 \\ (a_1 b_1 + a_2 b_2 + a_3 b_3)\theta^j \end{pmatrix} , \tag{2.5}$$

and from the fact that the second component of $\underline{x}_{3j+1}$ tends to zero as $j \to \infty$, we deduce the vector

$$\underline{x}_{3j+1} = \begin{pmatrix} 0 \\ -[3a_1 + 2a_2 - (2+\sqrt{2})(a_1 + a_2)]\theta^j/(1-\theta) \end{pmatrix} , \tag{2.6}$$

assuming without loss of generality that $(\underline{x}_{3j+1})_1 = 0$. Similarly $\underline{x}_{3j+2}$ and $\underline{x}_{3j+3}$ can be calculated, and we note the value

$$(\underline{x}_{3j+2})_2 = -[3\theta a_1 + 2a_2 - (2+\sqrt{2})(a_1+a_2)]\theta^j/(1-\theta) \ . \tag{2.7}$$

Letting a_1 and a_2 be any positive constants, and letting ℓ be any positive integer, one can apply this procedure to find sequences $\{\underline{x}_k;\ k=1,2,\ldots,\ell\}$ and $\{\underline{g}_k;\ k=1,2,\ldots,\ell\}$ that are consistent with the conjugate gradient method, that satisfy condition (1.9) for $m=3$, and there exist upper bounds on $\|\underline{g}_k\|$ and $\|\underline{g}_k\|^{-1}$ that are independent of k and ℓ. However, if we let $\ell\to\infty$, we violate the continuity of $\{\underline{\nabla}F(\underline{x});\ \underline{x}\in\mathbb{R}^2\}$.

Specifically, if we have first derivative continuity, and if we let $c_{3j+1}=-1$ for definiteness, its sign being determined by condition (2.3), then equation (2.6) and the form of $\underline{g}_{3j+1}$ imply the relation

$$F(\underline{x}_{3j+1}) - F(\underline{x}^*) = [\,(\sqrt{2}-1)a_1 + \sqrt{2}a_2]\theta^j/(1-\theta) + o(\theta^j) \ , \tag{2.8}$$

where $F(\underline{x}^*)$ is the limit of the sequence $\{F(\underline{x}_k);\ k=1,2,3,\ldots\}$. However, expression (2.7) and the ratio

$$c_{3j+2}/c_{3j+1} = (b_3 - b_2\theta)/(b_1\theta - b_2\theta) = 4 + 3\sqrt{2} \tag{2.9}$$

show that the dominant part of $F(\underline{x}_{3j+2}) - F(\underline{x}^*)$ has the value

$$(4+3\sqrt{2})\,[\,(\sqrt{2}+1)a_1 + \sqrt{2}a_2]\theta^j/(1-\theta) \ , \tag{2.10}$$

which is greater than expression (2.8) for sufficiently large j. Thus we violate the condition that the sequence of function values $\{F(\underline{x}_k);\ k=1,2,3,\ldots\}$ must decrease monotonically. It can be shown that it is not possible to overcome this defect by using other feasible values of b_1, b_2 and b_3.

This example illustrates quite well the method that is used to construct all the given examples. In particular we note that the parameters θ and $\{b_i;\ i=1,2,\ldots,m\}$ in the differences $\{\underline{\delta}_k = \underline{x}_{k+1} - \underline{x}_k;\ k=1,2,3,\ldots\}$ define the gradients $\{\underline{g}_k;\ k=1,2,3,\ldots\}$, except for the scaling factors $\{c_k;\ k=1,2,3,\ldots\}$, whose ratios are fixed by the conjugacy condition (1.8). For any choice of θ and $\{b_i;\ i=1,2,\ldots,m\}$, the points $\{\underline{x}_k;\ k=1,2,3,\ldots\}$ are found by the construction that gives the vector (2.6). At this stage the parameters $\{a_i;\ i=1,2,\ldots,m\}$ are available to help the sequence $\{F(\underline{x}_k);\ k=1,2,3,\ldots\}$ to decrease monotonically.

The following example with $n=2$ and $m=8$ shows that decreasing function values do not always conflict with second derivative continuity. Let the steps of the algorithm have the form

$$\underline{\delta}_{8j+i} = a_i\begin{pmatrix} 1 \\ b_i\phi^{2j}\end{pmatrix}, \quad \underline{\delta}_{8j+i+4} = a_i\begin{pmatrix} -1 \\ b_i\phi^{2j+1}\end{pmatrix}, \quad i=1,2,3,4, \tag{2.11}$$

where the numbers $\{a_i;\ i=1,2,3,4\}$ are all positive, and consider the values $\phi = \frac{1}{2}$, $b_1 = -2$, $b_2 = 1$, $b_3 = -1$ and $b_4 = -2$. Thus, if equation (1.9) holds, we have $\theta = \phi^2 = 1/4$. The gradients

$$\left.\begin{aligned} \underline{g}_{8j+1} &= \begin{pmatrix} -8\phi^{2j} \\ 2 \end{pmatrix}, \quad & \underline{g}_{8j+2} &= \begin{pmatrix} -4\phi^{2j} \\ -2 \end{pmatrix} \\ \underline{g}_{8j+3} &= \begin{pmatrix} -\phi^{2j} \\ 1 \end{pmatrix}, \quad & \underline{g}_{8j+4} &= \begin{pmatrix} 3\phi^{2j} \\ 3 \end{pmatrix} \end{aligned}\right\} \tag{2.12}$$

and the relations

$$\left.\begin{aligned} (\underline{g}_{k+4})_1 &= -\phi(\underline{g}_k)_1 \\ (\underline{g}_{k+4})_2 &= (\underline{g}_k)_2 \end{aligned}\right\} \quad k = 1,2,3,\ldots, \tag{2.13}$$

satisfy all the line search and conjugacy conditions.

Due to symmetry, we can reduce the objective function on every iteration if we achieve the relations

$$F(\underline{x}_{8j+1}) > F(\underline{x}_{8j+2}) > F(\underline{x}_{8j+3}) > F(\underline{x}_{8j+4}) > F(\underline{x}_{8j+5}) \ . \tag{2.14}$$

Now, when the first component of $\underline{x}$ is equal to the first component of $\underline{x}_k$, where k is any positive integer, then the values (2.12) allow the second component of $\nabla F(\underline{x})$ to be constant, provided that the first components of the points $\{\underline{x}_{8j+i};\ i=1,2,\ldots 8\}$ are all different. Thus we satisfy the equation

$$F(\underline{x}_k) - F(\underline{x}^*) = (\underline{x}_k)_2 (\underline{g}_k)_2, \quad k=1,2,3,\ldots \ . \tag{2.15}$$

It follows that expression (2.14) is equivalent to the inequalities

$$\begin{aligned} &-2(a_1b_1 + a_2b_2 + a_3b_3 + a_4b_4) \\ &\quad > 2(\phi a_1b_1 + a_2b_2 + a_3b_3 + a_4b_4) \\ &\quad > -(\phi a_1b_1 + \phi a_2b_2 + a_3b_3 + a_4b_4) \\ &\quad > -3(\phi a_1b_1 + \phi a_2b_2 + \phi a_3b_3 + a_4b_4) \\ &\quad > -2\phi(a_1b_1 + a_2b_2 + a_3b_3 + a_4b_4) \ . \end{aligned} \tag{2.16}$$

These inequalities are consistent because, if $a_1 = 336$, $a_2 = 864$, $a_3 = 364$ and $a_4 = 1$, then the five lines of this expression take the values 348, 324, 270, 264 and 174 respectively. Further, if we let $(\underline{x}_1)_1 = 0$, then the first components of the vectors $\{\underline{x}_{8j+i};\ i=1,2,\ldots,8\}$ have the values 0, 336, 1200, 1564, 1565, 1229, 365 and 1, which are all different.

It is straightforward to construct a twice continuously differentiable function $\{F(\underline{x});\ \underline{x}\in\mathbb{R}^2\}$ that satisfies the gradient conditions (2.12) and (2.13). In order to describe a suitable construction, we note that the required value

$$\underline{g}_{8j+2} = \underline{\nabla}F\begin{pmatrix} 336 \\ -324\phi^{2j} \end{pmatrix} = \begin{pmatrix} -4\phi^{2j} \\ -2 \end{pmatrix} \tag{2.17}$$

is obtained if $F(.)$ has the form

$$F(x_1,x_2) = \frac{1}{81}(x_1 - 336)x_2 - 2x_2 + e_2(x_1 - 336)^2 \tag{2.18}$$

whenever $|x_1 - 336| \le 0.1$, where e_2 is any constant. By choosing e_2 to be sufficiently large, we ensure that $\underline{x}_{8j+2}$ is a local minimum of the objective function on the line through $\underline{x}_{8j+1}$ and $\underline{x}_{8j+2}$ for all j. The remaining gradient conditions are satisfied similarly, which defines $F(x_1,x_2)$ whenever $|x_1 - (\underline{x}_k)_1| \le 0.1$ for some integer k. Intermediate values of $F(\underline{x})$ may then be chosen in a way that gives second derivative continuity.

To show the main disadvantage of this example we consider the line search from

$$\underline{x}_{8j+7} = \begin{pmatrix} 365 \\ 540\phi^{2j+1} \end{pmatrix} \quad \text{to} \quad \underline{x}_{8j+8} = \begin{pmatrix} 1 \\ 176\phi^{2j+1} \end{pmatrix}.$$

The point $(336,\ 511\phi^{2j+1})$ is on this line segment, and, due to expression (2.18), we have the value

$$F(336,\ 511\phi^{2j+1}) = F(\underline{x}^*) - 1022\phi^{2j+1}. \tag{2.19}$$

Therefore a smaller step-length in the line search would make the objective function much less than $F(\underline{x}^*)$, which shows that we have not satisfied the conditions of Step 4 of the algorithm of Section 1. Perhaps this algorithm always gives the limit (1.4) when $F(.)$ is twice continuously differentiable, when L is bounded, and when there are only two variables.

3. Three variable examples

First we let $n=3$ and $m=4$, and we consider a cycle of the form

$$\underline{\delta}_{4j+1} = a_1 \begin{pmatrix} 1 \\ b_1\phi^{2j} \\ h_1\phi^{2j} \end{pmatrix}, \qquad \underline{\delta}_{4j+2} = a_2 \begin{pmatrix} 1 \\ b_2\phi^{2j} \\ h_2\phi^{2j} \end{pmatrix},$$

$$\underline{\delta}_{4j+3} = a_1 \begin{pmatrix} -1 \\ b_1\phi^{2j+1} \\ h_1\phi^{2j+1} \end{pmatrix}, \qquad \underline{\delta}_{4j+4} = a_2 \begin{pmatrix} -1 \\ b_2\phi^{2j+1} \\ h_2\phi^{2j+1} \end{pmatrix},$$

where $j=0,1,2,\ldots$. In this case, applying the construction that is described in Section 1, we find that $\underline{g}_{4j+2}$, for example, is a multiple of the vector

$$\begin{pmatrix} b_1(b_1-b_2)\phi^{2j} + h_1(h_1-h_2)\phi^{2j} \\ (b_2-b_1) + h_1(b_2h_1-b_1h_2)\phi^{4j} \\ (h_2-h_1) + b_1(b_1h_2-b_2h_1)\phi^{4j} \end{pmatrix}.$$

We avoid complicated expressions by substituting the parameter values $b_1=9$, $b_2=55$ $h_1=h_2=45$ and $\phi=5/18$, which gives the gradients

$$\underline{g}_{4j+1} = c_{4j+1} \begin{pmatrix} 360\phi^{2j} \\ 1-5832\phi^{4j} \\ 1+7128\phi^{4j} \end{pmatrix}, \qquad \underline{g}_{4j+2} = c_{4j+2} \begin{pmatrix} -9\phi^{2j} \\ 1+2025\phi^{4j} \\ -405\phi^{4j} \end{pmatrix},$$

$$\underline{g}_{4j+3} = c_{4j+3} \begin{pmatrix} -360\phi^{2j+1} \\ 1-5832\phi^{4j+2} \\ 1+7128\phi^{4j+2} \end{pmatrix}, \qquad \underline{g}_{4j+4} = c_{4j+4} \begin{pmatrix} 9\phi^{2j+1} \\ 1+2025\phi^{4j+2} \\ -405\phi^{4j+2} \end{pmatrix},$$

where the conjugacy condition (1.8) is obtained by forcing the multipliers $\{c_k;\ k=2,3,4,\ldots\}$ to satisfy the equations

$$\left.\begin{aligned} c_{4j}(9+9\phi) &= c_{4j+1}(414\phi + 268272\phi^{4j+1}) \\ c_{4j+1}(460) &= c_{4j+2}(46 + 93150\phi^{4j}) \\ c_{4j+2}(9+9\phi) &= c_{4j+3}(414\phi + 268272\phi^{4j+3}) \\ c_{4j+3}(460) &= c_{4j+4}(46 + 93150\phi^{4j+2}) \end{aligned}\right\}. \tag{3.1}$$

Thus we deduce the ratios

$$\left.\begin{aligned} c_k/c_{k+2} &= (1+648\phi^k)(1+2025\phi^k),\ k \text{ even} \\ c_k/c_{k+2} &= (1+648\phi^{k+1})(1+2025\phi^{k-1}),\ k \text{ odd} \end{aligned}\right\}, \qquad (3.2)$$

so the gradient norms $\{\|\underline{g}_k\|\,;\ k=1,2,3,\ldots\}$ are uniformly bounded and bounded away from zero. Moreover, equations (3.1)-(3.2) allow $c_k < 0$ for all k, so we can satisfy the descent conditions $\{\underline{\delta}_k^T\underline{g}_k < 0;\ k=1,2,3,\ldots\}$.

Because terms of order ϕ^{4j} may be ignored, the condition for monotonically decreasing function values that corresponds to expression (2.16) is the relation

$$\begin{aligned} -c_{4j+1}[(a_1b_1+a_2b_2)+(a_1h_1+a_2h_2)] &> -c_{4j+2}(\phi a_1b_1+a_2b_2) \\ &> -c_{4j+3}\phi[(a_1b_1+a_2b_2)+(a_1h_1+a_2h_2)] \quad . \end{aligned} \qquad (3.3)$$

Therefore, remembering $c_k < 0$, and noting that expression (3.1) gives the approximate ratios

$$c_{4j+1} : c_{4j+2} : c_{4j+3} \simeq 1 : 10 : 1 \ , \qquad (3.4)$$

we require a_1 and a_2 to satisfy the inequalities

$$\begin{aligned} 54a_1+100a_2 &> 25a_1+550a_2 \\ &> \phi(54a_1+100a_2) \ . \end{aligned} \qquad (3.5)$$

The values $a_1 = 20$ and $a_2 = 1$ are suitable.

In this case, however, if we let $(\underline{x}_{4j+1})_1 = 0$, then we have the points

$$\underline{x}_{4j+1} = \begin{pmatrix} 0 \\ -235\phi^{2j}/(1-\phi) \\ -945\phi^{2j}/(1-\phi) \end{pmatrix}, \qquad \underline{x}_{4j+2} = \begin{pmatrix} 20 \\ -105\phi^{2j}/(1-\phi) \\ -295\phi^{2j}/(1-\phi) \end{pmatrix}.$$

Thus the vector

$$\hat{\underline{x}}_j = \begin{pmatrix} 1 \\ -228.5\phi^{2j}/(1-\phi) \\ -912.5\phi^{2j}/(1-\phi) \end{pmatrix} \qquad (3.6)$$

is the point on the line segment from $\underline{x}_{4j+1}$ to $\underline{x}_{4j+2}$ whose first component is $(\underline{x}_{4j+4})_1$. Thus, using the given gradients, we deduce the values

$$\left.\begin{aligned} F(\underline{x}_{4j+1}) &= F(\underline{x}^*) - 1180c_{4j+1}\phi^{2j}/(1-\phi) + o(\phi^{2j}) \\ F(\hat{\underline{x}}_j) &= F(\underline{x}^*) - 228.5c_{4j+4}\phi^{2j}/(1-\phi) + o(\phi^{2j}) \end{aligned}\right\} . \tag{3.7}$$

Because $c_{4j+1} < 0$ and because $c_{4j+4} \approx 10c_{4j+1}$, it follows that $F(\hat{\underline{x}}_j)$ is larger than $F(\underline{x}_{4j+1})$. Therefore again the conditions of Step 4 of the algorithm of Section 1 are not satisfied. It seems that it is not possible to satisfy these conditions by adjusting the parameters a_i, b_i and h_i $(i=1,2)$.

For $n=3$ and $m=6$ we consider steps of the form

$$\underline{\delta}_{6j+i} = a_i \begin{pmatrix} 1 \\ b_i\phi^{2j} \\ h_i\phi^{2j} \end{pmatrix}, \qquad \underline{\delta}_{6j+i+3} = a_i \begin{pmatrix} -1 \\ b_i\phi^{2j+1} \\ h_i\phi^{2j+1} \end{pmatrix}, \qquad i=1,2,3. \tag{3.8}$$

In fact it is suitable if the parameters have the values $b_1=1$, $b_2=-12$, $b_3=-0.3$, $h_1=h_2=117$, $h_3=120.9$, $\phi=0.3$, $a_1=42$, $a_2=21$, $a_3=28$ and $(\underline{x}_1)_1=0$, which gives the points and gradients

$$\underline{x}_{6j+1} = \begin{pmatrix} 0 \\ 312\phi^{2j} \\ -15366\phi^{2j} \end{pmatrix}, \qquad \underline{g}_{6j+1} = c_{6j+1} \begin{pmatrix} -403\phi^{2j} \\ -403\phi^{4j} \\ -1-\phi^{4j} \end{pmatrix},$$

$$\underline{x}_{6j+2} = \begin{pmatrix} 42 \\ 354\phi^{2j} \\ -104520\phi^{2j} \end{pmatrix}, \qquad \underline{g}_{6j+2} = c_{6j+2} \begin{pmatrix} -\phi^{2j} \\ 1+13689\phi^{4j} \\ -117\phi^{4j} \end{pmatrix},$$

$$\underline{x}_{6j+3} = \begin{pmatrix} 63 \\ 102\phi^{2j} \\ -7995\phi^{2j} \end{pmatrix}, \qquad \underline{g}_{6j+3} = c_{6j+3} \begin{pmatrix} 81\phi^{2j} \\ -3-42471\phi^{4j} \\ -1-4356\phi^{4j} \end{pmatrix},$$

$$\underline{x}_{6j+4} = \begin{pmatrix} 91 \\ 312\phi^{2j+1} \\ -15366\phi^{2j+1} \end{pmatrix}, \qquad \underline{g}_{6j+4} = c_{6j+4} \begin{pmatrix} 403\phi^{2j+1} \\ -403\phi^{4j+2} \\ -1-\phi^{4j+2} \end{pmatrix},$$

$$\underline{x}_{6j+5} = \begin{pmatrix} 49 \\ 354\phi^{2j+1} \\ -10452\phi^{2j+1} \end{pmatrix}, \qquad \underline{g}_{6j+5} = c_{6j+5} \begin{pmatrix} \phi^{2j+1} \\ 1+13689\phi^{4j+2} \\ -117\phi^{4j+2} \end{pmatrix},$$

$$\underline{x}_{6j+6} = \begin{pmatrix} 28 \\ 102\phi^{2j+1} \\ -7995\phi^{2j+1} \end{pmatrix}, \qquad \underline{g}_{6j+6} = c_{6j+6}\begin{pmatrix} -81\phi^{2j+1} \\ -3-42471\phi^{4j+2} \\ -1-4356\phi^{4j+2} \end{pmatrix}.$$

The signs of the gradients have been chosen so that the downhill conditions $\{\underline{\delta}_k^T \underline{g}_k < 0\,;\, k=2,3,4,\ldots\}$ are satisfied if and only if $\{c_k > 0\,;\, k=2,3,4,\ldots\}$. Further, the conjugacy condition (1.8) implies the equations

$$\left.\begin{aligned} c_{6j}(3+471900\phi^{4j}) &= c_{6j+1}(4+4\phi^{4j}) \\ c_{6j+1}(40-363\phi^{4j}) &= c_{6j+2}(1+136890\phi^{4j}) \\ c_{6j+2}(1+140400\phi^{4j}) &= c_{6j+3}(30+3953070\phi^{4j}) \end{aligned}\right\} . \qquad (3.9)$$

Thus as $k \to \infty$ we have $c_{k+3}/c_k = 1 + O(\phi^{2k/3})$, which ensures that the sequences $\{\|\underline{g}_k\|\,,\, k=1,2,3,\ldots\}$ and $\{\|\underline{g}_k\|^{-1}\,;\, k=1,2,3,\ldots\}$ are uniformly bounded. We show that the required gradients allow $\{F(\underline{x})\,;\, \underline{x}\in\mathbb{R}^3\}$ to be twice continuously differentiable, and to decrease monotonically on the line segment from $\underline{x}_k$ to $\underline{x}_{k+1}$ for each k.

We begin by assuming that all terms of order ϕ^{4j} are negligible. Therefore we seek a function $\{\bar{F}(\underline{x})\,;\, \underline{x}\in\mathbb{R}^3\}$ that satisfies the required monotonicity conditions and that has the derivative values

$$\underline{\nabla}\bar{F}\begin{pmatrix} 0 \\ 312\psi \\ -15366\psi \end{pmatrix} = \begin{pmatrix} -1209\psi \\ 0 \\ -3 \end{pmatrix}, \qquad \underline{\nabla}\bar{F}\begin{pmatrix} 42 \\ 354\psi \\ -10452\psi \end{pmatrix} = \begin{pmatrix} -120\psi \\ 120 \\ 0 \end{pmatrix},$$

$$\underline{\nabla}\bar{F}\begin{pmatrix} 63 \\ 102\psi \\ -7995\psi \end{pmatrix} = \begin{pmatrix} 324\psi \\ -12 \\ -4 \end{pmatrix}, \qquad \underline{\nabla}\bar{F}\begin{pmatrix} 91 \\ 312\psi \\ -15366\psi \end{pmatrix} = \begin{pmatrix} 1209\psi \\ 0 \\ -3 \end{pmatrix},$$

$$\underline{\nabla}\bar{F}\begin{pmatrix} 49 \\ 354\psi \\ -10452\psi \end{pmatrix} = \begin{pmatrix} 120\psi \\ 120 \\ 0 \end{pmatrix}, \qquad \underline{\nabla}\bar{F}\begin{pmatrix} 28 \\ 102\psi \\ -7995\psi \end{pmatrix} = \begin{pmatrix} -324\psi \\ -12 \\ -4 \end{pmatrix}.$$

We let $\bar{F}(.)$ have the form

$$\bar{F}(x_1,x_2,x_3) = \lambda(x_1)x_2 + \mu(x_1)x_3 , \qquad (3.10)$$

where x_1, x_2 and x_3 are the components of $\underline{x}$, and where $\lambda(.)$ and $\mu(.)$ are twice continuously differentiable functions of one variable such that $\lambda(x_1) = \lambda(91-x_1)$ and $\mu(x_1) = \mu(91-x_1)$ for $0 \le x_1 \le 91$. Further, for $0 \le x_1 \le 91$, we let $\sigma(x_1)\psi$ be the value of $\bar{F}(\underline{x})$ at the point $\underline{x}$ whose first component is x_1, and whose last two components are defined by the property that $\underline{x}$ is on the piecewise linear curve that joins

$$\begin{pmatrix} 0 \\ 312\psi \\ -15366\psi \end{pmatrix}, \begin{pmatrix} 42 \\ 354\psi \\ -10452\psi \end{pmatrix}, \begin{pmatrix} 63 \\ 102\psi \\ -7995\psi \end{pmatrix} \text{ and } \begin{pmatrix} 91 \\ 312\phi\psi \\ -15366\phi\psi \end{pmatrix}.$$

Therefore we require $\{\sigma(x_1);\ 0 \le x_1 \le 91\}$ to decrease strictly monotonically.

We consider in sequence the values of $\underline{\nabla}\bar{F}(.)$ that are given in the previous paragraph. The first value implies $\sigma(0) = 46098$, because $\lambda(0) = 0$ and $\mu(0) = -3$, and $\sigma'(0+) = -1560$. Similarly we find $\sigma(42) = 42480$, $\sigma'(42-) = 0$, $\sigma'(42+) = -1560$, $\sigma(63) = 30756$, $\sigma'(63-) = 0$, $\sigma'(63+) = -156$, $\sigma(91) = 13829.4$, $\sigma'(91-) = 0$, $\sigma(49) = 32400$ and $\sigma(28) = 44280$, so our parameters allow $\{\sigma(x_1);\ 0 \le x_1 \le 91\}$ to be monotonic.

We let $\{\sigma(x);\ 0 \le x \le 91\}$ be any real valued function that is three times continuously differentiable on each of the intervals $[0,42]$, $[42,63]$ and $[63,91]$, that satisfies the conditions of the previous paragraph, whose first derivative is negative on $(0,42)$, $(42,63)$ and $(63,91)$, that has the second derivative values $\sigma''(42-) = \sigma''(63-) = \sigma''(91-) = M$, where M is a parameter, and, in order that $\{\bar{F}(\underline{x});\ \underline{x}\in\mathbb{R}^3\}$ is twice continuously differentiable when $x_1 = 42$ and 63, we require particular values of $\sigma''(42+)$ and $\sigma''(63+)$ that are discussed later. Of course M is positive, in order that, for $i = 1,2,3$, $\underline{x}_{6j+i+1}$ is a local minimum of $\bar{F}(.)$ on the line through $\underline{x}_{6j+i}$ and $\underline{x}_{6j+i+1}$.

The choice of $\sigma(.)$ imposes a constraint on the functions $\lambda(.)$ and $\mu(.)$ of expression (3.10). Further, due to the symmetry of $\lambda(x_1)$ and $\mu(x_1)$ about $x_1 = 45.5$, we must have that $\{\tau(x_1) \equiv \sigma(91-x_1);\ 0 \le x_1 \le 91\}$ is the value of $\bar{F}(\underline{x})/\psi$ at the point $\underline{x}$ whose first component is x_1, and whose last two components are defined by the property that $\underline{x}$ is on the piecewise linear curve that joins

$$\begin{pmatrix} 91 \\ 312\psi \\ -15366\psi \end{pmatrix}, \begin{pmatrix} 49 \\ 354\psi \\ -10452\psi \end{pmatrix}, \begin{pmatrix} 28 \\ 102\psi \\ -7995\psi \end{pmatrix} \text{ and } \begin{pmatrix} 0 \\ 312\phi\psi \\ -15366\phi\psi \end{pmatrix}.$$

Thus, for each x_1, $\lambda(x_1)$ and $\mu(x_1)$ have to satisfy two equations. We claim that, having chosen $\sigma(.)$, these equations define the twice continuously differentiable functions $\lambda(.)$ and $\mu(.)$ uniquely.

For example, on the interval [63,91] we have the equations

$$\left.\begin{aligned}\sigma(x_1) &= (120.9 - 0.3x_1)\lambda(x_1) + (-15611.7 + 120.9x_1)\mu(x_1) \\ \tau(x_1) &= (403 - x_1)\lambda(x_1) + (-4719 - 117x_1)\mu(x_1)\end{aligned}\right\} , \qquad (3.11)$$

which imply the values

$$\left.\begin{aligned}\lambda(x_1) &= \frac{-(121 + 3x_1)\sigma(x_1) + (400.3 - 3.1x_1)\tau(x_1)}{4(x_1 - 91)(x_1 - 403)} \\ \mu(x_1) &= \frac{(-403 + x_1)\sigma(x_1) + (120.9 - 0.3x_1)\tau(x_1)}{156(x_1 - 91)(x_1 - 403)}\end{aligned}\right\} \qquad (3.12)$$

for $63 \leq x_1 < 91$. Both $\lambda(.)$ and $\mu(.)$ are bounded near $x_1 = 91$ because, due to $\sigma(91) = 0.3\tau(91)$, the numerators of expression (3.12) are zero at $x_1 = 91$. Further, the required value of $\underline{\nabla}\bar{F}(\underline{x}_{6j+4})$ is given by continuity and by the conditions $\sigma(91) = 13829.4$, $\sigma'(91-) = 0$, $\tau(91) = 46098$ and $\tau'(91-) = 1560$. Both $\lambda(.)$ and $\mu(.)$ are twice continuously differentiable at $x_1 = 91$ because we have chosen each piece of $\sigma(.)$ to be three times continuously differentiable.

By considering the equations that correspond to expression (3.11) on each of the intervals [0,28], [28,42], [42,49] and [49,63], one can verify that $\lambda(.)$ and $\mu(.)$ are well-defined by $\sigma(.)$ and $\tau(.)$ throughout the range $0 \leq x_1 \leq 91$, but one has to give careful attention to the boundedness of $\lambda(.)$ and $\mu(.)$ near $x_1 = 45.5$, because it depends on the condition $\tau(45.5) = \sigma(45.5)$.

We now turn to the question of choosing $\sigma''(63+)$, for example, so that $\bar{F}(\underline{x})$ is twice continuously differentiable at $x_1 = 63$. Because $\tau(.)$ has no derivative discontinuities at $x_1 = 63$, it is sufficient if the value of $\sigma''(63+)$ allows $\lambda''(.)$ and $\mu''(.)$ to be continuous at $x_1 = 63$. Therefore we note that expression (3.11) implies the second derivative

$$\begin{aligned}\sigma''(63+) &= -0.6\lambda'(63) + 102\lambda''(63+) \\ &\quad + 241.8\mu'(63) - 7995\mu''(63+) .\end{aligned} \qquad (3.13)$$

A similar calculation on the interval [42,63] gives the identity

$$\begin{aligned}\sigma''(63-) &= -24\lambda'(63) + 102\lambda''(63-) \\ &\quad + 234\mu'(63) - 7995\mu''(63-) .\end{aligned} \qquad (3.14)$$

It follows that we must choose the value

$$\sigma''(63+) = M + 23.4\lambda'(63) + 7.8\mu'(63) \quad . \qquad (3.15)$$

The derivatives $\lambda'(63)$ and $\mu'(63)$ may be determined from equation (3.11) if

$\sigma'(63+)$ and $\tau'(63)$ are known. The derivative $\sigma'(63+) = -156$ has been specified already, and $\tau'(63) = -\sigma'(28)$ may be set to any positive number before $\sigma''(63+)$ is calculated. Thus it is straightforward to achieve second derivative continuity and the value $\sigma''(63-) = M$, where M is a positive parameter. Of course a similar technique may be used at $x_1 = 42$.

Having established the existence of $\bar{F}(.)$, we now modify it in order to take account of the $O(\phi^{4j})$ terms that have been ignored so far. For each integer k the required derivative $\underline{g}_k = \underline{\nabla}F(\underline{x}_k)$ is obtained by altering $\bar{F}(\underline{x})$ only for values of $\underline{x}$ that satisfy $|x_1 - (\underline{x}_k)_1| < 1$, where x_1 and $(\underline{x}_k)_1$ are still the first components of $\underline{x}$ and $\underline{x}_k$ respectively. Therefore it is sufficient to describe the modification for only one of the points $\{\underline{x}_{6j+i};\ i=1,2,\ldots,6\}$. We give our attention to the vectors

$$\underline{x}_{6j+3} = \begin{pmatrix} 63 \\ 102\phi^{2j} \\ -7995\phi^{2j} \end{pmatrix}, \qquad \underline{g}_{6j+3} = c_{6j+3}\begin{pmatrix} 81\phi^{2j} \\ -3-42471\phi^{4j} \\ -1-4356\phi^{4j} \end{pmatrix},$$

remembering the equation

$$\underline{\nabla}\bar{F}(\underline{x}_{6j+3}) = 4\begin{pmatrix} 81\phi^{2j} \\ -3 \\ -1 \end{pmatrix},$$

and that the overall scaling of the gradients provides $c_{6j+3} = 4 + O(\phi^{4j})$.

First we seek a function $\{\hat{F}(\underline{x});\ \underline{x}\in\mathbb{R}^3\}$ that has the gradient

$$\underline{\nabla}\hat{F}(\underline{x}_{6j+3}) = 4\begin{pmatrix} 81\phi^{2j} \\ -3-42471\phi^{4j} \\ -1-4356\phi^{4j} \end{pmatrix}, \tag{3.16}$$

and that satisfies all the monotonicity and continuity conditions. Therefore we let $\{\eta(x_1);\ x_1\in\mathbb{R}\}$ be a twice continuously differentiable function that has the properties

$$\left.\begin{array}{ll} \eta(x_1) = 0, & |x_1| \geq 1 \\ \eta(0) = 1, & \\ \eta(-x_1) = \eta(x_1), & \text{all } x_1 \\ \eta'(x_1) \geq 0, & x_1 \leq 0 \\ \eta'(0) = 0 & \end{array}\right\}, \tag{3.17}$$

and we consider the form

$$\hat{F}(x_1,x_2,x_3) = \bar{F}(x_1,x_2,x_3) + 4\,\eta(x_1-63)\,\hat{G}(x_2,x_3) \quad , \tag{3.18}$$

where $\hat{G}(.)$ is a homogeneous cubic function of two variables that satisfies the equation

$$\begin{pmatrix} \partial\hat{G}/\partial x_2 \\ \partial\hat{G}/\partial x_3 \end{pmatrix} = \begin{pmatrix} -42471\psi^2 \\ -4356\psi^2 \end{pmatrix} \text{ at } \begin{pmatrix} x_2 \\ x_3 \end{pmatrix} = \begin{pmatrix} 102\psi \\ -7995\psi \end{pmatrix}. \tag{3.19}$$

Thus the derivative (3.16) is obtained, and continuous second derivatives are preserved.

Let $\{\hat{\sigma}_j(x_1)\,;\; 42 \leq x_1 \leq 63\}$ be the value of $\hat{F}(\underline{x})$ at the point on the line segment from $\underline{x}_{6j+2}$ to $\underline{x}_{6j+3}$ whose first component is x_1. We have $\hat{\sigma}_j{}'(63) = 0$, and, because the contributions from $\bar{F}(.)$ and $\hat{G}(.)$ to $\sigma_j{}''(63)$ are both $O(\phi^{2j})$, we may choose the parameter M of $\bar{F}(.)$ so that $\hat{\sigma}_j{}''(63)$ is positive. Thus, not only is $\underline{x}_{6j+3}$ a local minimum on the line through $\underline{x}_{6j+2}$ and $\underline{x}_{6j+3}$, but also there exists a positive constant ε such that $\hat{\sigma}_j(x_1)$ decreases strictly monotonically for $63-\varepsilon \leq x_1 \leq 63$. Now the contribution from $\bar{F}(.)$ to $\{\sigma_j{}'(x_1)\,;\; 42 \leq x_1 \leq 63-\varepsilon\}$ is bounded above by a negative multiple of ϕ^{2j}, while the contribution to this derivative from $\hat{G}(.)$ is only $O(\phi^{4j})$. It follows that $\{\hat{\sigma}_j(x_1)\,;\; 42 \leq x_1 \leq 63\}$ decreases strictly monotonically for all sufficiently large values of j. Therefore, by arranging for the iterations of the conjugate gradient method to start at a later stage if necessary, we preserve the required conditions on the line search from $\underline{x}_{6j+2}$ to $\underline{x}_{6j+3}$. The same technique is also used to ensure that $\hat{F}(.)$ decreases monotonically from $\underline{x}_{6j+3}$ to $\underline{x}_{6j+4}$ and from $\underline{x}_{6j+4}$ to $\underline{x}_{6j+5}$.

In order to make a further modification to the objective function to allow for the factor c_{6j+3} in $\underline{g}_{6j+3}$, we apply the following construction. Let $u(.)$ be a twice continuously differentiable function from $\mathbb{R}$ to $\mathbb{R}$ such that $|u(t)| = O(|t|^3)$, and, in addition to the conditions (3.17), let $\eta(.)$ satisfy the equation $\{\eta(x_1)=1\,;\; |x_1| \leq \frac{1}{2}\}$. We choose $F(.)$ to have the form

$$F(\underline{x}) = \hat{F}(\underline{x}) + \eta(x_1-63)\,u(\hat{F}(\underline{x})), \quad \underline{x}\in\mathbb{R}^3 \quad . \tag{3.20}$$

Then $\underline{\nabla}F(\underline{x}) = \underline{\nabla}\hat{F}(\underline{x})$ for $|x_1-63| \geq 1$, and we have the gradient

$$\underline{\nabla}F(\underline{x}) = [1+u'(\hat{F}(\underline{x}))]\underline{\nabla}\hat{F}(\underline{x}) \quad , \quad |x_1-63| \leq \tfrac{1}{2} \; . \tag{3.21}$$

Further, for $\frac{1}{2} \leq |x_1-63| \leq 1$, the directional derivatives of $[F(.)-\hat{F}(.)]$ along the lines from $\underline{x}_{6j+2}$ to $\underline{x}_{6j+3}$, from $\underline{x}_{6j+3}$ to $\underline{x}_{6j+4}$, and from $\underline{x}_{6j+4}$ to $\underline{x}_{6j+5}$ are only $O(\phi^{4j})$. Therefore we apply the technique of the previous paragraph to ensure that $F(.)$ inherits the required monotonicity properties of $\hat{F}(.)$ for

$|x_1 - 63| \geq \frac{1}{4}$. Because the remaining monotonicity conditions are satisfied if the factor $[1 + u'(\hat{F}(\underline{x}))]$ in equation (3.21) is positive, we make a further postponement of the start of the conjugate gradient iterations if necessary to ensure $u'(\hat{F}(\underline{x})) < 1$ for all $\hat{F}(\underline{x})$ in $[0, \hat{F}(\underline{x}_1)]$.

Expressions (3.16) and (3.21) imply that we require $u(.)$ to satisfy the conditi

$$1 + u'(\hat{F}(\underline{x}_{6j+3})) = \tfrac{1}{4}c_{6j+3} , \qquad (3.22)$$

and equation (3.9) and the scaling of gradients give the value

$$\tfrac{1}{4}c_{6j+3} = \prod_{\ell=1}^{\infty} \frac{(4 + 4\phi^{4j+2\ell})(1 + 13689\phi^{4j+2\ell})(30 + 395307\phi^{4j+2\ell})}{(3 + 47190\phi^{4j+2\ell})(40 - 363\phi^{4j+2\ell})(1 + 14040\phi^{4j+2\ell})} . \qquad (3.23)$$

Thus $\tfrac{1}{4}c_{6j+3} = v(\phi^{2j})$, where $v(.)$ is the function

$$v(t) = \prod_{\ell=1}^{\infty} \frac{(1 + t^2\phi^{2\ell})(1 + 13689t^2\phi^{2\ell})(1 + 13176.9t^2\phi^{2\ell})}{(1 + 15730t^2\phi^{2\ell})(1 - 9.075t^2\phi^{2\ell})(1 + 14040t^2\phi^{2\ell})} , \qquad (3.24)$$

which is analytic for small t and satisfies $v(t) = 1 + O(t^2)$. Moreover, by integrating $\underline{\nabla}\hat{F}(.)$ along the straight line from $(63,0,0)$ to $\underline{x}_{6j+3}$, we find the value

$$\hat{F}(\underline{x}_{6j+3}) = 30756\phi^{2j} + 40658904\phi^{6j} . \qquad (3.25)$$

Therefore we define $u(.)$ by the equations

$$\left.\begin{array}{l} u'(30756t + 40658904t^3) = v(t) - 1 \\ u(0) = 0 \end{array}\right\} , \qquad (3.26)$$

which gives $|u(t)| = O(|t^3|)$. The definition of $F(\underline{x})$ for $|x_1 - 63| \leq 1$ is now complete. Similar constructions are used to form $F(\underline{x})$ when x_1 is close to other values of $(\underline{x}_k)_1$.

4. Discussion

Because Sections 2 and 3 only describe successful or partially successful attempts to construct counter-examples to the global convergence of the conjugate gradient method, we mention now some other sequences that were tried that satisfy the periodicity conditions (1.9).

When $m=2$, $\underline{d}_k^T\underline{d}_{k-1}$ must be negative for all sufficiently large k, and in this case, by Step 3 of the algorithm of Section 1, we have $\beta_k<0$. However, $\beta_k<0$ and equation (1.1) imply the condition

$$\begin{aligned}|\beta_k| &= (\underline{g}_k^T\underline{g}_{k-1} - \|\underline{g}_k\|^2)/\|\underline{g}_{k-1}\|^2 \\ &\leq (\|\underline{g}_k\|/\|\underline{g}_{k-1}\|) - (\|\underline{g}_k\|/\|\underline{g}_{k-1}\|)^2 \\ &\leq \tfrac{1}{4}, \qquad\qquad \beta_k<0, \end{aligned} \tag{4.1}$$

where the last line is just the maximum value of the function $\{t-t^2;\ t\in\mathbb{R}\}$. Therefore, for $m=2$, the numbers $\{\|\underline{d}_k\|;\ k=1,2,3,\dots\}$ remain bounded, which implies the limit (1.4).

Further, in every complete cycle of the form (1.9) we have $\underline{d}_k^T\underline{d}_{k-1}<0$ at least twice for sufficiently large k, so, if $\|\underline{d}_k\|\to\infty$, which is necessary in a counter-example, some mechanism has to combat the regular small values of the ratio $\|\underline{d}_k\|/\|\underline{d}_{k-1}\|$ that are implied by inequality (4.1). It is therefore a little surprising that the first example of Section 2 shows that a suitable mechanism can exist when $m=3$. The mechanism is that, due to equation (1.1), a large positive value of β_k occurs when $\|\underline{g}_k\| >> \|\underline{g}_{k-1}\|$, but in this case for $n=2$, due to the fact that $(\underline{g}_k)_1\to 0$, we have $F(\underline{x}^*)<F(\underline{x}_k)<F(\underline{x}_{k-1})$ only if $|(\underline{x}_k)_2|$ is much smaller than $|(\underline{x}_{k-1})_2|$, which is why the first example of Section 2 is unsatisfactory. In three dimensions, however, the relations $\|\underline{g}_k\| >> \|\underline{g}_{k-1}\|$ and $F(\underline{x}^*)<F(\underline{x}_k)<F(\underline{x}_{k-1})$ imply only that the projection of $(\underline{x}_k-\underline{x}_{k-m})$ in the direction of $\underline{g}_k$ must be small, which allows enough freedom for the first example of Section 3, but it seems that the nice properties of the given $n=3$, $m=4$ case cannot be achieved when $n=m=3$.

The success of the $n=2$, $m=8$ example of Section 2 was not expected by the author, because attempts to make the function values $\{F(\underline{x}_k);\ k=1,2,3,\dots\}$ decrease strictly monotonically, $F(.)$ being twice continuously differentiable, had failed for the form of symmetry of expression (2.11) using $n=2$, $m=4$ and $n=2$, $m=6$. The numbers that occur in the second example of Section 3 were found in the way that is suggested in Section 1, namely the conditions for consistency with the conjugate gradient method were expressed as inequality constraints on the parameters, and then a search was made for feasible values.

The examples of Section 2 are relevant not only to the conjugate gradient method, but also to all variable metric algorithms in Broyden's linear family that make exact line searches (see Fletcher, 1980, for instance). The reason is well-known, namely that condition (1.8) and $\underline{d}_k^T \underline{g}_k < 0$ define the direction of $(\underline{x}_{k+1} - \underline{x}_k)$ when there are only two variables. Therefore the DFP and BFGS algorithms may fail to converge, if the condition on the step-length α_k is only that it be a local minimum of the function $\{F(\underline{x}_k + \alpha \underline{d}_k)\ ;\ \alpha \geq 0\}$ that satisfies $F(\underline{x}_k + \alpha_k \underline{d}_k) < F(\underline{x}_k)$. An important consequence of this remark is that if a proof of convergence of one of these algorithms for general twice continuously differentiable objective functions could be found, which now seems unlikely, then the proof would depend on line search conditions that are stronger than one usually assumes. Examples where the DFP algorithm fails to converge are also given by Thompson (1977), but he allows the objective function to have first derivative discontinuities.

The last example of Section 3 shows that the Polak-Ribière version of the conjugate gradient algorithm without restarts, described in Section 1, may fail to find small values of $\|\underline{\nabla} F(\underline{x})\|$ in exact arithmetic. The given objective function may be modified so that it is ℓ times continuously differentiable, where ℓ is any positive integer.

Finally we turn our attention to the Fletcher-Reeves version of the conjugate gradient algorithm, although it is sometimes much less efficient than the Polak-Ribière version (Powell, 1977). Therefore we let the parameter β_k of the algorithm of Section 1 have the value

$$\beta_k = \|\underline{g}_k\|^2 / \|\underline{g}_{k-1}\|^2 \ , \tag{4.2}$$

instead of expression (1.1). Because β_k is now positive for all k, it is no longer possible for $\|\underline{d}_{k-1}\|$ and $\|\underline{d}_k\|$ to be very large and for $\underline{d}_k^T \underline{d}_{k-1}$ to be negative, which rules out examples of the form that are given in Sections 2 and 3. In fact, assuming the conditions that are stated in Section 1, it can be proved as follows that the Fletcher-Reeves algorithm gives the limit (1.4).

From the definition of $\underline{d}_k$ and exact line searches we deduce the equation

$$\begin{aligned} \|\underline{d}_k\|^2 &= \|\underline{g}_k\|^2 + \beta_k^2 \|\underline{d}_{k-1}\|^2 \\ &= \|\underline{g}_k\|^2 + \beta_k^2 [\|\underline{g}_{k-1}\|^2 + \beta_{k-1}^2 \|\underline{d}_{k-2}\|^2] = \ldots \\ &= \sum_{\ell=1}^{k} \{ \prod_{j=\ell+1}^{k} \beta_j^2 \} \|\underline{g}_\ell\|^2 = \sum_{\ell=1}^{k} \|\underline{g}_k\|^4 / \|\underline{g}_\ell\|^2 \ , \end{aligned} \tag{4.3}$$

where the product is defined to be one if $\ell = k$, and where the last line depends on the value (4.2). Further, we recall that $\|\underline{g}_k\|$ is bounded above because $\underline{x}_k$ is in the level set (1.3). Therefore, if the gradients were bounded away from zero, we would have $\|\underline{d}_k\|^2 \leq kc$, where c is a positive constant. Thus the sum

$$\sum_k \rho_k^2 = \sum_k \|\underline{g}_k\|^2 / \|\underline{d}_k\|^2 \tag{4.4}$$

would be divergent, which would contradict the remarks that follow inequality (1.7). Hence the Fletcher-Reeves version of the conjugate gradient algorithm achieves the limit (1.4).

Our results suggest that further attention should be given to the choice of the parameters $\{\beta_k;\ k=1,2,3,\ldots\}$ in the conjugate gradient method.

References

A. Buckley (1982), "Conjugate gradient methods", in Nonlinear Optimization 1981, ed. M.J.D. Powell, Academic Press (London).

R. Fletcher (1980), Practical Methods of Optimization, Vol. I: Unconstrained Optimization, John Wiley & Sons (Chichester).

R. Fletcher and C.M. Reeves (1964), "Function minimization by conjugate gradients", The Computer Journal, Vol. 7, pp. 149-154.

M.J.D. Powell (1977), "Restart procedures for the conjugate gradient method", Math. Programming, Vol. 12, pp. 241-254.

J.R. Thompson (1977), "Examples of non-convergence of conjugate descent algorithms with exact line searches", Math. Programming, Vol. 12, pp. 356-360.

G. Zoutendijk (1970), "Nonlinear programming, computational methods", in Integer and Nonlinear Programming, ed. J. Abadie, North-Holland Publishing Co. (Amsterdam).

PARTICLE APPROXIMATION OF LINEAR HYPERBOLIC EQUATIONS OF THE FIRST ORDER

P.-A. Raviart

1. INTRODUCTION.

Among the numerical practical methods of approximation of P.D.E.'s, the particle methods are frequently ignored although they are constantly used in a number of specific applications. In Physics, these methods seem fairly well adapted to the numerical solution of kinetic equations of Boltzmann type and are often based on a Monte-Carlo methodology. For applications to Physics and in particular to plasma simulations, we refer to the book of Hockney and Eastwood [7] . In Fluid Mechanics, particle methods have been introduced by Harlow [6] for the computation of multi-fluid compressible flows. On the other hand, vortex methods are currently used in the numerical simulation of incompressible fluid flows at large Reynolds numbers ; in that direction, see the survey of Leonard [8] .

In fact, to the author's best knowledge, it is only recently that a precise mathematical analysis of the particle method has been obtained : see the pioneering work of Hald [5] and the results of Beale & Majda [1] , [2] concerning the vortex method.

The purpose of this paper is twofold. We intend on the one hand to give a short mathematical introduction to particle methods and on the other hand to present some extensions and improvements of the results of [9] concerning the particle approximation of hyperbolic equations. For the sake of brevity, we have only given the essential points of the proofs of our main new results.

For a mathematical discussion of the convergence of the vortex method, we refer again to [1] , [2] , [9] . For an analysis of the particle approximation of Vlasov-Poisson equations arising in Plasma Physics, see [3] .

2. MEASURE SOLUTIONS OF FIRST ORDER HYPERBOLIC EQUATIONS.

Consider the Cauchy problem for a linear hyperbolic equation of the first order written in conservation form :

$$(2.1) \qquad \begin{cases} Lu \equiv \dfrac{\partial u}{\partial t} + \displaystyle\sum_{i=1}^{n} \dfrac{\partial}{\partial x_i}(a_i\, u) + a_0\, u = f \quad , \quad x \in \mathbb{R}^n \quad , \; t > 0 \; , \\ u(x,o) = u_0(x). \end{cases}$$

For simplicity, we shall assume that the coefficients a_i , $0 \leq i \leq n$, are sufficiently smooth and bounded together with the first derivatives $\dfrac{\partial a_i}{\partial x_j}$, $1 \leq i,j \leq n$.

Define the characteristic curves associated with the first order differential operator $\frac{\partial}{\partial t} + \sum_{i=1}^{n} a_i \frac{\partial}{\partial x_i}$ to be the solutions of the differential system

(2.2) $$\frac{dX}{dt} = a(X,t),$$

where $a = (a_1,\ldots,a_n)$ and $X = (X_1,\ldots,X_n)$. For all $x \in \mathbb{R}^n$ and all $s \geq 0$, we denote by $t \to X(t;x,s)$ the solution of (2.2) which satisfies the initial condition

(2.3) $$X(s) = x$$

and we set

(2.4) $$J(t;x,s) = \det(\frac{\partial X_i}{\partial x_j}(t;x,s)).$$

Then, if the data u_0 and f are sufficiently smooth, it is a simple matter to check that Problem (2.1) has a unique classical solution given by

(2.5) $$\begin{cases} u(x,t) = u_0(X(0;x,t))\, J(0;x,t) \exp(-\int_0^t a_0(X(s;x,t),s)ds) + \\ + \int_0^t f(X(s;x,t),s)\, J(s;x,t) \exp(-\int_s^t a_0(X(\sigma;x,t),\sigma)d\sigma)ds . \end{cases}$$

In fact, (2.5) makes sense even if the data are not smooth. We thus introduce the notion of weak solution of Problem (2.1). Denote by L^* the formal adjoint of the differential operator L, i.e.,

$$L^* v = -\frac{\partial v}{\partial t} - \sum_{i=1}^{n} a_i \frac{\partial v}{\partial x_i} + a_0 v .$$

We notice that, if u is a classical solution of (2.1), we have for all function $\varphi \in C_0^1(\mathbb{R}^n \times [0,T[)$ (i.e., φ is a C^1 function which vanishes for $|x|$ large enough and $t > T - \varepsilon$ for some $\varepsilon > 0$)

(2.6) $$\int_0^T \int_{\mathbb{R}^n} u\, L^* \varphi\, dx\, dt = \int_0^T \int_{\mathbb{R}^n} f \varphi\, dx\, dt + \int_{\mathbb{R}^n} u_0\, \varphi(.,0)dx .$$

Now, a function $u \in L^1_{loc}(\mathbb{R}^n \times (0,T))$ is called a weak solution of Problem (2.1) if (2.6) holds for all $\varphi \in C_0^1(\mathbb{R}^n \times [0,T[)$. One can prove

Theorem 1. Assume that $u_0 \in L^1_{loc}(\mathbb{R}^n)$ and $f \in L^1_{loc}(\mathbb{R}^n \times (0,T))$. Then, Problem (2.1) has a unique weak solution $u \in L^1_{loc}(\mathbb{R}^n \times (0,T))$ given by (2.5).

Next, for defining the particle method of approximation of weak solutions of (2.1), we need to consider still weaker solutions of (2.1) : the measure solutions. Given a subset S of $\mathbb{R}^d$, we denote by $M(S)$ the space of measures defined on S (i.e., the dual space of the space $C_0^0(S)$ of all continuous functions $S \to \mathbb{R}$ with compact support) and by $< \cdot,\cdot >$ the duality pairing between $M(S)$ and $C_0^0(S)$. Then, a measure $u \in M(\mathbb{R}^n \times [0,T])$ is called a measure solution of Problem (2.1) if

$$(2.7)\qquad \langle u, L^*\varphi \rangle = \langle f,\varphi \rangle + \langle u_0, \varphi(\cdot,0) \rangle \quad \forall\, \varphi \in C_0^1(\mathbb{R}^n \times [0,T[\,).$$

The following result is an easy generalization of Theorem 1.

Theorem 2. Assume that $u_0 \in M(\mathbb{R}^n)$ and $f \in M(\mathbb{R}^n \times [0,T])$. Then, Problem (2.1) has a unique measure solution $u \in M(\mathbb{R}^n \times [0,T])$.

Let us consider examples of measure solutions which will play a crucial role in the numerical approximation.

Example 1. Denote by $\delta(x-x_0)$ the Dirac measure located at the point $x_0 \in \mathbb{R}^n$. If

$$u_0 = \delta(x-x_0) \quad , \quad f = 0 \quad ,$$

it is readily seen that the corresponding measure solution of (2.1) is given by

$$(2.8)\qquad u(x,t) = \alpha_0(t)\ \delta(x-X(t;x_0,0))$$

where

$$\alpha_0(t) = \exp(-\int_0^t a_0(X(s;x_0,0),s)ds)\ .$$

In other words, $u \in M(\mathbb{R}^n \times [0,T])$ is defined by

$$\langle u,\varphi \rangle = \int_0^T \alpha_0(t)\ \varphi(X(t;x_0,0),t)dt \quad \forall\, \varphi \in C_0^0(\mathbb{R}^n \times [0,T]).$$

Hence, for all $t \in [0,T]$, $u(\cdot,t)$ is proportional to a Dirac measure whose trajectory in the (x,t)-space coincides with the characteristic curve passing through the point $(x_0,0)$. ∎

Example 2. If

$$u_0 = 0 \ ,\ f = \delta(x-x_0) \otimes \delta(t-t_0) \ ,\ x_0 \in \mathbb{R}^n \ ,\ t_0 \in [0,T] \quad ,$$

the corresponding measure solution of (2.1) is given by

$$(2.9)\qquad u(x,t) = \beta_0(t)\ \delta(x-X(t;x_0,t_0))$$

where

$$\beta_0(t) = H(t-t_0)\ \exp(-\int_{t_0}^t a_0(X(s;x_0,t_0),s)ds)$$

and

$$H(t) = \begin{cases} 0 & \text{for } t < 0 \\ 1 & \text{for } t \geqslant 0\ . \end{cases}$$

This means that

$$\langle u,\varphi \rangle = \int_{t_0}^T \beta_0(t)\ \varphi(X(t;x_0,t_0),t)dt \qquad \forall\, \varphi \in C_0^0(\mathbb{R}^n \times [0,T])\ .$$

Again, for $t \in [t_0,T]$, $u(\cdot,t)$ is proportional to a Dirac measure whose trajectory is the characteristic curve passing through the point (x_0,t_0). ∎

We shall say that (2.8) and (2.9) are the fundamental particle solutions associated with the first order differential operator L. The physical meaning of these solutions is clear.

3. THE PARTICLE METHOD.

Let us now introduce a general method of approximation of weak solutions of Problem (2.1). We begin by approximating the initial condition u_0 by a linear combination of Dirac measures on $\mathbb{R}^n$

$$u_h^0 = \sum_{j\in J} \alpha_j \, \delta(x-x_j)$$

for some set $(x_j,\alpha_j)_{j\in J}$ of points $x_j \in \mathbb{R}^n$ and weights $\alpha_j \in \mathbb{R}$. Then, by Example 1, the unique measure solution of

$$\begin{cases} L\, u_h = 0 \\ u_h(\cdot,0) = u_h^0 \end{cases}$$

is given by

$$u_h(x,t) = \sum_{j\in J} \alpha_j(t) \, \delta(x-X(t;x_j,0))$$

where

$$\alpha_j(t) = \alpha_j \exp(-\int_0^t a_0(X(s;x_j,0),s)ds) .$$

In practice, except in a few simple cases, the functions $t \to X(t;x_j,0)$ and $t \to \alpha_j(t)$ need to be determined numerically. In fact, one solves the differential system

$$\begin{cases} \frac{d}{dt} X(t;x_j,0) = a(X(t;x_j,0),t) \\ X(0;x_j,0) = x_j \end{cases}$$

and the differential equation

$$\begin{cases} \frac{d}{dt} \alpha_j(t) + a_0(X(t;x_j,0),t)\, \alpha_j(t) = 0 \\ \alpha_j(0) = \alpha_j \end{cases}$$

by means of a classical numerical method (Runge-Kutta, linear multistep,...).

Next, we approximate the function f by a linear combination of Dirac measures on $\mathbb{R}^n \times [0,T]$

$$f_h = \sum_{k\in K} \beta_k \, \delta(x-x_k) \otimes \delta(t-t_k)$$

for some set of points $(x_k,t_k)_{k\in K}$ of $\mathbb{R}^n \times [0,T]$ and some weights $(\beta_k)_{k\in K}$. It follows from Example 2 that the unique measure solution of

$$\begin{cases} L\, u_h = f_h \\ u_h(\cdot,0) = 0 \end{cases}$$

is given by

$$u_h(x,t) = \sum_{k\in K} \beta_k(t) \, \delta(x-X(t;x_k,t_k))$$

where

$$\beta_k(t) = \beta_k\ H(t-t_k)\ \exp(-\int_{t_k}^{t} a_o(X(s;x_k,t_k),s)ds)\ .$$

Again the trajectories $t \to X(t;x_k,t_k)$ and the weight functions $t \to \beta_k(t)$ are determined numerically by solving one differential system and one differential equation per particle.

Now, by the superposition principle, a particle approximation of Problem (2.1) is obtained as the unique measure solution of

$$(3.1)\qquad \begin{cases} L\ u_h = f_h \\ u_h(\cdot,0) = u_h^o \end{cases}$$

and is given explicitely by

$$(3.2)\qquad u_h(x,t) = \sum_{j\in J} \alpha_j(t)\ \delta(x-X(t;x_j,0)) + \sum_{k\in K} \beta_k(t)\ \delta(x-X(t;x_k,t_k))\ .$$

At that stage, it is worthwile to notice that the particle method may be viewed as a variant of the classical method of characteristics which satisfies the two following properties :

(i) it is a conservative method, i.e., we have if $a_o \equiv 0$

$$< u_h(\cdot,t),1 > = < u_h(\cdot,0),1 > + \int_0^t < f_h(\cdot,s),1 > ds\ ;$$

(ii) the method does not introduce any numerical diffusion.

4. CONVERGENCE OF THE PARTICLE METHOD I : PRELIMINARY RESULTS.

It remains to make more precise the choices of the approximate data u_h^o, f_h and to study the properties of the particle approximation (3.2) of the solution of Problem (2.1). We begin by considering the approximation of the initial conditior u_o by u_h^o in the sense of measures on $\mathbb{R}^n$. Thus, let φ be in $C_0^0(\mathbb{R}^n)$; we have to compare

$$< u_o,\varphi > = \int_{\mathbb{R}^n} u_o\ \varphi\ dx \quad \text{and} \quad < u_h^o,\varphi > = \sum_{j\in J} \alpha_j\ \varphi(x_j)\ .$$

We reckognize here the classical problem of numerical quadrature. Let us describe a simple procedure (but not necessarily the most effective in practice) for constructing such quadrature formulae. Given a parameter $\Delta x > 0$, we cover $\mathbb{R}^n$ with a uniform mesh with meshsize Δx : for all $j = (j_1,\dots,j_n) \in Z^n$, we denote by B_j the cell

$$B_j = \{x \in \mathbb{R}^n\ ;\ (j_i - \tfrac{1}{2})\ \Delta x \leq x_i \leq (j_i + \tfrac{1}{2})\ \Delta x\ ,\ 1 \leq i \leq n\}$$

and by $x_j = (j_i\ \Delta x)_{1\leq i\leq n}$ the center of B_j. Then, we set

$$(4.1)\qquad u_h^o = \sum_{j\in\mathbb{Z}^n} \alpha_j\ \delta(x-x_j)$$

where α_j is some approximation of $\int_{B_j} u_o \, dx$. Assuming that $u_o \in C^0(\mathbb{R}^n)$, we choose

(4.2) $$\alpha_j = \Delta x^n \, u_o(x_j)$$

so that

$$< u_o - u_h^o, \varphi > = \sum_{j \in \mathbb{Z}^n} E_j(u_o \, \varphi)$$

where

(4.3) $$E_j(g) = \int_{B_j} g \, dx - \Delta x^n \, g(x_j) \quad , \quad g \in C^0(B_j) \, .$$

In a similar way, for approximating the function f , we introduce a time-step $\Delta t > 0$ and we set

$$t_\ell = \ell \, \Delta t \quad , \quad t_{\ell+1/2} = (\ell + \tfrac{1}{2})\Delta t \quad , \quad \ell \in \mathbb{Z} \, .$$

Then, we define

(4.4) $$f_h = \sum_{\ell \geq 0} \sum_{j \in \mathbb{Z}^n} \beta_j^{\ell+1/2} \, \delta(x-x_j) \otimes \delta(t-t_{\ell+1/2}) \, .$$

Assuming that the function f is continuous, we take

(4.5) $$\beta_j^{\ell+1/2} = \Delta x^n \, \Delta t \, f(x_j, t_{\ell+1/2}) \, .$$

With the choices (4.1), (4.2), (4.4), (4.5) and setting $h = (\Delta x \, , \Delta t)$, the particle approximation u_h of the solution u of Problem (2.1) becomes

(4.6) $$\left\{ \begin{aligned} u_h(x,t) = & \sum_{j \in \mathbb{Z}^n} \alpha_j(t) \, \delta(x - X(t;x_j,0)) + \\ & + \sum_{\ell \geq 0} \sum_{j \in \mathbb{Z}^n} \beta_j^{\ell+1/2}(t) \, \delta(x - X(t;x_j,t_{\ell+1/2})) \end{aligned} \right.$$

with

(4.7) $$\left\{ \begin{aligned} & \alpha_j(t) = \Delta x^n \, u_o(x_j) \exp(- \int_0^t a_o(X(s;x_j,0),s)ds) \, , \\ & \beta_j^{\ell+1/2}(t) = \Delta x^n \, \Delta t \, f(x_j,t_{\ell+1/2}) \, H(t-t_{\ell+1/2}) \times \\ & \qquad \times \exp(- \int_{t_{\ell+1/2}}^t a_o(X(s;x_j,t_{\ell+1/2}),s)ds) \, . \end{aligned} \right.$$

Remark 1. In practice, the functions u_o and f tend to zero rapidly as $|x|$ tends to zero (or even vanish for $|x|$ large enough) so that we can neglect α_j and $\beta_j^{\ell+1/2}$ for $|j|$ large enough and consider only a finite number of particles. ∎

Let us next compare $u(\cdot,t)$ and $u_h(\cdot,t)$ in the sense of measures on $\mathbb{R}^n$. We can state

Lemma 1. We have for all function $\varphi \in C_0^0(\mathbb{R}^n)$

(4.8) $$\left\{\begin{aligned} &< u(\cdot,t) - u_h(\cdot,t),\varphi > = \sum_{j\in\mathbb{Z}^n} E_j(v(\cdot,t)) + \\ &+ \int_0^t \int_{\mathbb{R}^n} g(y,s,t)dy\, ds - \Delta t \sum_{\ell\geq 0} H(t-t_{\ell+1/2}) \int_{\mathbb{R}^n} g(y,t_{\ell+1/2},t)dy + \\ &+ \Delta t \sum_{\ell\geq 0} H(t-t_{\ell+1/2}) \sum_{j\in\mathbb{Z}^n} E_j(g(\cdot,t_{\ell+1/2},t)) \end{aligned}\right.$$

where

(4.9) $$\left\{\begin{aligned} &v(y,t) = u_0(y) \exp(- \int_0^t a_0(X(s;y,0),s)ds)\, \varphi(X(t;y,0)) \\ &g(y,s,t) = f(y,s) \exp(- \int_s^t a_0(X(\sigma;y,s),s)ds)\, \varphi(X(t;y,s)) \end{aligned}\right.$$

Proof. Let φ be in $C_0^0(\mathbb{R}^n)$; using the explicit form (2.5) of the solution u and making simple changes of variables, we obtain

$$< u(\ ,t),\varphi > = \int_{\mathbb{R}^n} u(x,t)\, \varphi(x)dx =$$

$$= \int_{\mathbb{R}^n} u_0(y) \exp(- \int_0^t a_0(X(s;y,0),s)ds)\, \varphi(X(t;y,0))dy +$$

$$+ \int_0^t \int_{\mathbb{R}^n} f(y,s) \exp(- \int_s^t a_0(X(\sigma;y,s),\sigma)d\sigma)\, \varphi(X(t;y,s))dy\, ds$$

so that (4.8) follows easily from (4.6) and (4.7). ∎

Now, in order to obtain convergence results, we need to recall some results concerning the error functional E_j associated with the generalized midpoint quadrature rule. If Ω is an open subset of $\mathbb{R}^d$, we introduce the Sobolev space

$$W^{m,\infty}(\Omega) = \{\varphi \in L^\infty(\Omega) \ ;\ \partial^\alpha \varphi = \frac{\partial^{|\alpha|}\varphi}{\partial x_1^{\alpha_1} \dots \partial x_n^{\alpha_n}} \in L^\infty(\Omega) \ ,\ |\alpha| \leq m\}$$

provided with the norm and semi-norm

$$\|\varphi\|_{m,\infty,\Omega} = \max_{|\alpha|\leq m} \|\partial^\alpha \varphi\|_{L^\infty(\Omega)} \ ,\ |\varphi|_{m,\infty,\Omega} = \max_{|\alpha|=m} \|\partial^\alpha \varphi\|_{L^\infty(\Omega)} .$$

Then,we may state

Lemma 2. We have for $m = 1,2$

(4.10) $$|E_j(g)| \leq C\, \Delta x^{m+n} |g|_{m,\infty,B_j} \qquad \forall\, g \in W^{m,\infty}(B_j) .$$

More generally, there exist constants d_α , $\alpha \in \mathbb{N}^n$ with $|\alpha| \geq 2$, independant of Δx and j such that for all integer $m \geq 3$

(4.11) $$\left\{\begin{aligned} &|E_j(g) - \sum_{2\leq|\alpha|\leq m-1} d_\alpha\, \Delta x^{|\alpha|} \int_{B_j} \partial^\alpha g\, dx| \leq \\ &\qquad \leq C\, \Delta x^{m+n} |g|_{m,\infty,B_j} \qquad \forall\, g \in W^{m,\infty}(B_j) . \end{aligned}\right.$$

Here and in all the sequel, C, c_1, c_2,... denote various positive constants independent of Δx , Δt (and ε to be introduced later).

The first part of the lemma is standard. The asymptotic expansion (4.11) is proved in [9]. As an immediate consequence, we obtain

Lemma 3. *Let $m \geqslant 1$ be an integer. Then, we have for all function $g \in W^{m,\infty}(\mathbb{R}^n) \cap L^1(\mathbb{R}^n)$ if $m \leqslant 2$ and all function $g \in W^{m,\infty}(\mathbb{R}^n) \cap W^{m-1,1}(\mathbb{R}^n)$ (1) if $m \geqslant 3$*

$$(4.12) \qquad \Big|\sum_{j\in\mathbb{Z}^n} E_j(g)\Big| \leqslant C\, \Delta x^{m+n} \sum_{j\in\mathbb{Z}^n} |g|_{m,\infty,\mathbb{R}^n} .$$

Finally, let us state a first convergence result.

Theorem 3. *Let $u_0 \in C^0(\mathbb{R}^n)$ and $f \in C^0(\mathbb{R}^n \times [0,T])$. Then, we have for all function $\varphi \in C_0^0(\mathbb{R}^n)$*

$$\lim_{h\to 0} < u(\cdot,t) - u_h(\cdot,t),\varphi > = 0 \quad \text{uniformly in } t \in [0,T] .$$

Proof. The conclusion follows easily from Lemma 1 and

$$\lim_{\Delta x\to 0} \Big|\sum_{j\in\mathbb{Z}^n} E_j(g)\Big| = 0 \qquad \forall\, g \in C_0^0(\mathbb{R}^n) ,$$

$$\lim_{\Delta t\to 0} \Big|\int_0^t \psi(s)ds - \Delta t \sum_{\ell\geqslant 0} H(t-t_{\ell+1/2})\, \psi(t_{\ell+1/2})\Big| = 0 \qquad \forall\, \psi \in C^0([0,T]) . \qquad ∎$$

5. CONVERGENCE OF THE PARTICLE METHOD II : ERROR ESTIMATES.

In fact, for obtaining a practically useful approximation of the function $u(\cdot,t)$ at each time t , we need to define from the particle solution $u_h(\cdot,t)$ a continuous function $u_h^\varepsilon(\cdot,t)$ which will approximate $u(\cdot,t)$ in the L^∞ norm. To do that, we introduce a cut-off function $\zeta \in C^0(\mathbb{R}^n) \cap L^1(\mathbb{R}^n)$ such that $\int_{\mathbb{R}^n} \zeta\, dx = 1$. For specificity, we shall assume in the sequel that the function ζ has a compact support. Next, we set for all $\varepsilon > 0$

$$\zeta_\varepsilon(x) = \frac{1}{\varepsilon^n}\, \zeta(\frac{x}{\varepsilon})$$

and

$$(5.1) \qquad u_h^\varepsilon(\cdot,t) = u_h(\cdot,t) \star \zeta_\varepsilon$$

i.e.,

$$u_h^\varepsilon(x,t) = \sum_{j\in\mathbb{Z}^n} \alpha_j(t)\, \zeta_\varepsilon(x-X(t;x_j,0)) + \sum_{\ell\geqslant 0} \sum_{j\in\mathbb{Z}^n} \beta_j^{\ell+1/2}\, \zeta_\varepsilon(x-X(t;x_j,t_{\ell+1/2})) .$$

---

(1) i.e., $g \in W^{m,\infty}(\mathbb{R}^n)$ and $\partial^\alpha g \in L^1(\mathbb{R}^n)$ for $|\alpha| \leqslant m-1$.

In order to derive a uniform bound for $u(\cdot,t) - u_h^\varepsilon(\cdot,t)$, we write

$$u(\cdot,t) - u_h^\varepsilon(\cdot,t) = (u(\cdot,t) - u(\cdot,t) \star \zeta_\varepsilon) + (u(\cdot,t) - u_h(\cdot,t)) \star \zeta_\varepsilon .$$

Using Taylor's formula with integral remainder, the first term in the right hand side is easily estimated by means of

Lemma 4. (cf. [9]) Assuming that for some integer $k \geqslant 1$

(5.2) $$\int_{\mathbb{R}^n} \zeta \, dx = 1 , \int_{\mathbb{R}^n} x^\alpha \zeta \, dx = 0 \quad (^1) , \forall \alpha \in \mathbb{N}^n , 1 \leqslant |\alpha| \leqslant k-1 .$$

Then, we have for all function $v \in W^{k,\infty}(\mathbb{R}^n)$

(5.3) $$\|v \star \zeta_\varepsilon - v\|_{L^\infty(\mathbb{R}^n)} \leqslant C \, \varepsilon^k \, |v|_{k,\infty,\mathbb{R}^n} .$$

The crucial point lies in estimating

$$\|(u(\cdot,t) - u_h(\cdot,t)) \star \zeta_\varepsilon\|_{L^\infty(\mathbb{R}^n)} .$$

Such a bound depends of the smoothness properties of the function ζ . Thus, we introduce a class of cut-off functions. Given an integer $m \geqslant 1$, we denote by C_m the set of all functions $\zeta \in W^{m-1,\infty}(\mathbb{R}^n)$ with compact support which satisfy the following property : there exists a decomposition of the support of ζ of the form

(5.4) $$\mathrm{supp}(\zeta) = \bigcup_{r=1}^{R} \bar{\Omega}_r$$

such that

(i) the Ω_r's are open disjoint subsets of $\mathbb{R}^n$ with Lipschitz continuous boundaries $\partial \Omega_r$,

(ii) $\zeta_{|\Omega_r} \in W^{m,\infty}(\Omega_r)$, $1 \leqslant r \leqslant R$.

Now, given $x \in \mathbb{R}^n$ and using Lemma 1 with the function $\varphi : y \to \zeta_\varepsilon(x-y)$, we may write

(5.5) $$\left\{ \begin{aligned} &((u(\cdot,t) - u_h(\cdot,t)) \star \zeta_\varepsilon)(x) = \sum_{j \in \mathbb{Z}^n} E_j(v(x,\cdot,t)) + \\ &+ \int_0^t \int_{\mathbb{R}^n} g(x,y,s,t)dy\,ds - \Delta t \sum_{\ell \geqslant 0} H(t-t_{\ell+1/2}) \int_{\mathbb{R}^n} g(x,y,t_{\ell+1/2},t)dy + \\ &+ \Delta t \sum_{\ell \geqslant 0} H(t-t_{\ell+1/2}) \sum_{j \in \mathbb{Z}^n} E_j(g(x,\cdot,t_{\ell+1/2},t)) \end{aligned} \right.$$

where

(5.6) $$\left\{ \begin{aligned} &v(x,y,t) = u_o(y) \exp(- \int_0^t a_o(X(s;y,0),s)ds) \, \zeta_\varepsilon(x-X(t;y,0)) \\ &g(x,y,s,t) = f(y,s) \exp(- \int_s^t a_o(X(\sigma;y,s),\sigma)d\sigma) \, \zeta_\varepsilon(x-X(t;y,s)) . \end{aligned} \right.$$

---

(1) i.e., $\int_{\mathbb{R}^n} x_1^{\alpha_1} \ldots x_n^{\alpha_n} \zeta(x)dx = 0$, $\alpha = (\alpha_1,\ldots,\alpha_n)$.

We begin by proving

Lemma 5. Let $m = 1$ or 2. If ζ belongs to the class C_m and $f(\cdot,s)$ belongs to $W^{m,\infty}(\mathbb{R}^n)$, we have for some constant $C = C(T)$

(5.7) $$\left|\sum_{j\in\mathbb{Z}^n} E_j(g(x,\cdot,s,t))\right| \leq C(1+\frac{\Delta x}{\varepsilon})^{n-1}(\frac{\Delta x}{\varepsilon})^m \|f(\cdot,s)\|_{m,\infty,\mathbb{R}^n}\ , \quad 0 \leq s,t \leq T.$$

Proof. Let us introduce the mapping $\Phi_s^t : x \in \mathbb{R}^n \to \Phi_s^t(x) = X(t;x,s) \in \mathbb{R}^n$. In the su

$$\sum_{j\in\mathbb{Z}^n} E_j(g(x,\cdot,s,t))\ ,$$

we have only to take into account the indices $j \in J$ where

$$J = J(\varepsilon,x,s,t) = \{j \in \mathbb{Z}^n\ ;\ \Phi_s^t(B_j) \cap \operatorname{supp}(y \to \zeta_\varepsilon(x-y)) \neq \phi\}\ .$$

Using the decomposition (5.4) of the support of the cut-off function ζ, we set

$$\Omega_r^\varepsilon = \{y \in \mathbb{R}^n\ ;\ \frac{y}{\varepsilon} \in \Omega_r\}\quad ,\quad r = 1,\dots,R$$

and we consider the subset J_1 of J defined by

$$J_1 = \{j \in J\ ;\ \Phi_s^t(B_j) \subset x - \overline{\Omega}_r^\varepsilon \text{ for at least one } r \in \{1,\dots,R\}\}\ .$$

Since

$$\operatorname{meas}(\Omega_r^\varepsilon) \leq c_1\ \varepsilon^n$$

and

$$\operatorname{meas}(\Phi_s^t(B_j)) \geq c_2\ \Delta x^n\quad ,\quad 0 \leq s,t \leq T\quad ,\quad c_2 = c_2(T)\quad ,$$

we obtain that

$$\operatorname{card}(J_1) \leq c_3(\frac{\varepsilon}{\Delta x})^n$$

with $c_3 = c_1\ c_2^{-1}$. Moreover, since $\zeta \in C_m$, the function $g(x,\cdot,s,t)$ belongs to $W^{m,\infty}(B_j)$ for all $j \in J_1$ so that by Lemma 2

$$\sum_{j\in J_1} E_j(g(x,\cdot,s,t)) \leq c_4\ \Delta x^{m+n} \sum_{j\in J_1} |g(x,\cdot,s,t)|_{m,\infty,B_j}\ .$$

But, if $j \in J_1$, we have

$$|g(x,\cdot,s,t)|_{m,\infty,B_j} \leq \frac{c_5}{\varepsilon^{m+n}} \|f(\cdot,s)\|_{m,\infty,B_j}\quad ,\quad 0 \leq s,t \leq T$$

for some constant $c_5 = c_5(T)$ and therefore

$$\sum_{j\in J_1} |g(x,\cdot,s,t)|_{m,\infty,B_j} \leq \frac{c_5}{\varepsilon^{m+n}}\ \operatorname{card}(J_1)\ \|f(\cdot,s)\|_{m,\infty,B_j} \leq$$

$$\leq \frac{c_6}{\varepsilon^m\ \Delta x^n} \|f(\cdot,s)\|_{m,\infty,\mathbb{R}^n}$$

with $c_6 = c_3\ c_5$. Hence, we obtain for some constant $c_7 = c_7(T)$

(5.8) $$\sum_{j\in J_1} E_j(g(x,\cdot,s,t)) \leq c_7(\frac{\Delta x}{\varepsilon})^m \|f(\cdot,s)\|_{m,\infty,\mathbb{R}^n}\quad ,\quad 0 \leq s,t \leq T\ .$$

Next, let $J_2 = J \setminus J_1$. If $j \in J_2$, $\Phi_s^t(B_j)$ intersects $\partial\,\Omega_r^\varepsilon$ for at least one $r = 1,\ldots,R$. Since

$$\operatorname{diam}(\Phi_s^t(B_j)) \leqslant c_8\ \Delta x \quad , \quad 0 \leqslant s,t \leqslant T \quad , \quad c_8 = c_8(T) \quad ,$$

there exists a constant $c_9 = c_9(T)$ such that

$$j \in J_2 \Rightarrow \Phi_s^t(B_j) \subset D = \bigcup_{r=1}^{R} \Big(\bigcup_{x \in \partial\Omega_r^\varepsilon} S(x, c_9\ \Delta x)\Big)$$

where $S(x,\rho)$ denotes the ball of center x and radius ρ. But

$$\operatorname{meas}(D) \leqslant c_{10}\ \Delta x(\varepsilon + \Delta x)^{n-1} \quad , \quad c_{10} = c_{10}(T)$$

so that

$$\operatorname{card}(J_2) \leqslant c_{11}\ \frac{\Delta x(\varepsilon + \Delta x)^{n-1}}{\Delta x^n} = c_{11}(1 + \frac{\varepsilon}{\Delta x})^{n-1}$$

with $c_{11} = c_{10}\ c_2^{-1}$. Since $g(x,\cdot,s,t) \in W^{m-1,\infty}(\mathbb{R}^n)$, we have by Lemma 2

$$\Big|\sum_{j\in J_2} E_j(g(x,\cdot,s,t))\Big| \leqslant c_{12}\ \Delta x^{m+n-1} \sum_{j\in J_2} |g(x,\cdot,s,t)|_{m-1,\infty,B_j} \ .$$

Now

$$|g(x,\ ,s,t)|_{m-1,\infty,B_j} \leqslant \frac{c_{13}}{\varepsilon^{m+n-1}}\ \| f(\cdot,s)\|_{m-1,\infty,B_j} \quad , \quad 0 \leqslant s,t \leqslant T$$

with $c_{13} = c_{13}(T)$. Hence, we find for some constant $c_{14} = c_{14}(T)$

$$(5.9) \qquad \Big|\sum_{j\in J_2} E_j(g(x,\cdot,s,t))\Big| \leqslant c_{14}(1 + \frac{\Delta x}{\varepsilon})^{n-1} (\frac{\Delta x}{\varepsilon})^m\ \| f(\cdot,s)\|_{m-1,\infty,\mathbb{R}^n} \ .$$

The desired estimate (5.7) follows from (5.8) and (5.9). ■

When the cut-off function ζ is sufficiently smooth, a direct application of Lemma 3 gives with a simpler proof.

Lemma 6. Let $m \geqslant 1$ be an integer. Assume that the cut-off function ζ belongs to the space $W^{m,\infty}(\mathbb{R}^n)$ and has a compact support. Then, if $f(\cdot,s)$ belongs to $W^{m,\infty}(\mathbb{R}^n)$, we have for some constant $C = C(T)$ and all $s,t \in [\,0,T\,]$

$$(5.10) \qquad \Big|\sum_{j\in \mathbb{Z}^n} E_j(g(x,\cdot,s,t))\Big| \quad C(1 + \frac{\Delta x}{\varepsilon})^n (\frac{\Delta x}{\varepsilon})^m\ \| f(\cdot,s)\|_{m,\infty,\mathbb{R}^n}$$

We are now able to state the main result of this Section which extends some of the results of [9] by taking into account a source term f and relaxing the smoothness of the cut-off function ζ.

Theorem 4. Assume the following hypotheses :

(i) there exists an integer $k \geqslant 1$ such that (5.2) holds ;

(ii) we have

$$(5.11) \qquad \Delta x + \Delta t \leqslant c\ \varepsilon \ .$$

Then, if $\zeta \in C_m$ with $m = 1$ or 2, we obtain for sufficiently smooth data u_o and f and for all $t_p = p\,\Delta t \in [0,T]$

$$\|u(\cdot,t_p) - u_h^\varepsilon(\cdot,t_p)\|_{L^\infty(\mathbb{R}^n)} \leq C(T,u_o,f)\left(\varepsilon^k + \left(\frac{\Delta x}{\varepsilon}\right)^m + \left(\frac{\Delta t}{\varepsilon}\right)^m\right). \tag{5.12}$$

If $\zeta \in W^{m,\infty}(\mathbb{R}^n)$ for some integer $m \geq 2$ and has a compact support, we obtain again for sufficiently smooth data u_o and f

$$\|u(\cdot,t_p) - u_h^\varepsilon(\cdot,t_p)\|_{L^\infty(\mathbb{R}^n)} \leq C(T,u_o,f)\left(\varepsilon^k+\left(\frac{\Delta x}{\varepsilon}\right)^m+\left(\frac{\Delta t}{\varepsilon}\right)^2\right), \quad 0 \leq t_p \leq T. \tag{5.13}$$

Proof. Assume that the data u_o and f are sufficiently smooth. Then, using (5.2) and Lemma 4 yields

$$\|u(\cdot,t_p) - u(\cdot,t_p) * \zeta_\varepsilon\|_{L^\infty(\mathbb{R}^n)} \leq c_1\, \varepsilon^k |u(\cdot,t_p)|_{k,\infty,\mathbb{R}^n} \leq$$

$$\leq c_2\, \varepsilon^k \qquad , \qquad c_2 = c_2(T,u_o,f).$$

On the other hand, for estimating

$$\|(u(\cdot,t_p) - u_h(\cdot,t_p) * \zeta_\varepsilon\|_{L^\infty(\mathbb{R}^n)}$$

we start from (5.5) and (5.6). Assume first $\zeta \in C_m$ for $m = 1$ or 2. Then, it follows from Lemma 5 (with the function $g(x,\cdot,s,t)$ replaced by $v(x,\cdot,t)$) and (5.11) that

$$\Big|\sum_{j\in\mathbb{Z}^n} E_j(v(x,\cdot,t))\Big| \leq c_3\left(1 + \frac{\Delta x}{\varepsilon}\right)^{n-1}\left(\frac{\Delta x}{\varepsilon}\right)^m \|u_o\|_{m,\infty,\mathbb{R}^n} \leq$$

$$\leq c_4\left(\frac{\Delta x}{\varepsilon}\right)^m, \qquad c_4 = c_4(T,u_o).$$

Similarly, we have again by Lemma 5

$$\Big|\Delta t \sum_{\ell\geq 0} H(t_p - t_{\ell+1/2}) \sum_{j\in\mathbb{Z}^n} E_j(g(x,\cdot,t_{\ell+1/2},t_p))\Big| =$$

$$= \Big|\Delta t \sum_{\ell=0}^{p-1} \sum_{j\in\mathbb{Z}^n} E_j(g(x,\cdot,t_{\ell+1/2},t_p))\Big| \leq$$

$$\leq c_5\left(\frac{\Delta x}{\varepsilon}\right)^m \max_{0\leq s\leq t_p} \|f(\cdot,s)\|_{m,\infty,\mathbb{R}^n} \leq c_6\left(\frac{\Delta x}{\varepsilon}\right)^m$$

with $c_6 = c_6(T,f)$.

Finally, it remains to show that

$$\Big|\int_0^{t_p}\int_{\mathbb{R}^n} g(x,y,s,t_p)\,dy\,ds - \Delta t \sum_{\ell=0}^{p-1}\int_{\mathbb{R}^n} g(x,y,t_{\ell+1/2},t_p)\,dy\Big| \leq$$

$$\leq c_7\left(\frac{\Delta t}{\varepsilon}\right)^m \sum_{i=0}^{m}\int_0^{t_p} \Big\|\frac{\partial^i f}{\partial t^i}(\cdot,s)\Big\|_{L^\infty(\mathbb{R}^n)}\,ds.$$

By using similar techniques to that of the proof of Lemma 5, this follows from the standard estimate

$$\left|\int_0^{t_p} \psi(s)ds - \Delta t \sum_{\ell=0}^{p-1} \psi(t_{\ell+1/2})\right| \leqslant c_8 \, \Delta t^i \int_0^{t_p} \left|\frac{d^i\psi}{dt^i}(s)\right| ds \quad , \ i = 1, 2.$$

The estimate (5.12) is thus proved.

When $\zeta \in W^{m,\infty}(\mathbb{R}^n)$ for some integer $m \geqslant 2$ and has a compact support, the estimate (5.13) is obtained in a similar way by using Lemma 6 instead of Lemma 5. ■

Let us briefly discuss the choice of the cut-off function ζ. In dimension $n = 1$, it is often convenient to take for ζ a B-spline of the form

$$\zeta = \chi^{\star m} = \underbrace{\chi \star \chi \star \cdots \star \chi}_{m \text{ times}}$$

where χ is the characteristic function of the interval $[-\frac{1}{2}, \frac{1}{2}]$. It is readily seen that such a function ζ belongs to the class C_m (and therefore to $W^{m-1,\infty}(\mathbb{R})$).

In the general n-dimensional case, we may choose

$$\zeta(x) = \prod_{i=1}^{n} \chi^{\star m}(x_i) \ .$$

Again, this cut-off function ζ belongs to C_m or to $W^{m-1,\infty}(\mathbb{R}^n)$ and satisfies (5.2) with $k = 2$. Moreover, this function ζ is $\geqslant 0$ so that the conditions $u_0 \geqslant 0$ and $f \geqslant 0$ imply $u_h^\varepsilon \geqslant 0$, a property which is often required in practice.

For other choices of cut-off functions, see [1], [5], [9].

6. REINITIALIZATION OF PARTICLES.

When the source term f is $\neq 0$, we have seen that we neeed to introduce at each time $t_{\ell+1/2}$ a new set of particles. Clearly, after some time-steps, the total number of particles can become too large. Hence, for computational economy, we have to reinitialize our whole set of particles.

The mathematical problem is as follows. Given a distribution of particles

(6.1) $$u_h = \sum_{k\in K} \alpha_k \, \delta(x-y_k) \ ,$$

we want to approximate u_h by a measure $v_h \in M(\mathbb{R}^n)$ of the form

$$v_h = \sum_{j\in\mathbb{Z}^n} \beta_j \, \delta(x-x_j) \qquad , \quad x_j = (j_i \, \Delta x)_{1\leqslant i\leqslant n} \ .$$

Then, we introduce a function $\hat{w} \in C_0^0(\mathbb{R}^n)$; we set

$$w_h(x) = \hat{w}\left(\frac{x}{\Delta x}\right)$$

and

(6.2) $$v_h = \sum_{j\in\mathbb{Z}^n} (w_h \star u_h)(x_j)\ \delta(x - x_j)$$

or equivalently

$$v_h = \sum_{j\in\mathbb{Z}^n} \Big(\sum_{k\in K} \alpha_k\, w_h(x_j - y_k)\Big)\ \delta(x - x_j)$$

Let us compare the measures u_h and v_h. Given $\varphi \in C_0^0(\mathbb{R}^n)$, we have

$$< v_h, \varphi > = \sum_{j\in\mathbb{Z}^n} \Big(\sum_{k\in K} \alpha_k\, w_h(x_j - y_k)\Big)\ \varphi(x_j) = < u_h, \tilde{\varphi} >$$

where

(6.3) $$\tilde{\varphi}(x) = \sum_{j\in\mathbb{Z}^n} w_h(x_j - x)\ \varphi(x_j)\ .$$

Hence, we obtain

(6.4) $$< u_h - v_h, \varphi > = < u_h, \varphi - \tilde{\varphi} >$$

Given $x \in \mathbb{R}^n$, we define

$$\mathcal{O}_h(x) = x + \mathrm{supp}(w_h)$$

For convenience, we assume that $\mathrm{supp}(\hat{w})$ is star-shaped with respect to the origi so that $\mathcal{O}_h(x)$ is star-shaped with respect to x. We are now able to state

Lemma 7. Assume that there exists an integer $\ell \geqslant 1$ such that

(6.5) $$\sum_{j\in\mathbb{Z}^n} j^\alpha\, \hat{w}(j - \xi) = \xi^\alpha \qquad \forall\ \xi \in \mathbb{R}^n\ ,\ \forall\ \alpha \in \mathbb{N}^n \ \text{with}\ |\alpha| \leqslant \ell - 1.$$

Then, if $\varphi \in W^{\ell,\infty}(\mathcal{O}_h(x))$, we have

(6.6) $$|(\varphi - \tilde{\varphi})(x)| \leqslant C\ \Delta x^\ell\ |\varphi|_{\ell,\infty,\mathcal{O}_h(x)}\ .$$

Proof. First, it is a simple matter to check that the relations (6.5) imply

(6.7) $$\sum_{j\in\mathbb{Z}^n} (x_j - x)^\alpha\, w_h(x_j) = x^\alpha \qquad \forall\ x \in \mathbb{R}^n \quad , \quad |\alpha| \leqslant \ell - 1.$$

Next, using (6.7) and Taylor's formula with integral remainder at the point x yields (6.6). ∎

For examples of functions $\hat{w}$ which satisfy the relations (6.5), we refer to [4].

Now, we want to estimate the error at time t due to a reinitialization of the set of particles at time s. To this purpose, we introduce the operator solution $S(s,t) : v_0 \to v(\cdot,t) = S(s,t)v_0$ of the Cauchy problem

$$\begin{cases} L\, v = 0\ , \\ v(\cdot,s) = v_0\ . \end{cases}$$

We have to bound $\|(S(s,t)(u_h - v_h)) \star \zeta_\varepsilon\|_{L^\infty(\mathbb{R}^n)}$

Theorem 5. Assume the following hypotheses :

(i) there exists an integer $\ell \geq 1$ such that (6.5) holds ;

(ii) the cut-off function ζ belongs to the class C_ℓ ;

(iii) there exists a constant $\delta > 0$ such that

(6.7) $$|y_k - y_{k'}| \geq \delta \qquad \forall\, k,k' \in K \,,\quad k \neq k' \,.$$

Then, we have for some constant $C(T)$ and for all $s < t < T$

(6.8) $$\|(S(s,t)(u_h - v_h)) * \zeta_\varepsilon\|_{L^\infty(\mathbb{R}^n)} \leq C(T)\frac{1}{\delta^n}(\sup_{k\in K} |\alpha_k|)(1 + \frac{\Delta x}{\varepsilon})^n (\frac{\Delta x}{\varepsilon})^\ell \,.$$

Proof. First, we observe that for all $v_0 \in M(\mathbb{R}^n)$ and all $\varphi \in C_0^0(\mathbb{R}^n)$

$$< S(s,t)v_0, \varphi > \; = \; < v_0, S(s,t)^\star \varphi >$$

where

$$(S(s,t)^\star \varphi)(x) = \varphi(X(t;x,s)) \exp(-\int_s^t a_0(X(\sigma;x,s),\sigma)d\sigma) \,.$$

Hence, using (6.3) and (6.4), we have

$$((S(s,t)(u_h - v_h)) * \zeta_\varepsilon)(x) = \sum_{k\in K} \alpha_k(\varphi_\varepsilon(x,y_k) - \tilde{\varphi}_\varepsilon(x,y_k))$$

where

$$\varphi_\varepsilon(x,y) = \zeta_\varepsilon(x - X(t;y,s)) \exp(-\int_s^t a_0(X(\sigma;x,s),\sigma)d\sigma)$$

and

$$\tilde{\varphi}_\varepsilon(x,y) = \sum_{j\in\mathbb{Z}^n} w_h(x_j,y)\, \varphi_\varepsilon(x,x_j) \,.$$

Given a point $x \in \mathbb{R}^n$, we denote by $K(x)$ the set of indices $k \in K$ such that $y_k \in \text{supp}(\varphi_\varepsilon(x,\cdot)) \cup \text{supp}(\tilde{\varphi}_\varepsilon(x,\cdot))$. Then, we consider the set

$$K_1(x) = \{k \in K(x);\, x - X(t;\mathcal{O}_h(y_k),s) \subset \bar{\Omega}_r^\varepsilon \text{ for at least one } r \in \{1,\cdot,R\}\} \,,$$

where Ω_r^ε is defined as in the proof of Lemma 5. Using (6.7), it is easy to check that

$$\text{card}(K_1(x)) \leq c_1 (\frac{\varepsilon + \Delta x}{\delta})^n \,.$$

Moreover, for $k \in K_1(x)$, the function $y \to \varphi_\varepsilon(x,y)$ belongs to $W^{\ell,\infty}(\mathcal{O}_h(y_k))$. Hence, applying Lemma 7 gives

$$|\sum_{k\in K_1(x)} \alpha_k(\varphi_\varepsilon(x-y_k) - \tilde{\varphi}_\varepsilon(x,y_k))| \leq$$

$$\leq c_2 \, \Delta x^\ell (\sup_{k\in K} |\alpha_k|)\, \text{card}(K_1(x))\, |\varphi_\varepsilon(x,\cdot)|_{\ell,\infty,\mathcal{O}_h(y_k)} \,.$$

But

$$|\varphi_\varepsilon(x,\cdot)|_{\ell,\infty,\mathcal{O}_h(y_k)} \leq \frac{c_3}{\varepsilon^{n+\ell}}$$

so that

$$\Big|\sum_{k\in K_1(x)} \alpha_k(\varphi_\varepsilon(x,y_k) - \tilde{\varphi}_\varepsilon(x,y_k))\Big| \leq$$

$$\leq c_4 \frac{1}{\delta^n} (\sup_{k\in K} |\alpha_k|) (1 + \frac{\Delta x}{\varepsilon})^n (\frac{\Delta x}{\varepsilon})^\ell .$$

Consider next the set $K_2(x) = K(x) \setminus K_1(x)$. If $k \in K_2(x)$. We have

$$(x - X(t;0_h(y_k),s)) \cap \partial\Omega_r^\varepsilon \neq \phi \quad \text{for at least one} \quad r \in \{1,\ldots,R\} .$$

Using again (6.7), one can check that

$$\mathrm{card}(K_2(x)) \leq c_5 \frac{\Delta x \ \varepsilon^{n-1}}{\delta^n}$$

For $k \in K_2(x)$, we note that the function $y \to \varphi_\varepsilon(x,y)$ belongs only to $W^{\ell-1,\infty}(\mathbb{R}^n)$. Hence

$$\Big|\sum_{k\in K_2(x)} \alpha_k(\varphi_\varepsilon(x,y_k) - \tilde{\varphi}_\varepsilon(x,y_k))\Big| \leq$$

$$\leq c_6 \, \Delta x^{\ell-1}(\sup_{k\in K} |\alpha_k|) \ \mathrm{card}(K_2(x)) \ |\varphi_\varepsilon(x,\cdot)|_{\ell,\infty,\mathbb{R}^n} \leq$$

$$\leq c_7 \frac{1}{\delta^n} (\sup_{k\in K} |\alpha_k|) (\frac{\Delta x}{\varepsilon})^\ell .$$

Thus, we obtain

$$\Big|\sum_{k\in K(x)} \alpha_k(\varphi_\varepsilon(x,y_k) - \tilde{\varphi}_\varepsilon(x,y_k))\Big| \leq$$

$$\leq c_8 \frac{1}{\delta^n} (\sup_{k\in K} |\alpha_k|) (1 + \frac{\Delta x}{\varepsilon})^n (\frac{\Delta x}{\varepsilon})^\ell$$

and (6.8) follows. ∎

Let us give an immediate consequence of the above theorem. Assume that we need to reinitialize at time s the set of particles produced by the particle approximation (4.6), (4.7) of the solution u of Problem (2.1). Denote by v_h this new particle distribution. Then, applying the estimate (6.8) successively to the measures

$$\sum_{j\in\mathbb{Z}^n} \alpha_j(s) \ \delta(x - X(s;x_j,0))$$

and

$$\sum_{j\in\mathbb{Z}^n} \beta_j^{\ell+1/2}(s) \ \delta(x - X(s;x_j,t_{\ell+1/2})) \quad , \quad t_{\ell+1/2} \leq s$$

and noticing that in each case $\delta \geq c \ \Delta x$, we obtain

$$\|(S(s,t)(u_h(\cdot,s)-v_h)) * \zeta_\varepsilon\|_{L^\infty(\mathbb{R}^n)} \leq C(T,u_0,f)(\frac{\Delta x}{\varepsilon}) \quad , \ s \leq t \leq T ,$$

provides that (6.5) holds and ζ belongs to the class C_ℓ .

Finally, as a simple example of application of the results of this paper, we remark that the standard choice

$$\zeta = \hat{w} = \chi^{\star 2}$$

(i.e., ζ and $\hat{w}$ are the usual hat function) leads to the estimates

$$\| u(\cdot,t_p) - u_h(\cdot,t_p) \|_{L^\infty(\mathbb{R}^n)} \leq C(\varepsilon^2 + (\frac{\Delta x}{\varepsilon})^2 + (\frac{\Delta t}{\varepsilon})^2)$$

and

$$\| (S(s,t)(u_h(\cdot,s) - v_h)) \star \zeta_\varepsilon \|_{L^\infty(\mathbb{R}^n)} \leq C(\frac{\Delta x}{\varepsilon})^2$$

if we assume that the exact solution u is smooth enough and (5.11) holds.

REFERENCES.

1. Beale, J.T., and Majda, A. "Vortex methods I : Convergence in three dimensions", Math. Comp., 32, 1-27 (1982).
2. Beale, J.T., and Majda, A. "Vortex methods II : Higher order accuracy in two and three dimensions", Math. Comp., 32, 29-56 (1982).
3. Cottet, G.H., and Raviart, P.A., "Particle methods for the one-dimensional Vlasov-Poisson equations", SIAM J. Numer. Anal. (1983) (to appear).
4. Denavit, J. "Numerical simulation of plasmas with periodic smoothing in phase space", J. Comput. Phys., 9, 75-98 (1972).
5. Hald, O. "Convergence of Vortex methods II", SIAM J. Numer. Anal., 16, 726-755 (1979).
6. Harlow, F.H. "The particle in cell computing method for fluid dynamics", Methods in Computational Physics (B. Alder, S. Fernbach and M. Rotenberg ed.), Vol. 3, Academic Press, New-York 1964.
7. Hockney, R.W., and Eastwood, J.W. "Computer Simulation Using Particles", Mc Graw Hill, New-York 1981.
8. Leonard, A. "Vortex methods for flow simulations", J. Comput. Phys., 37, 283-335 (1980).
9. Raviart, P.A. "An analysis of particle methods", CIME Course in Numerical Methods in Fluid Dynamics, Como, July 1983 (to be published in Lectures Notes in Mathematics, Springer Verlag).

Global Error Estimation for Stiff ODEs†

Lawrence F. Shampine

1. Introduction

The typical code for the numerical solution of the initial value problem for a system of ordinary differential equations (ODEs),

$$y' = f(x,y) \quad , a \leqslant x \leqslant b \quad , \qquad y(a) \text{ given} \quad ,$$

steps through the interval [a,b] producing approximations $y_m \doteq y(x_m)$ on a mesh $a = x_0 < x_1 < \ldots$. Each step size $h = x_{n+1} - x_n$ is chosen so as to control the error made in the step from x_n to x_{n+1}. Although this is a natural thing to do in the code, it is unfortunately only indirectly related to the error which really concerns the user of the code, namely the true or global error $y(x_{n+1}) - y_{n+1}$. There are fundamental difficulties with the direct control of global error, but in recent years quite a lot of progress has been made with the estimation of global error. Excellent surveys by Stetter [7] and Prothero [4] describe this activity.

For a variety of theoretical and practical reasons, stiff ODEs are much harder to solve than non-stiff ODEs. On the other hand, estimation of the global error is in some respects easier. Formation of an approximation to the Jacobian $f_y(x,y(x))$ seems essential to the numerical solution of stiff ODEs. It provides valuable information not usually available when solving non-stiff problems. Approaches to global error estimation which involve the integration of a differential equation for the global error may benefit from the stability properties of the original problem; the effect of an error made in the approximation of the global error at one step may be strongly damped at succeeding steps.

Some years ago [6] we devised a way to estimate the global error when solving stiff ODEs which is interesting because it is very simple and very cheap. Undoubtedly the most popular codes for the solution of stiff ODEs are based on the backward differentiation formulas (BDF). The scheme described in [6] applies to the BDF of orders 1-3 but not to orders 4-6. Because we were unable to handle the popular codes which all vary the order of the BDF used between one and five, we gave the scheme only the limited circulation of a report. On returning to the matter recently we saw how to broaden the applicability of the method greatly, in particular to include all the BDF, and to understand much better an important approximation made. In the next section we explain this new treatment of our approach to global error estimation.

---

† This work performed at Sandia National Laboratories supported by the U. S. Department of Energy under contract number DE-AC04-76DP00789.

Since the report [6], Prothero [4] has given an approach to estimation of global error which is closely related to the way we now view our approach. We shall review his approach in section 3, where we contrast his assumptions with the ones we make and point out an important restriction. In a final section we draw upon an approximation made by Prothero to extend the applicability of the procedure developed in section 2 and in some cases to improve the accuracy of the procedure.

2. Estimation of Global Error

The initial value problem for the system of ODEs

$$y' = f(x,y) \quad , \qquad a \leqslant x \leqslant b \quad , \qquad y(a) \text{ given} \tag{2.1}$$

is solved by some discrete-valued method yielding vectors y_n which approximate the solution values $y(x_n)$ on a mesh $a = x_0 < x_1 < \ldots$. The true or global error at x_n is

$$E_n = y(x_n) - y_n \quad .$$

At each $\{x_n, y_n\}$ there is a local solution $u(x)$ of the differential equation defined by

$$u' = f(x,u) \quad , \qquad u(x_n) = y_n \quad . \tag{2.2}$$

The local error of the numerical method made in stepping from x_n to $x_{n+1} = x_n + h$ is defined to be

$$u(x_{n+1}) - y_{n+1} \quad .$$

A fundamental decomposition of the global error is

$$E_{n+1} = y(x_{n+1}) - y_{n+1} = \left(y(x_{n+1}) - u(x_{n+1})\right) + \left(u(x_{n+1}) - y_{n+1}\right) \quad . \tag{2.3}$$

We suppose that the code used produces a reasonably good estimate of the local error,

$$le_{n+1} \doteq u(x_{n+1}) - y_{n+1} \quad . \tag{2.4}$$

In our approach it does not matter <u>how</u> this is accomplished. Indeed, it does not even matter what kind of method is used for advancing a step. Merely to give the reader a feeling for the issue, we comment that with one-step methods it is customary to compute two approximations to $y(x_{n+1})$; let us call them y_{n+1} and y^*_{n+1}. If y^*_{n+1} is a higher order approximation than y_{n+1}, we first write

$$u(x_{n+1}) - y_{n+1} = (y^*_{n+1} - y_{n+1}) + \left(u(x_{n+1}) - y^*_{n+1}\right) \quad .$$

On dropping the higher order term, we obtain a computable approximation

$$le_{n+1} = y^{*}_{n+1} - y_{n+1} \doteq u(x_{n+1}) - y_{n+1}$$

which can be fully justified as $h \to 0$. It is tempting here to advance the integration with the higher order result, and there are various arguments for and against this "local extrapolation." In the present context local extrapolation is excluded because we require an estimate of the local error of y_{n+1} in order to deduce an estimate of its global error.

An estimate of the local error is fundamental because it is used by the code to decide whether to accept the step and to select an efficient step size. We can presume then that an adequate estimate le_{n+1} is available at each step and so devote our attention to estimating the other term in (2.3). This other term is the difference of two solutions of the same differential equation which differ at x_n by $y(x_n) - y_n = E_n$. A standard result from the theory of differential equations says that

$$y(x_{n+1}) - u(x_{n+1}) = v(x_{n+1}) + O(\|E_n\|^2) \ ,$$

where $v(x)$ is the solution of the equation of first variation

$$v' = f_y(x,y(x))v \quad , \qquad v(x_n) = E_n \quad . \tag{2.5}$$

We shall approximate

$$y(x_{n+1}) - u(x_{n+1}) \doteq v(x_{n+1}) \tag{2.6}$$

and further consider how to approximate $v(x_{n+1})$ numerically.

A numerical approximation J to $f_y(x,y(x))$ plays several roles in a code for the solution of stiff ODEs. In each role it is necessary that J be a "reasonably" good approximation, at least on $[x_n, x_{n+1}]$. Otherwise, a new J is formed and/or h is reduced. We shall approximate the solution $v(x)$ of (2.5) on $[x_n, x_{n+1}]$ by the solution $w(x)$ of

$$w' = Jw \quad , \qquad w(x_n) = E_n \quad . \tag{2.7}$$

It is possible to quantify in some measure the standard assumption that $w(x)$ is a "reasonable" approximation to $v(x)$.

Let

$$|||f_y - J||| = \max_{[x_n, x_{n+1}]} \|f_y(x,y(x)) - J\| \quad ,$$

$$|||w||| = \max_{[x_n, x_{n+1}]} \|w(x)\| \quad .$$

Then

$$\|w'(x) - f_y(x,y(x))w(x)\| = \|(J-f_y)w\| \leq |||f_y - J|||\ |||w||| , \quad x_n \leq x \leq x_{n+1} .$$

Further let

$$|f_y| \geq \mu[f_y(x,y(x))] \qquad \text{for} \qquad x_n \leq x \leq x_{n+1} ,$$

where $\mu[M]$ is the logarithmic "norm" of the matrix M with respect to the norm $\|\cdot\|$. Theorem 1.1 of Dahlquist [3] implies that

$$\|v(x) - w(x)\| \leq [\exp(h|f_y|) - 1] \frac{|||f_y - J|||}{|f_y|} |||w||| , \qquad x_n \leq x \leq x_{n+1} .$$

(The result has a slightly different form if $|f_y| = 0$.) Then

$$\frac{|||v - w|||}{|||w|||} \leq [\exp(h|f_y|) - 1] \frac{|||f_y - J|||}{|f_y|} .$$

The left hand side is a kind of relative error of the approximation of $v(x)$ by $w(x)$ on $[x_n,x_{n+1}]$. The term $|||f_y - J|||/|f_y|$ is a kind of relative error of the approximation of f_y by J. As noted earlier the code varies h and forms J as often as needed to keep this error reasonably small. Also, for reasons of stability and accuracy, the code will keep h small enough that $h|f_y|$ is not large. (This is no restriction on h at all if $|f_y| < 0$, which is a common assumption.) This implies that the factor $\mathrm{sgn}(|f_y|)[\exp(h|f_y|) - 1]$ is not large. Finally then we conclude that with the usual expectation that the code provide a reasonable approximation to the Jacobian and control its step size appropriately, $w(x_{n+1})$ will be a reasonable approximation to $v(x_{n+1})$ and we can use

$$y(x_{n+1}) - u(x_{n+1}) \doteq w(x_{n+1}) \tag{2.8}$$

in place of (2.6).

The analytical solution $w(x_{n+1}) = \exp(hJ)E_n$ is impractical. In the primary integration of $y(x)$, the code will form and factor

$$M = I - h\beta J = I - \frac{h}{\alpha} J \tag{2.9}$$

where β (or equivalently α) is a constant characteristic of the method. Indeed some methods carry this out for several different β at each step. Our aim is to compute a practical approximation to $w(x_{n+1})$ by exploiting the fact that linear systems with the matrix M of (2.9) can be solved cheaply.

It is not at all necessary to approximate $w(x_{n+1})$ very accurately, but it is necessary to accomplish this in a stable way. We propose to integrate (2.7) by s steps of length h/s taken with the well-known θ-method:

$$w_{n,i+1} = w_{n,i} + \frac{h}{s}\left[(1-\theta)Jw_{n,i+1} + \theta Jw_{n,i}\right] \qquad i = 0,1,\ldots,s-1 \quad .$$

Here

$$w_{n,0} = w(x_n) = E_n \qquad \text{and} \qquad w_{n,s} \doteq w(x_{n+1}) \quad .$$

It is better to rewrite the formula as

$$w_{n,i+1} = \left[-\frac{\theta}{1-\theta} I + \frac{1}{1-\theta}\left(I - \frac{h}{s}(1-\theta)J\right)^{-1}\right]w_{n,i} \quad .$$

(Of course a linear system is to be solved rather than a matrix inverted.) This trick, which we learned from [5], not only avoids the storage of J, it also results in a somewhat cheaper step. The whole process amounts to a way of computing a rational approximation to exp(hJ), but it is convenient to view the matter as a numerical integration because the θ-method is well understood: The method is accurate of order 1 if $\theta \neq 1/2$ and of order 2 if $\theta = 1/2$. If $0 \leq \theta \leq 1/2$, the method is A-stable.

Because the step size is chosen by the primary integration, we must assure ourselves that this secondary integration is stable. One way to do this is by using only A-stable formulas. We shall restrict ourselves to methods for the primary integration such that we can select a stable method for the secondary integration based on the factored matrix $I-h\beta J$ provided by the primary integration. Thus we want to define θ and s by

$$\beta = \frac{1-\theta}{s} \qquad \text{with} \qquad 0 \leq \theta \leq 1/2 \quad .$$

In principle, any $0 < \beta \leq 1$ can be accommodated, but in practice we need a small integer s. The main possibilities are then

$$1 \geq \beta \geq 1/2 \qquad \text{with} \qquad s = 1 \quad ,$$

$$1/2 \geq \beta \geq 1/4 \qquad \text{with} \qquad s = 2 \quad .$$

Our method is really very simple. It approximates (2.3) by

$$E_{n+1} \doteq le_{n+1} + w_{n+1} \quad . \tag{2.10}$$

Here le_{n+1} is the approximation (2.4) to the local error provided by the code and

$$w_{n+1} = [(\frac{\beta s-1}{\beta s})I + (\frac{1}{\beta s})M^{-1}]^s E_n \quad . \tag{2.11}$$

This procedure is applicable to any method for the primary integration which produces a factored matrix M of (2.9) with

$$0 < \theta = 1 - \beta s < 1/2 \quad .$$

If, say, $1 > \beta > 1/4$, the estimator requires remarkably little extra computation and storage.

With the sole exception of the constraint on β, the method used for the primary integration is irrelevant. It appears in the global error estimate via the local error estimate, which we presume supplied by the code. Our approach to global error estimation is completely local. It does not matter if the step size and/or method is altered at every step. As an example we cite the popular codes based on the BDF which do vary step size and order. For the popular quasi fixed step size implementations, one has

order	1	2	3	4	5	6
β	1	2/3	6/11	12/25	60/137	60/147

Evidently our estimator is applicable with s = 1 if the current order is in the range 1-3 and with s = 2 if the order is in the range 4-6.

3. Prothero's Method

Here we sketch the derivation of a method presented by Prothero in section 6 of [4]. His estimator is based on the local discretization (truncation) error rather than the local error we use. For a general class of k-step formulas of the form

$$\sum_{i=0}^{k} \alpha_i y_{n+1-i} - h\phi(x_n; y_{n+1}, \ldots, y_{n+1-k}; h) = 0 \quad ,$$

the local discretization error d_{n+1} is defined by

$$d_{n+1} = -\frac{1}{\rho'(1)h} \{\sum_{i=0}^{k} \alpha_i y(x_{n+1-i}) - h\phi(x_n; y(x_{n+1}), \ldots, y(x_{n+1-k}); h)\}$$

where

$$\rho(\theta) = \sum_{i=0}^{k} \alpha_i \theta^i \quad .$$

This is essentially the residual of the true solution y(x) in the formula. The local error we use is defined in terms of a local solution. The quantities are related, and some codes estimate one and some the other.

If the starting values $y_0,\ldots,y_{k-1}$ are provided in a suitable way and there is a principal error function $\Psi(x)$ such that

$$d_n = h^p\Psi(x_n) + O(h^{p+1}) \quad ,$$

then it can be shown that as the (constant) step size h tends to zero

$$y_n - y(x_n) = h^p e(x_n) + O(h^{p+1}) \quad ,$$

where

$$e' = f_y(x,y(x))e + \Psi(x) \quad .$$

This linearization is in some respects less satisfactory than what we did earlier. For one thing, it is less accurate since we dropped terms of order h^{2p} and here terms of order h^{p+1} are dropped. More important is the fact that our analysis is completely local, and this standard asymptotic error analysis is global. The standard analysis does not accommodate a change of formula (order) in the course of the integration, and it is not clear how to justify its use to describe variable order BDF codes which are, in fact, the main object of Prothero's attention. His numerical results do suggest that a justification is possible.

It is to be noted that the roles of the local errors are quite different in the two approaches to global error estimation. In particular, the nature of the local discretization error is critical to the linearization Prothero uses. In our approach there is no interaction of the local error with the linearization, and its nature is irrelevant.

Prothero integrates the equation

$$E'(x) = f_y(x,y(x))E(x) + h^p\Psi(x)$$

for the principal part of the global error by the BDF of order 1:

$$E_{n+1} = E_n + h[f_y(x_{n+1},y(x_{n+1}))E_{n+1} + h^p\Psi(x_{n+1})] \quad .$$

Of course the Jacobian needed here is not available, so it is approximated by a matrix J produced by the code, just as we did. The implications of this approximation were discussed in the last section. The term $h^{p+1}\Psi(x_{n+1})$ is approximated by a local truncation error estimate lte_{n+1} produced by the code. This is the analog of our use of an estimate of the local error. Note that the BDF1 is a good numerical method in any case, but it is useful here that one does not have to approximate $\Psi(x)$ at off-mesh points. The result of the approximations is a formula

$$E_{n+1} = [I-hJ]^{-1}[E_n + lte_{n+1}]$$

which bears a considerable resemblance to ours.

As in our approach, it is not practical to form and factor I-hJ, so an approximation

$$[I-hJ]^{-1} \doteq M^{-1}[(2-\alpha)I + (\alpha-1)M^{-1}] \tag{3.1}$$

is made which is based on the matrix M of (2.9) available in the code in factored form. The procedure is finally

$$E_{n+1} = M^{-1}[(2-\alpha)I + (\alpha-1)M^{-1}][E_n + lte_{n+1}] \quad .$$

Prothero says that the right hand side of (3.1) is a stable, first order approximation. Just as in our analysis it is important that the approximation be A-stable. However, it is easy to see that the approximation is not A-stable for all $\alpha > 0$. Suppose that λ is any eigenvalue of J with $Re(\lambda) < 0$. If we let

$$\zeta = h\lambda/\alpha \quad , \qquad \mu = 2 - \alpha \quad ,$$

the corresponding eigenvalue of the right hand side of (3.1) is

$$\frac{1 - \mu\zeta}{(1-\zeta)^2} \quad .$$

Letting

$$\zeta = a + ib$$

we see that

$$\frac{|1-\mu\zeta|^2}{|1-\zeta|^4} = \frac{1 + \mu^2 b^2 + O(a)}{1 + 2b^2 + b^4 + O(a)} \quad .$$

If $\mu^2 > 2$, one can first choose b and then $a < 0$ sufficiently small in magnitude that for this ζ with $Re(\zeta) < 0$,

$$\frac{|1-\mu\zeta|}{|(1-\zeta)^2|} > 1 \quad .$$

Thus it is necessary that $\mu^2 < 2$ for the approximation to be A-stable. A little calculation shows the condition to be sufficient as well. As a consequence, Prothero's scheme should be restricted to

$$2 - \sqrt{2} < \alpha < 2 + \sqrt{2}$$

or equivalently,

$$0.3 \doteq \frac{2-\sqrt{2}}{2} < \beta < \frac{2+\sqrt{2}}{2} \doteq 1.7 \quad .$$

Prothero was most concerned with the BDF, and on inspection of the corresponding β given in the last section, we see that his method is justified for them. This is confirmed by his good numerical results.

4. An Improved Estimator

The matrix approximation (3.1) used by Prothero is not only a first order approximation to $(I-hJ)^{-1}$ but also a first order approximation to $\exp(hJ)$. Consequently it is an alternative to (2.11) for producing the approximation to $w(x_{n+1})$ needed in our global error estimate (2.10). Specifically we could use (2.10) and

$$w_{n+1} = M^{-1}[(\frac{2\beta-1}{\beta})I + (\frac{1-\beta}{\beta})M^{-1}]E_n \tag{4.1}$$

whenever the matrix here is stable. There are various arguments for and against this choice which we now take up.

Case 1. If $1 < \beta < \frac{2+\sqrt{2}}{2} \doteq 1.7$, use (2.10) and (4.1).

In this range of β the approximation of section 2 is not applicable, so by using (4.1) we extend the applicability of the global error estimate. There are semi-implicit [1] and Rosenbrock [2] formulas which have been seriously considered for the solution of stiff problems which do have $\beta > 1$, e.g., a scheme of Crouzeix has $\beta \doteq 1.06$, so that the extension is useful.

Case 2. If $1/2 < \beta < 1$, use (2.10) and (2.11) with $s = 1$.

In this range of β the approximation (2.11) is more attractive than (4.1) because it requires half as many solutions of linear systems per step. Also in the special case of $\beta = 1/2$, the approximation (2.11) is of second order of accuracy.

Case 3. If $0.3 \doteq \frac{2-\sqrt{2}}{2} < \beta < 1/2$, use (2.10) and (4.1).

In this range of β the approximation (2.11) requires $s = 2$ so that both require two solutions of linear equations per step. Both are of first order of accuracy, but their accuracy at infinity is different. The θ-method is only damped at infinity for $\theta \neq 0$, whereas the approximation (4.1) is strongly damped. Other things being equal, we prefer (4.1) for this reason.

Case 4. If $1/4 < \beta < \frac{2-\sqrt{2}}{2}$, use (2.10) and (2.11) with $s = 2$.

In this range of β the approximation (4.1) is not stable, so we must use the other approximation.

For $\beta < 1/4$ the approximation (2.11) is applicable with $s > 2$, but it is not clear that the work involved is acceptable. The four cases given cover all the common formulas and involve remarkably little extra work or storage in the code. In view of this fact it would seem worth adding the estimate to existing codes even if the estimated global errors produced are rather crude. The principal constraints on the applicability arise in our use of an estimate of the local error. Because of it the code must estimate the local error, as opposed to the local discretization (truncation) error, and the code must not do local extrapolation.

References

1. R. Alexander, Diagonally implicit Runge-Kutta methods for stiff O.D.E.'s, SIAM J. Numer. Anal. 6 (1977) 1006-1021.

2. T. D. Bui, Some A-stable and L-stable methods for the numerical integration of stiff ordinary differential equations, J. ACM 26 (1979) 483-493.

3. G. Dahlquist, Stability and error bounds in the numerical integration of ordinary differential equations, Trans. Royal Inst. Technology, Stockholm, 130 (1959).

4. A. Prothero, Estimating the accuracy of numerical solutions to ordinary differential equations, pp. 103-128 in I. Gladwell and D. K. Sayers, eds., Computational Techniques for Ordinary Differ- ential Equations, Academic, London, 1980.

5. A. Robinson and A. Prothero, Global error estimates for solutions to stiff systems of ordinary differential equations, contributed paper, Dundee Numerical Analysis Conference, 1977.

6. L. F. Shampine, Global error estimation for stiff ODEs, Rept. SAND79-1587, Sandia National Laboratories, Albuquerque, NM, 1979.

7. H. J. Stetter, Global error estimation in ODE-solvers, pp. 179-189 in Lecture Notes in Mathematics, 630, Springer, Berlin, 1978.

NUMERICAL TECHNIQUES FOR NONLINEAR MULTI-PARAMETER PROBLEMS

A. Spence* and A. Jepson**

* School of Mathematics, University of Bath, Bath, UK, BA2 7AY

** Department of Computer Science, University of Toronto, Toronto, Canada, M5S 1A7

1. INTRODUCTION

Many physical systems have multiple equilibria for some values of the control parameters. For example, certain simple models of chemical reactors are well understood and exhibit a wealth of such phenomena [11],[22]. Other well-known examples occur in fluid flow problems [3], [15], [19], in elasticity theory [1], and in biological models [4],[9]. A major reason for the study of steady state problems is that a complete understanding of the equilibria of a given system may lead to an understanding of dynamical behaviour. In this respect we mention again the models of chemical reactors [11], [22], where stable (and unstable) periodic solutions may bifurcate from stable steady state solutions, and the theory of thermal ignition [2], [5].

We consider a general nonlinear problem of the form

$$f(\underline{x},\lambda,\underline{\alpha}) = 0 \qquad f : \mathbb{R}^n \times \mathbb{R} \times \mathbb{R}^p \to \mathbb{R}^n \tag{1.1}$$

where $\underline{x} \in \mathbb{R}^n$ is the state variable; $\lambda \in \mathbb{R}$ is the bifurcation parameter; $\underline{\alpha} \in \mathbb{R}^p$ is a vector of control parameters; and f is a smooth function. The distinction between the bifurcation parameter λ and the control parameters $\underline{\alpha}$ arises for two reasons. First, it is often the case in experiments that there is one parameter which can be easily changed and this parameter is varied quasi-statically while the other parameters remain fixed. Second, this distinction makes the mathematics more tractable and once the results have been obtained they can be readily reinterpreted in terms of other choices of the bifurcation parameter. Problems like (1.1) may arise naturally, for example, in the analysis of aircraft in flight [17], or as the result of the discretization of a differential or integral equation [11].

A basic task is to be able to compute solutions of (1.1) for a fixed value of $\underline{\alpha}$ i.e. for $\underline{\alpha} = \underline{\alpha}_o$ we need to be able to solve

$$g(\underline{x},\lambda) := f(\underline{x},\lambda,\underline{\alpha}_o) = 0 . \qquad (1.2)$$

We discuss the theory of one parameter problem like (1.2) and some numerical techniques for the computation of their solutions in Section 2. In this Section we merely note that in (1.2) it is common to regard $\underline{x}$ as a function of λ and to describe the solutions of (1.2) using a *bifurcation diagram*, $D(g) = \{(\underline{x},\lambda),\ \underline{x}\in\mathbb{R}^n,\ \lambda\in\mathbb{R} : g(\underline{x},\lambda) = 0\}$. For $n > 1$ it is usual to consider the set $\{(k(\underline{x}),\lambda),\ k: \mathbb{R}^n\to\mathbb{R} : g(\underline{x},\lambda) = 0\}$ (see Figure 1).

A simple example for the case $x \in \mathbb{R}$, henceforth known as the *scalar case*, helps to illustrate many important concepts. Consider

$$f(x,\lambda,\alpha) := x^3 - \lambda x + \alpha_1 + \alpha_2 x^2 . \qquad (1.3)$$

Solution (bifurcation) diagrams for three choices of $\underline{\alpha}$ are given, qualitatively, in Figures 1.1 (a), (b) and (c).

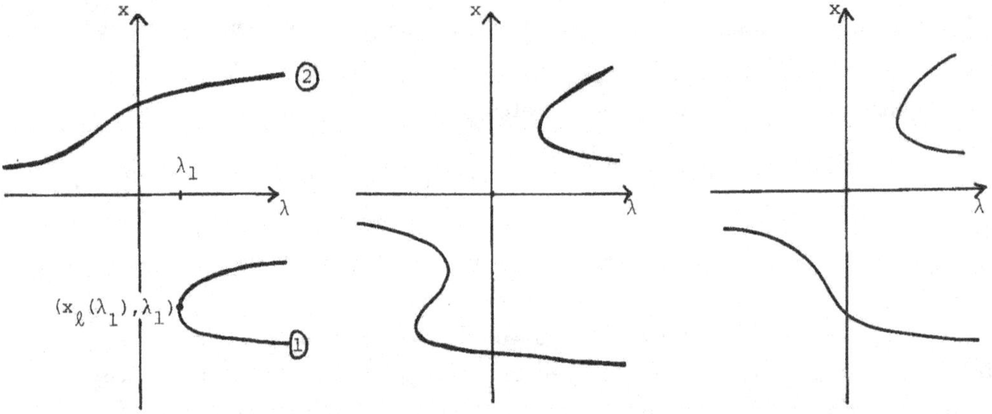

(a) $\underline{\alpha} = (-2,2)$ (b) $\underline{\alpha} = (.1,2)$ (c) $\underline{\alpha} = (2,2)$

Figure 1.1

In Figure 1.1(a) we see that there is a unique solution x for any $\lambda < \lambda_1$, whereas, for $\lambda > \lambda_1$, there are three solutions. The value λ_1 will often be of interest in physical situations. The point $(x_\ell(\lambda_1),\lambda_1)$ is a *singular point* (in fact, a turning point, see Section 2) in the sense that,

$$(f,f_x) = (0,0) \qquad (1.4)$$

and it is precisely at such points that a dynamical system $x_t = f(x,\lambda,\alpha)$ might lose stability. A complete qualitative picture of the solutions of (1.3) is given in Figure 1.2 where the control parameter space is divided into open regions R_i, $i = 1,\ldots,4$, in which the bifurcation diagrams are *qualitatively similar*, by which we mean essentially that all bifurcation diagrams in a region have the same shape and number of each type of singular point.

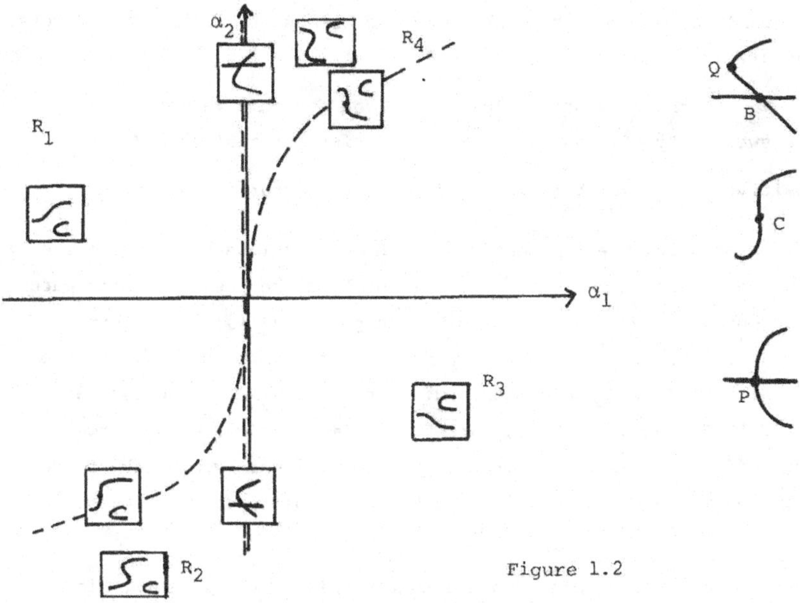

Figure 1.2

The regions, R_i, are separated by the curves $\alpha_1 = 0$ and $\alpha_1 = \alpha_2^3/27$. On these lines the solutions of (1.3) have higher order singularities than the solutions of (1.3) for values of $\underline{\alpha}$ lying inside the regions R_i. In particular on $\alpha_1 = 0$ we have

$$(f, f_x, f_\lambda) = (0,0,0) \tag{1.5}$$

and on $\alpha_1 = \alpha_2^3/27$ we have

$$(f, f_x, f_{xx}) = (0,0,0). \tag{1.6}$$

The curves intersect at $\underline{\alpha} = (0,0)$, at which point we have

$$(f, f_x, f_{xx}, f_\lambda) = (0,0,0,0). \tag{1.7}$$

One might ask what happens if a given general $f(x,\lambda,\underline{\alpha})$ satisfies (1.7). In particular, will it exhibit solution behaviour like that in Figure 1.1 for some values of $\underline{\alpha}$? Also, can a complete picture of its solutions be given by a diagram like Figure 1.2? These questions are discussed by the authors in [13] for the case $x \in \mathbb{R}$, where the singularity theory of [10] is used together with the continuation techniques of [14] to produce a technique for the numerical solution of problems like $f(x,\lambda,\underline{\alpha}) = 0$. This work is also discussed briefly in Section 4.

Now consider the general case (1.1) i.e. $\underline{x} \in \mathbb{R}^n$. For many problems one will be interested in qualitative information about the solutions of (1.1). For example, typical questions are as follows:

i) for $\underline{\alpha} = \underline{\alpha}_0$ do there exist multiple solutions for some values of λ ? If the answer is "yes", how many solutions exist for a given λ ?

ii) how sensitive is a given bifurcation diagram to perturbations in $\underline{\alpha}$?

To help in understanding the implications of these questions consider Figure 1.1 If we know a point on the solution curve ① it will probably be possible to compute the whole of that connected component, including the point $(x_\ell(\lambda_1),\lambda_1)$ by some continuation method (see Section 2). However, for a general problem, how can the existence of the solution curve ② be recognised? Also, if $\alpha_2 = 2$ one can easily see from Figure 1.2 that the solution diagrams for $\alpha_1 = .1$ and $\alpha_1 = -.1$ are qualitatively very different. Ideally for the problem (1.1) one would like to be able to do the same as was done in example 1.3. In particular, it would be nice if one were able

i) to reduce the control parameter space to the union of regions in which the solutions of (1.1) have qualitatively similar bifurcation diagrams. (This would involve computing all the manifolds of higher order singular points, e.g. the curves $\alpha_1 = 0$, $\alpha_1 = \alpha_2^3/27$ in Figure 1.2)

ii) to compute the complete bifurcation diagram for any given $\underline{\alpha}$. (This would involve the computation of all the connected components of the bifurcation diagram, e.g. curves ① and ② in Figure 1.1(a)).

In this paper we indicate an approach to help accomplish these goals. For the case $\alpha \in \mathbb{R}$ the approach has proved viable in several examples [6] [12] [21]. The techniques presented in this paper generalise to $\underline{\alpha} \in \mathbb{R}^p$ $(p > 1)$ though for a given physical problem the approach may not be practicable (for reasons we give later). For problems with $\underline{\alpha} \in \mathbb{R}^p$ $(p > 1)$ it may be that a different approach is needed but the principles of such an approach will probably be the same as those outlined here.

The plan of the paper is as follows. In Section 2 we discuss one parameter problems, in Section 3 two parameter problems, and in Section 4 multiparameter problems. Section 5 contains a brief discussion of some computational aspects.

2. ONE PARAMETER PROBLEMS

As was mentioned in the introduction we need to be able to compute the bifurcation diagrams for a one parameter problem which we write as

$$g(\underline{x},\lambda) = 0 \quad \underline{x} \in \mathbb{R}^n, \ \lambda \in \mathbb{R}, \quad g: \mathbb{R}^n \times \mathbb{R} \to \mathbb{R}^n. \tag{2.1}$$

Here, as in the rest of this paper, we assume g is smooth. In this section we give an outline of the basic theoretical results for (2.1) and describe briefly the algorithm used to compute the solutions of (2.1).

First we give some definitions. Assume $g(\underline{x}_o,\lambda_o) = 0$. If $g_x^o := g_x(\underline{x}_o,\lambda_o)$, the Jacobian of g, is nonsingular, then $(\underline{x}_o,\lambda_o)$ is a *regular point*, and the Implicit Function Theorem ensures the existence and uniqueness of a solution of (2.1) in a neighbourhood of $(\underline{x}_o,\lambda_o)$. If g_x^o is singular, then various types of behaviour can occur. In this case let us assume

(a) 0 is an algebraically simple eigenvalue of g_x^o,

(b) $\text{Null}\,(g_x^o) = \text{span}\,\{\underline{\phi}_o\} \qquad \underline{\phi}_o \in \mathbb{R}^n\setminus\{\underline{0}\}$

(c) $\text{Range}\,(g_x^o) = \{\underline{y} \in \mathbb{R}^n : \underline{\psi}_o^T\underline{y} = 0\} \qquad \underline{\psi}_o \in \mathbb{R}^n\setminus\{\underline{0}\}$. (2.2)

Following [7] it is convenient to introduce another variable to represent the solution curve, say $(\underline{x},\lambda) = (\underline{x}(s), \lambda(s))$ and seek conditions that ensure the existence of solutions of

$$g(\underline{x}(s),\lambda(s)) = 0$$

near $(\underline{x}_o,\lambda_o) = (\underline{x}(s_o),\lambda(s_o))$. The various conditions are found by successive differentiation of the above equations and subsequent evaluation at $s = s_o$. It is usual to distinguish the cases

$$\underline{\psi}_o^T g_\lambda^o \neq 0 \tag{2.3}$$

and

$$\underline{\psi}_o^T g_\lambda^o = 0 \tag{2.4}$$

and we briefly outline the main results (see [12], [21]).

First introduce $\underline{z}_o \in \mathbb{R}^n$ defined by

$$\underline{z}_o = 0 \qquad \text{if} \qquad \underline{\psi}_o^T g_\lambda^o \neq 0 \tag{2.5a}$$

$$g_x^o\underline{z}_o = -g_\lambda^o, \ \underline{\psi}_o^T\underline{z}_o = 0, \quad \text{if} \quad \underline{\psi}_o^T g_\lambda^o = 0. \tag{2.5b}$$

If $(\underline{x}_o,\lambda_o)$ satisfies (2.1), (2.2) and (2.3) then it is called a *simple turning point*. If, in addition,

$$a_g := \underline{\psi}_o \, g^o_{xx} \, \underline{\phi}_o \underline{\phi}_o \neq 0 \tag{2.5}$$

then it is called a simple *quadratic* turning point (see Q in Figure 1.2). If

$$a_g = 0 \; , \quad d_g := \underline{\psi}_o \, g^o_{xxx} \, \underline{\phi}_o \underline{\phi}_o \underline{\phi}_o + 3\underline{\psi}_o \, g^o_{xx} \, \underline{\phi}_o \underline{v}_o \neq 0 \tag{2.6}$$

where

$$g^o_x \, \underline{v}_o = - g^o_{xx} \, \underline{\phi}_o \underline{\phi}_o \; , \qquad \underline{\psi}_o^T \underline{v}_o = 0 \; , \tag{2.7}$$

then it is called a simple *cubic* turning point (or *hysteresis point*) and clearly this terminology extends to higher degeneracies (see C in Figure 1.2).

If (x_o,λ_o) satisfies (2.1), (2.2) and (2.4) then bifurcation is possible. To describe the various types of behaviour we need some definitions. Define $b_{g\lambda}$, $c_{g\lambda}$ and D_1 by

$$b_{g\lambda} := \underline{\psi}_o^T \, g^o_{x\lambda} \, \underline{\phi}_o + \underline{\psi}_o^T \, g^o_{xx} \, \underline{\phi}_o \, \underline{z}_o \tag{2.8}$$

$$c_{g\lambda} := \underline{\psi}_o^T \, g^o_{\lambda\lambda} + 2 \, \underline{\psi}_o^T \, g^o_{x\lambda} \, \underline{z}_o + \underline{\psi}_o^T \, g^o_{xx} \, \underline{z}_o \underline{z}_o \tag{2.9}$$

$$D_1 := b^2_{g\lambda} - a_g \, c_{g\lambda} \, . \tag{2.10}$$

Now if $a_g \neq 0$, $D_1 > 0$, then $(\underline{x}_o,\lambda_o)$ is a *simple transcritical bifurcation point* (see B in Figure 1.2); if $a_g = 0$, $b_{g\lambda} \neq 0$, then $(\underline{x}_o,\lambda_o)$ is a *simple pitchfork bifurcation point* (see P in Figure 1.2); and if $D_1 < 0$ then $(\underline{x}_o,\lambda_o)$ is a point of *isola* formation, or an isolated conjugate point (see Figure 3.2b). If $D_1 = 0$ then further analysis is needed to determine the solution behaviour.

Following the introduction of another parameter s to help with the theory the same approach is used in numerical techniques for (2.1). Append to

$$g(\underline{x}(s),\lambda(s)) = 0 \tag{2.11a}$$

an extra normalising equation,

$$N(\underline{x}(s),\lambda(s),s) = 0 \tag{2.11b}$$

Equations (2.11a) and (2.11b) can be written

$$\hat{G}(\underline{y}(s),s) = 0, \qquad \underline{y}(s) = (\underline{x}(s),\lambda(s)),$$

and a *predictor-solver continuation* technique is used to find solutions of (2.12) for various values of s. Thus given $\underline{y}_o = \underline{y}(s_o)$ and a step size Δs, first *predict* an estimate of $\underline{y}(s_o + \Delta s)$, called $\underline{y}^p(s_o + \Delta s)$ say, and then use this predicted value as a starting value to solve $\hat{G}(\underline{y}(s_o + \Delta s), s_o + \Delta s_o) = 0$. A commonly used predictor-solver technique is the Euler Newton pseudo-arclength approach in [14]. The advantage of adding the extra condition (2.11b) can be easily explained. At simple turning points $\text{rank}[g_x^o, g_\lambda^o] = n$, using (2.3), and hence, provided $N_x \phi_o \neq 0$, $\hat{G}_y^o$ will be nonsingular even though g_x^o is singular. Essentially the turning point disappears. At simple bifurcation points $\text{rank}[g_x^o, g_\lambda^o] = n-1$ and so $\hat{G}_y^o$ will be singular. However techniques exist for dealing with such situations (see, for example, [8] when using the pseudo-arclength approach).

The main point to note is that all numerical techniques for finding the bifurcation diagram of (2.1) should be able to compute all parts of any connected component given a starting value $(\underline{x}_o, \lambda_o)$. As was mentioned in Section 1 with reference to Figure 1.1(a), given a starting value on curve ① the whole curve should be readily computed for any range of λ, however the curve ② will not be found. This is an obvious limitation of continuation processes and the techniques discussed in this paper help to overcome this limitation.

As was mentioned above G_y^o is nonsingular at simple turning points and the pseudo-arclength approach has no difficulty in computing the solution $(\underline{x}(s),\lambda(s))$ in the neighbourhood of a simple turning point. If the turning point was needed explicitly then it could be computed using some root finding algorithm on $\det(g_x(\underline{x}(s),\lambda(s)))$ (or the smallest pivot in the LU decomposition of $g_x(\underline{x}(s),\lambda(s))$) or, alternatively, on $\frac{d\lambda}{ds}$ (see [18]). An alternative technique is to set up an *extended system* which has an isolated solution at the simple turning point. Such a system is

$$G(\underline{y}) := \begin{bmatrix} g(\underline{x},\lambda) \\ \underline{\psi}(\underline{x},\lambda)^T\, g_x(\underline{x},\lambda)\, \underline{\phi}(\underline{x},\lambda) \end{bmatrix}, \quad \underline{y} = (\underline{x},\lambda), \tag{2.13}$$

where

$$\begin{aligned} g_x(\underline{x},\lambda)\, \underline{\phi}(\underline{x},\lambda) &= \sigma(\underline{x},\lambda)\, \underline{\phi}(\underline{x},\lambda) \\ \underline{\psi}^T(\underline{x},\lambda)\, g_x(\underline{x},\lambda) &= \sigma(\underline{x},\lambda)\, \underline{\psi}^T(\underline{x},\lambda)\,. \end{aligned} \tag{2.14}$$

To discuss the isolatedness of a solution of $G(\underline{y}) = 0$ we need to discuss the nonsingularity of the Jacobian

$$G_y(\underline{y}) = \begin{bmatrix} g_x & g_\lambda \\ \underline{\psi} g_{xx}\underline{\phi} + \underline{\psi}_x g_x \underline{\phi} + \underline{\psi} g_x \underline{\phi}_x & \underline{\psi} g_{x\lambda}\underline{\phi} + \underline{\psi}_\lambda g_x \underline{\phi} + \underline{\psi} g_x \underline{\phi}_\lambda \end{bmatrix} .$$

Thus at a turning point, since $g_x^o \underline{\phi}_o = 0$ and $\underline{\psi}_o^T g_x = 0$, we have

$$G_y(\underline{y}_o) = \begin{bmatrix} g_x^o & g_\lambda^o \\ \underline{\psi}_o^T g_{xx}^o \underline{\phi}_o & \underline{\psi}_o^T g_{x\lambda}^o \underline{\phi}_o \end{bmatrix} . \tag{2.15}$$

It is convenient to introduce the 2×2 matrix

$$E = \begin{bmatrix} 0 & \underline{\psi}_o^T g_\lambda^o \\ a_g & b_{g\lambda} \end{bmatrix} . \tag{2.16}$$

We now have

Theorem 2.1 Assume (2.2)and that $G(y)$ is defined by (2.13). Let $\underline{y}_o = (\underline{x}_o, \lambda_o)$ and write $G_y^o = G_y(\underline{y}_o)$. Then

$$\dim \text{Null } (G_y^o) = \dim \text{Null } (E) \tag{2.17}$$

$$\text{codim Range } (G_y^o) = \text{codim Range } (E) \tag{2.18}$$

Proof We prove only (2.17) here. The proof of (2.18) is similar. We seek a $\underline{\Phi} = (\underline{p}, r) \in \mathbb{R}^n \times \mathbb{R}$ such that $G_y^o \underline{\Phi} = 0$. From (2.15) we have

$$g_x^o \underline{p} + r g_\lambda^o = 0 \tag{2.19}$$

$$\underline{\psi}_o^T g_{xx}^o \underline{\phi}_o \underline{p} + r \underline{\psi}_o^T g_{x\lambda}^o \underline{\phi}_o = 0 \tag{2.20}$$

Now using (2.5) we have from (2.19)

$$\underline{p} = \gamma\underline{\phi}_o + r\underline{z}_o , \quad \text{for } \gamma \in \mathbb{R} ,$$

and substitution back into (2.19), (2.20) gives

$$E \begin{bmatrix} \gamma \\ r \end{bmatrix} = 0$$

The result (2.17) follows.

An immediate Corollary is:

Corollary 2.1 Under the assumptions in Theorem 2.1, $\underline{y}_o = (\underline{x}_o, \lambda_o)$ is an isolated solution of $G(\underline{y}) = 0$ if and only if $(\underline{x}_o, \lambda_o)$ is a quadratic turning point of (2.1) i.e. $a_g \neq 0$, $\underline{\psi}_o{}^T g_\lambda^o \neq 0$.

This result on the isolatedness of $\underline{y}_o$ has, of course, implications for the convergence of Newton's method applied to (2.13).

3. TWO PARAMETER PROBLEMS

In this section we consider the case

$$f(\underline{x}, \lambda, \alpha) = 0 \qquad f: \mathbb{R}^n \times \mathbb{R} \times \mathbb{R} \to \mathbb{R}^n \tag{3.1}$$

i.e. there is only one control parameter. In order to attain the objectives listed in Section 1 for this problem we must consider the singular points of (3.1).

First assume that for a fixed α ($=\alpha_o$ say) (3.1) has a simple quadratic turning point at $(\underline{x}_o, \lambda_o)$ i.e.

$$\underline{\psi}_o{}^T f_\lambda^o \neq 0 , \quad \underline{\psi}_o{}^T f_{xx}^o \underline{\phi}_o \underline{\phi}_o = a_f \neq 0 \tag{3.2}$$

Form the extended system

$$F(\underline{y}, \alpha) := \begin{bmatrix} f(\underline{x}, \lambda, \alpha) \\ \underline{\psi}^T f_x(\underline{x}, \lambda, \alpha)\underline{\phi} \end{bmatrix} \qquad \underline{y} = (\underline{x}, \lambda) \tag{3.3}$$

where $\underline{\psi}$ and $\underline{\phi}$ are defined as in (2.14). Under the conditions (3.2) Corollary 2.1 implies that $F(\underline{y},\alpha_o) = 0$ has an isolated solution at $\underline{y}_o = (\underline{x}_o,\lambda_o)$. For $\alpha \in [\alpha_o-\sigma,\ \alpha_o+\sigma]$, F_y remains nonsingular provided the conditions (3.2) hold, and the Implicit Function Theorem guarantees the existence of a neighbourhood of $(\underline{y}_o,\alpha_o)$ such that $\underline{y}$ is a smooth curve that can be parameterised by α. We call $\underline{y}(\alpha)$ the *fold curve* and the solution of (3.1) in the neighbourhood of the fold curve a *fold* (see Figure 3.1).

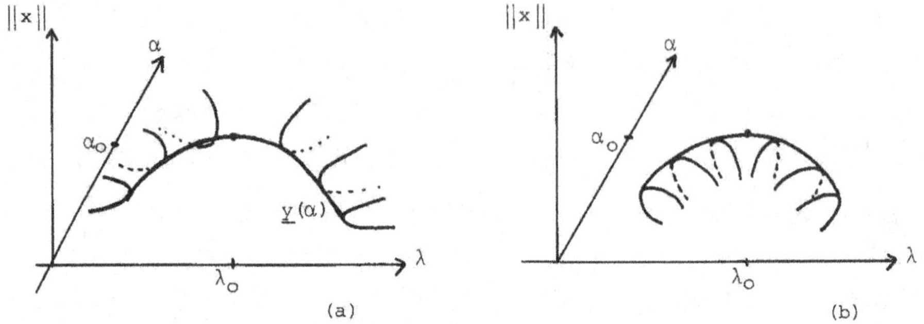

Figure 3.1

Clearly one can apply standard continuation techniques, for example, the pseudo-arc length approach, to compute the curve $\underline{y}(\alpha)$. This approach, using a different extended system for $\underline{y}_o$, is discussed in [6],[12].

In Figures 3.1(a) and 3.1(b) we note that the fold curve exists only for $\alpha < \alpha_o$ and that in both cases there are two turning points for each $\alpha < \alpha_o$. We now prove that, in fact, there is a turning point in $F(\underline{y},\alpha) = 0$ at $\alpha = \alpha_o$.

Following the analysis in Section 2, $(\underline{y}_o,\alpha_o)$ is a simple turning point of $F(\underline{y},\alpha) = 0$ if conditions equivalent to (2.2) and (2.3) hold. In this paper we discuss in detail the case

$$a_f \neq 0\ , \qquad \underline{\psi}_o^T f_\lambda = 0\ , \tag{3.4}$$

in which case, the corresponding E matrix (see (2.16)) clearly has dim Null (E) = rank (E) = 1. The case $a_f = 0$, $\underline{\psi}_o^T f_\lambda \neq 0$ is considered in [12] and [21] though we quote the results here. From (3.4), and using techniques similar to those in the proof of Theorem 2.1, we have that there exist $\underline{\Phi}_o, \underline{\Psi}_o$ such that

$$F_y^o \underline{\Phi}_o = 0\ , \quad \underline{\Psi}_o\, \underline{z} = 0\ , \quad \forall \underline{z} \in \text{Range}\ (F_y^o)\ . \tag{3.5}$$

In fact, it is easily shown that

$$\underline{\Phi}_o = \begin{pmatrix} \underline{z}_o + \gamma\underline{\phi}_o \\ 1 \end{pmatrix}\ , \qquad \underline{\Psi}_o = (\underline{\psi}_o, 0)\ , \tag{3.6}$$

where $\gamma = - b_{f\lambda}/a_f$. One easily has the result $\underset{\sim}{\Psi}_o F_\mu^o = \underset{\sim}{\psi}_o f_\mu^o$ and so we have proved half of the following theorem.

Theorem 3.1 Suppose $(\underset{\sim}{y}_o, \alpha_o)$ satisfies (3.3) and assume (2.2) holds for $f(\underset{\sim}{x}_o, \lambda_o, \alpha_o) = 0$. Then $(\underset{\sim}{y}_o, \alpha_o)$ is a simple turning point of $F(\underset{\sim}{y}, \alpha) = 0$ if and only if, *either* **A**:

$$a_f \neq 0 , \quad \underset{\sim}{\psi}_o^T f_\lambda = 0 ,$$

and

$$\underset{\sim}{\psi}_o^T f_\alpha^o \neq 0 , \tag{3.7}$$

or **B**:

$$a_f = 0 , \quad \text{rank } (E) = 1 , \tag{3.8}$$

and

$$\underset{\sim}{\psi}_o^T f_{\hat{\lambda}}^o \neq 0 , \quad b_{f\hat{\alpha}} := \underset{\sim}{\psi}_o^T f_{xx}^o \underset{\sim}{\phi}_o \underset{\sim}{w}_o + \underset{\sim}{\psi}_o^T f_{x\hat{\alpha}}^o \underset{\sim}{\phi}_o \neq 0. \tag{3.9}$$

In (3.9) we have used the definitions

$$\begin{pmatrix} \hat{\lambda} \\ \hat{\alpha} \end{pmatrix} = \begin{pmatrix} c_1 & c_2 \\ -c_2 & c_1 \end{pmatrix} \begin{pmatrix} \lambda \\ \alpha \end{pmatrix} , \qquad c_1 = \underset{\sim}{\psi}_o^T f_\lambda^o , \; c_2 = \underset{\sim}{\psi}_o^T f_\alpha^o ,$$

and

$$f_x^o \underset{\sim}{w}_o = - f_{\hat{\alpha}}^o , \quad \underset{\sim}{\psi}_o^T \underset{\sim}{w}_o = 0 .$$

One can prove the result that the turning point $(\underset{\sim}{y}_o, \alpha_o)$ in theorem (3.1) is *quadratic*, if, in case **A** ,

$$D_1 = b_{f\lambda}^2 - a_f \, c_{f\lambda} \neq 0 , \tag{3.11}$$

or, in case **B** ,

$$d_f \neq 0 \quad \text{and} \quad \underset{\sim}{\psi}_o f_{\hat{\lambda}}^o \neq 0 . \tag{3.12}$$

Note that case **A** corresponds to the Figure 3.1 and case **B**, the familar cusp catastrophe (see Figure 3 of [21] and (c) in Figure 3.2 below). Clearly for (3.1) the control parameter space is simply $\mathbb{R}$ and we can describe the solutions of

(3.1) using the diagrams in Figure 3.2.

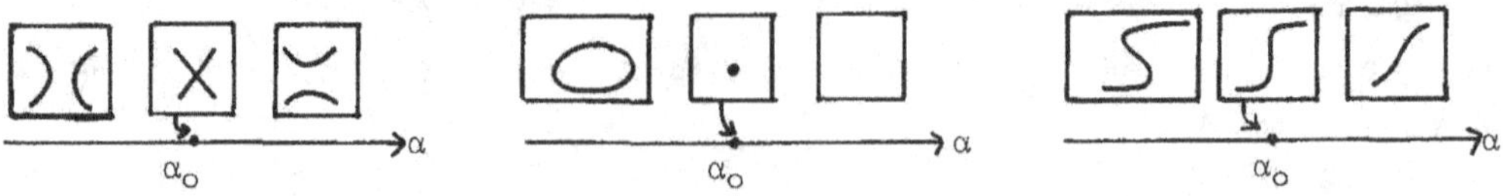

(a) case **A** $D_1 > 0$ (b) case **A** $D_1 < 0$ (c) case **B**

Figure 3.2

The co. ditions (3.7) and (3.9) are related to the "structural stability" of $f(\underline{x},\lambda,\alpha) = 0$, in the sense of catastrophe theory or singularity theory, and describe how α appears in f. We return to this topic in Section 4 but it is sufficient to say here that if the conditions of Theorem 3.1 hold with conditions (3.11) or (3.12), depending on whether case **A** or **B** is being considered, then $f(\underline{x},\lambda,\alpha) = 0$ is a universal unfolding of $f(\underline{x},\lambda,\alpha_o) = 0$.

The above results indicate that our objectives outlined in §1 can be attained for (3.1). Also, provided a_f and $\underline{\psi}_o^T f_\lambda$ are not both zero simultaneously then the singularities in the extended system (3.3) are only simple turning points (under assumption (2.2)). This gives us confidence that numerical techniques based on computing the fold curve will be viable.

The next question to consider is does this favourable situation for $\alpha \in \mathbb{R}$ extend to the case $\underline{\alpha} \in \mathbb{R}^p$ $p \geqslant 2$? This is discussed in the next section.

4. MULTIPARAMETER PROBLEMS

We consider the general problem introduced in Section 1, i.e.

$$f(\underline{x},\lambda,\underline{\alpha}) = 0 \qquad f\colon \mathbb{R}^n \times \mathbb{R} \times \mathbb{R}^p \to \mathbb{R}^n \qquad p > 1 . \tag{4.1}$$

In a recent paper [13] the authors discuss the case $n = 1$, the scalar problem and we first summarise the results of that paper.

4.1 The Scalar case, $x \in \mathbb{R}$

To discuss the approach in [13] for the problem

$$f(x,\lambda,\underline{\alpha}) = 0 \qquad f\colon \mathbb{R} \times \mathbb{R} \times \mathbb{R}^p \tag{4.2}$$

it is necessary to introduce some of the concepts of singularity theory discussed by

Golubitsky and Schaeffer. In [10] the possible singularities of (4.2) are grouped into equivalence classes using a certain notion of equivalence, called *contact equivalence*. Each equivalence class is referred to simply as a singularity. An important concept is the stability of a singularity of (4.2) with respect to small perturbations in f. A singularity at $(x_o, \lambda_o, \underline{\alpha}_o)$ is said to be *stable*, if, for any smooth g, for any neighbourhood U of $(x_o, \lambda_o, \underline{\alpha}_o)$ and for any positive ϵ sufficiently small, the perturbed problem

$$f_\epsilon(x,\lambda,\underline{\alpha}) := f(x,\lambda,\underline{\alpha}) + \epsilon g(x,\lambda,\underline{\alpha})$$

has a singularity of the same type at some point in U. With each particular singularity Golubitsky and Schaeffer associate an integer $q \geq 0$, called the *codimension*. Roughly speaking, the codimension is the minimum number of parameters besides λ, needed to make the singularity stable.

Using these concepts a singularity hierarchy was constructed in [13] for those singularities with codimension ≤ 3. Figure 4.1 gives the hierarchy for singularities with codimension ≤ 2. The hierarchy is used to construct systems of equations which characterise a given type of singularity.

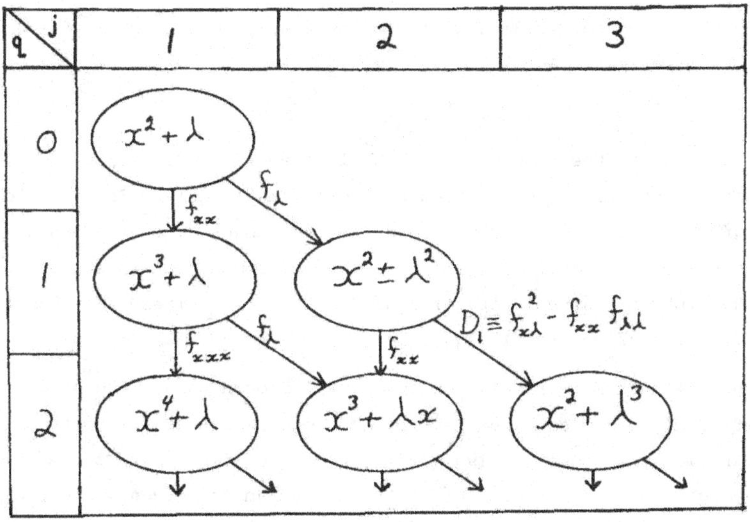

Figure 4.1 - The Singularity Hierarchy

The (q,j)-singularity is defined to be that type of singularity for the polynomial in the (q,j)-node of Figure 4.1. This singularity has codimension q. Moreover (4.2) has a (q,j)-singularity at $(x_o,\lambda_o,\underline{\alpha}_o)$ if and only if (i) all the functions labelling the branches of the hierarchy on a path up to the $(0,0)$-node vanish at $(x_o,\lambda_o,\underline{\alpha}_o)$ along with $f = 0$ and $f_x = 0$, and (ii) the two functions labelling branches leaving the (q,j) element are non-zero at $(x_o,\lambda_o,\underline{\alpha}_o)$.

An extended system for the (q,j)-singularity can be constructed by consideration of a path from the $(0,0)$-node to the (q,j)-node of the hierarchy. (See example (1.3) and the extended systems (1.4)-(1.7) in Section 1.) We define *side-constraints* for a (q,j)-extended system to be the conditions that the labels on the branches leaving the (q,j)-node of the hierarchy are non-zero.

The following result is proved in [13]. Assume $p-1 = q$. If at $(x_o,\lambda_o,\underline{\alpha}_o)$ a (q,j)-extended system for the (q,j)-singularity is

$$F_{q,j}(\underline{y},\alpha_p) = 0 \qquad \underline{y} = (x,\lambda,\alpha_1,\dots,\alpha_q) \tag{4.3}$$

with appropriate side constraints

$$C^{(1)}_{q,j} \neq 0, \qquad C^{(2)}_{q,j} \neq 0 \tag{4.4}$$

then $\underline{y}_o$ is a regular point of (4.3) and satisfies (4.4) precisely when $f(x_o,\lambda_o,\underline{\alpha}_o) = 0$ is universally unfolded by $f(x,\lambda,\underline{\alpha})\Big|_{\alpha_p=\alpha_{p,o}} = 0$.

This way of looking at the problem does highlight several nice features. The extended systems are specific; in the "generic" case the singularities are isolated solutions of these extended systems; and, as was indicated at the end of Section 3 (and is proved in [13]), a (q,j)-extended system has at worst, a simple turning point at a singularity of codimension $q+1$. These are important properties in the design of numerical algorithms.

In [13] numerical methods were discussed for moving down and moving up the hierarchy. Briefly, as one moves down the hierarchy one separates the control space $\mathbb{R}^p$ into regions with qualitatively similar bifurcation diagrams. As one moves up the hierarchy one can find the bifurcation diagram for given values of $\underline{\alpha}$ and hopefully, collect all the disconnected solution branches. It is instructive to reconsider example (1.3) in the light of these remarks.

4.2 The Vector Case, $\underline{x} \in \mathbb{R}^n$

As one can imagine the vector case is more complicated. The main problem is the appearance of the high order derivatives in the extended systems and the side constraints. For example the extended system for a cubic turning point is

$$F(\underline{y}) := \begin{bmatrix} f(\underline{x},\lambda,\alpha) \\ \underline{\psi}^T f_x(\underline{x},\lambda,\alpha)\underline{\phi} \\ \underline{\psi}^T f_{xx}(\underline{x},\lambda,\alpha)\underline{\phi}\underline{\phi} \end{bmatrix}$$

and the side constraints are

$$\underline{\psi}_o^T f_\lambda^o \neq 0 , \quad \underline{\psi}_o^T f_{xxx}^o \underline{\phi}_o\underline{\phi}_o\underline{\phi}_o + 3\underline{\psi}_o^T f_{xx} \underline{\phi}_o\underline{v}_o \neq 0$$

where $\underline{v}_o$ is given by the analogue of (2.7) for f (see[20]). Expressions for higher codimension singularities are even worse.

It seems clear that a more managable approach is needed when considering vector nonlinear problems depending on several parameters, since, in most cases, the computation of the high order derivatives would be prohibitively expensive. However any systematic approach will probably be based on ideas similar to those described in Section 4.1. Also there are a few problems, for example, mildly nonlinear elliptic boundary value problems, for which the high order derivatives can be computed. For such problems the direct generalisations to the case $\underline{x} \in \mathbb{R}^n$ might form the basis for a viable numerical approach.

5. NUMERICAL ASPECTS

We shall only give a brief indication of some numerical aspects here and refer the reader to existing literature. We mention some ideas and methods which have been used successfully in the past by the authors. There is other related work, based on using a different extended system for a quadratic turning point [6], [12], [21]. Also other approaches to such problems exist, notably [17], [18], and the final choice of algorithm will depend on various factors, for example, the cost of setting up and then factorising g_x^o, the ability to compute higher derivatives, the availability of existing software, etc.

As was mentioned in Sections 1 and 2 we would expect a numerical continuation algorithm to compute connected components of a solution diagram for a one parameter problem. At the very least it should detect the presence of quadratic turning points.

Given a starting value, close to $(\underline{x}_o, \lambda_o)$ a quadratic turning point, Newton's method (or a quasi-Newton method) can be applied directly to $G(\underline{y}) = 0$ (see (2.13)). Details of this approach are given in [20]. Note that the Jacobian G_y requires the eigenvectors $\underline{\psi}$ and $\underline{\phi}$ (see (2.14)). Approximations for $\underline{\psi}$ and $\underline{\phi}$ can be provided after g_x has been factorised, [16], and the factorisation of g_x is used again in a block factorisation of G_y. One nice feature of this approach is that an approximation to the E matrix (2.16) arises from the factorisation of G_y and so the quantities $\underline{\psi}_o^T g_\lambda$ and a_g (see (2.16)) can be checked to detect and locate singularities of higher codimension. Such singularities can also be computed using extensions of this approach and a discussion of how to compute a simple cubic turning point is given in [20], [21]. The computation of simple transcritical and pitchfork bifurcation points as defined in Section 2 can be carried out analogously.

Finally we mention that Euler-Newton continuation methods, sometimes using the pseudo-arclength continuation approach of [14], are used to compute paths of quadratic turning points in [12] and [21] and paths of pitchfork bifurcation points in [6].

It is well known that singular vectors and values are less sensitive to perturbation than eigenvectors and eigenvalues and we mention that the algorithms can be changed to use singular vectors.

ACKNOWLEDGEMENTS

Research supported by the University of Toronto, NSERC Canada, with travel funding under the British Council Academic Links and Interchange Scheme.

REFERENCES

[1] Bauer, L., Reiss, E.L. and Keller, H.B. Axisymmetric buckling of hollow spheres and hemispheres, Comm. Pure and Appl. Math. 23, (1970), pp 529-568.

[2] Bazley, N.W. and Wake, G.C. The disappearance of criticality in the theory of thermal ignition. ZAMP 29 (1979) pp 971-976.

[3] Benjamin, T.B. Bifurcation phenomena in steady flows of a visicous fluid I. Theory Proc. R. Soc. Lond. A, 359 (1977), pp 1-26: II. Experiments Proc. R. Soc. Lond. A. 359 (1977), pp 27-43

[4] Bigge, J. and Bohl, E. On the steady states of finitely many chemical cells. Preprint of the University of Konstanz (1982).

[5] Boddington, T, Gray, P. and Robinson, C. Thermal explosions and the disappear ance of criticality at small activation energies: exact results for the slab. Proc. R. Soc. Lond. A 368, (1979) pp 441-468.

[6] Cliffe, K.A. Numerical calculation of two-cell and single-cell Taylor flows, J. Fluid Mech. *(to appear)*.

[7] Crandall, M.G. and Rabinowitz, P.H. Bifurcation, perturbation of simple eigenvalues and linearised stability. Arch. Rat. Mech. Anal. 52 (1973) pp 161-180.

[8] Decker, D.W. and Keller, H.B. Path following near bifurcation. Comm. Pure Appl. Math. 34 (1981) pp 149-175.

[9] Ermentrout, G.B. and Cowen, J.D. Secondary bifurcation in neuronal nets. SIAM J. Appl Math. 39 (1980), pp 323-340.

[10] Golubitsky, M. and Schaeffer, D. A theory for imperfect bifurcation via singularity thoery. Comm Pure Appl. Math. 32 (1979), pp 21-98.

[11] Heinemann, R.F. and Poore, A.B. Multiplicity, stability and oscillatory dynamics of the tubular reactor. Chem. Eng. Sci. 36 (1981) pp 1411-1419.

[12] Jepson, A. and Spence, A. Folds in solutions of two-parameter systems: Part I, Tech. Rep. NA-82-02, Computer Science Department, Standford Univ. Stanford, CA, (1982), submitted to SIAM JNA

[13] Jepson, A. and Spence, A. The numerical solution of nonlinear equations having several parameters Part I: Scalar Equations (1983) (submitted)

[14] Keller, H.B. Numerical solution of bifurcation and nonlinear eigenvalue problems. In "Applications of Bifurcation Theory" ed½ P.H. Rabinowitz, Academic Press, New York, (1977), pp 359-384.

[15] Keller, H.B. and Szeto, R.K-H. Calculation of flows between rotating disks, in "Computing Methods in Applied Sciences and Engineering" ed. R. Glowinski and J.L. Lions (1980) pp 51-61 (North Holland).

[16] Keller, H.B. Singular systems, inverse iteration and least squares. (Private communication)

[17] Rheinboldt, W.C. Computation of critical boundaries on equilibrium manifolds SIAM J. Numer. Anal. 19 (1982) pp 653-669.

[18] Rheinboldt, W.C. and Burkardt, J.V. A locally parameterized continuation process ACM TOMS, 9 (1983) pp 215-235.

[19] Schaeffer. D.G. Qualitative analysis of a model for boundary effects in the Taylor problem. Math. Proc. Camb. Phil. Soc. 87, (1980) pp 307-337.

[20] Spence, A. and Jepson, A. The numerical computation of turning points of non-linear equations, in "Treatment of integral equations by Numerical Methods" ed. C.T.H. Baker, G.F. Miller, Academic Press, London, (1982) pp 169-183.

[21] Spence, A. and Werner, B. Non-simple turning points and cusps. IMA J. Num. Anal. 2, (1982) pp 413-427.

[22] Uppal, A., Ray, W.H. and Poore, A.B. The classification of the dynamic behaviour of continuous stirred tank reactors - influence of reactor residence time. Chem. Eng. Sci. 31 (1976) pp 205-214.

SEQUENTIAL DEFECT CORRECTION
FOR HIGH-ACCURACY FLOATING-POINT ALGORITHMS

by Hans J. Stetter, Vienna

1. Ideal Results in Floating-Point Arithmetic

During the past few years, numerical mathematics has gained a silent victory: After they had ignored the incessant request of numerical analysts for 25 years, hardware designers have finally accepted "clean" floating-point arithmetic as a natural standard. In a *clean arithmetic*, the computer-generated result $a \circledast b$ satisfies

$$(1) \qquad a \circledast b = \square (a * b)$$

for all rational operations $(+,-,\times,/)$ and all machine numbers a and b such that $a * b$ does not over/underflow. Here, $\square$ is a map from the reals into the set of machine numbers which satisfies the natural conditions for a rounding operation (cf. e.g. [1]).

The universal validity of (1.1) has become part of the IEEE Microprocessor Floating-Point Arithmetic Standard supported officially by IFIP; it has been implemented in a number of microprocessor chips (Intel, Motorola, etc.). Not a small part of this development is due to W. Kahan who has devoted his ingenuity and energy to this cause over a long period of time.

Yet the impact of this "victory" on scientific computing has remained rather unobtrusive: In spite of the shortcomings of actual computers, numerical analysts had assumed (1.1) anyway - and clean arithmetic in the sense of (1.1) has been found too low an aim: The results of trivial *sequences of operations* are not "clean"! E.g., in the multiplication of a 50×50 matrix A with a 50 component vector b, some 5000 rational operations are performed; the components of the computed vector $A \circledcirc b$ may deviate grossly from the rounded components of the accurate result vector Ab, but a realistic estimate of this deviation cannot be obtained in a practical way.

For a reasonably round-off error safe computation as well as for a realistic round-off error analysis of more complex algorithms, we need the extension of (1.1) to composite operations like matrix multiplication and general arithmetic (rational) expressions. The result $\boxed{F}(a_1,a_2,..,a_m)$ of the machine execution of such an operation $F(a_1,a_2,..,a_m)$ should satisfy

$$(1.2) \qquad \boxed{F}(a_1,a_2,..,a_m) = \square F(a_1,a_2,..,a_m)$$

for all floating-point arguments for which the exact result does not over/underflow. The right-hand side of (1.2) is often called the "ideal floating-point result" for obvious reasons.

The validity of (1.2) for rational expressions F would bring the associative and the distributive law back to numerical computation and thus constitute a step of a qualitative nature: The evaluation of algebraically equivalent expressions would lead to identical numerical results.

It is true that systems have been available for some time which permit the computation of ideal floating-point results for many composite operations: These systems, like SCRATCHPAD, MACSYMA, and others are able to use different formats for number representation (quotients of integers, multiple word representation or the like); none of them can be conveniently used within an ordinary Fortran program. Thus their impact on scientific computing has remained small.

2. High-accuracy algorithms

Recently, algorithms have been designed *and implemented* which have changed this situation fundamentally: Not only do they generate the ideal floating-point results for the evaluation of many kinds of rational expressions (including matrix products and polynomials), they also produce ideal results in the sense of (1.2) for solutions of systems of linear (algebraic) equations and for a wide class of non-linear problems: polynomial zeros, matrix eigenproblems, simple systems of non-linear algebraic equations, and for linear programming problems. These algorithms and their implementations have been developed at the University of Karlsruhe by a group of scientists led by Professor U. Kulisch; the development has been documented in the publications [1] - [4] (and many reports etc.). Implementations exist on some microprocessors like Z80 and Motorola 68000 as well as on Univac and IBM/370 computers; the latter have the form of Fortran subroutine libraries!

Due to their structure (see Appendix), these algorithms ordinarily generate a "last bit accurate" approximation of the exact result (i.e. the mathematically defined result of the specified operation with the specified data) of the following form:

$$(2.1) \qquad (\nabla F(a_1,a_2,\ldots,a_m),\ \Delta F(a_1,a_2,\ldots,a_m)) \ ,$$

where ∇ and Δ denote "downward" and "upward" roundings (see Fig. 1). It is clear that (2.1) locates the exact result $F(a_1,a_2,\ldots,a_m)$ as accurately as the nearest floating-point number $\square F(a_1,a_2,\ldots,a_m)$.

Fig. 1

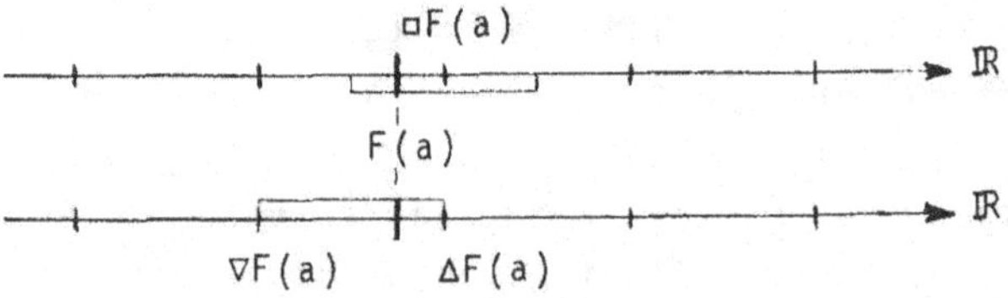

Furthermore, when the requested results are solutions of equations the generation of a computed result (2.1) by such an algorithm - without an error message - is equivalent to a *proof* of the fact that a (local) solution of the specified mathematical problem with the specified data exists and is unique within the interval (2.1). Thus, a linear equation solver of this type may be used to establish the non-singularity of a matrix on a floating-point computer.

Naturally, in the case of an *ill-conditioned problem* where the exact result changes very strongly with small changes of the data, the accuracy of (2.1) may not be meaningful if the data come from a real-life situation. But this is a feature of the mathematical problem and not of its approximative solution. Thus, the question to be posed and answered is: What it the meaningful accuracy for the solution of the mathematical problem, rather than for the result of the numerical computation.

Actually, it is just in the case of poor condition that conventional algorithms may produce "results" which deviate so strongly from the exact results that not even the meaningful accuracy may be recovered. If, on the other hand, the algorithm generates a tight approximation of the exact result which therefore reflects the true dependence of this result on data variations, the computer may even be used to analyze the sensitivity of the (exact) result experimentally and to establish its meaningful accuracy.

Besides, it often happens that a mathematical task is well-conditioned, but *numerical* ill-conditioning arises due to a representation of the problem which leads to a severe cancellation of leading digits. It is not always possible, without an excessive amount of analysis and computation, to change the representation to a less critical form. In these cases, the new high-accuracy algorithms remove the difficulty completely.

Example: The situation just mentioned typically arises in the evaluation of a function near a zero where normally the values become small because of a cancellation of additive terms. Thus, in the evaluation of a polynomial by Horner's algorithm, the "resolution" accuracy is determined by the largest intermediate result which need not be small at all. Fig. 2 shows floating-point values obtained for $p(t) = 2030\,t^4 - 5741\,t^3 - t^2 + 11842\,t - 8118$ in the interval (1.4142, 1.4143), by Horner's algorithm and by a corresponding high-accuracy algorithm (see [4] and the Appendix): While the values from the conventional computation permit no conclusion at all regarding the existence and number of zeros, the high-accuracy results give clear evidence of two zeros in the interval.

conventional algorithm

high-accuracy algorithm

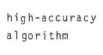

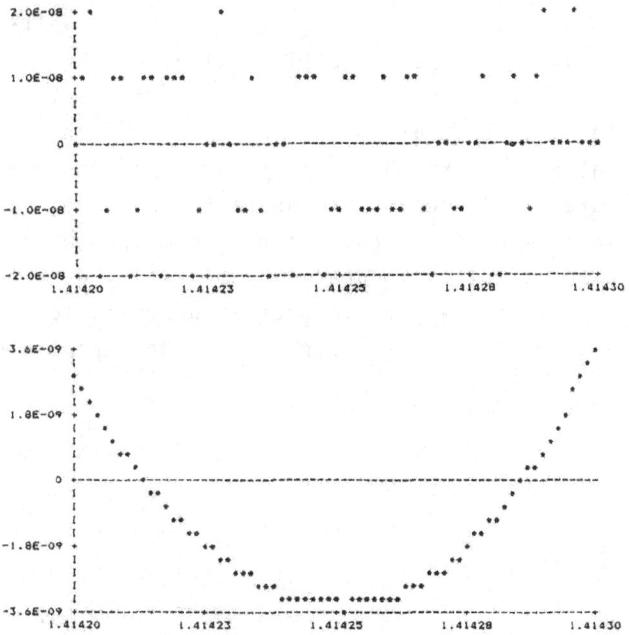

Fig. 2

3. Sequences of Problems

It is to be expected that the high-accuracy algorithms described in the previous section will soon become generally available, which will constitute a considerable relief in the programming of critical algebraic computations. However, a problem analogous to the one which has led us from (1.1) to (1.2) remains: Algebraic problems rarely appear isolated but rather in sequences. E.g., one may have to compute the largest zero of a polynomial whose coefficients are determined by a system of linear equations. In such sequences of computations, it may happen again that the sensitivity of the final result with respect to variations of the data is low but that one or more components of the sequence are much poorer conditioned.

In any case, if separate high-accuracy algorithms are used for the component problems, the following effect appears: The first algorithm produces (small) result intervals, e.g. for the coefficients of the polynomial. These intervals arise from the same set of original data and are thus highly correlated. When these correlated interval data are fed into further (interval) algorithms the computed result intervals may be unduly widened. This classical pitfall of interval mathematics is unavoidable if the information about the dependence on the primary data is lost on the way. It can only be excluded if the whole sequence is run as *one* high-accuracy computation.

This situation is analogous to what is well-known from the stepwise solution of systems of initial value problems for ordinary differential equations; see Fig. 3. Here, the (truncation) error in one step is called "local error" while the error of the value at the endpoint is the "global error". Ordinarily, it is not a priori known which local accuracy is necessary to guarantee a requested global accuracy. Furthermore, a naive transmission of componentwise interval bounds for the exact solution in one step as initial data bounds for the next step will normally lead to grossly unrealistic bounds for the error at the end, even when the bounding of the error in each step is rather tight.

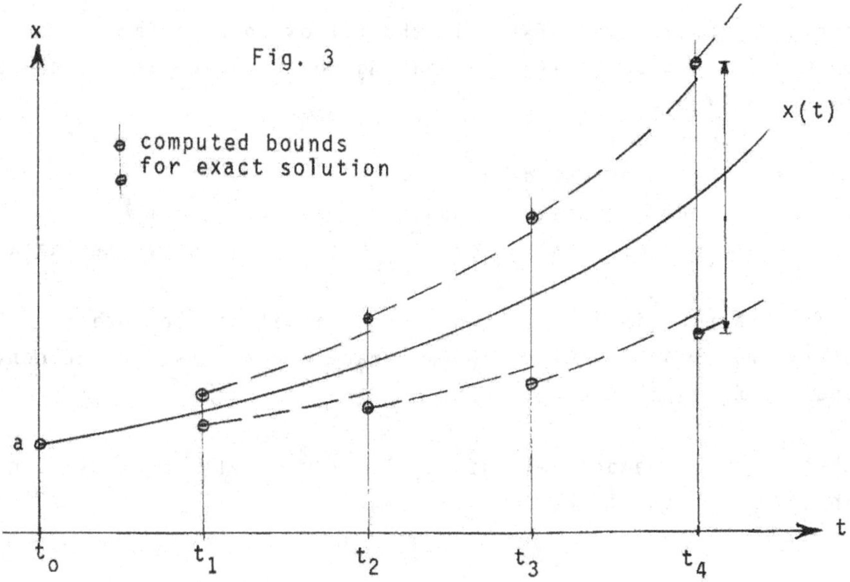

This leads us to the following <u>task</u> (S) which is the main object of this lecture:

<u>Given</u> a fixed finite sequence of algebraic computational processes:

$$
\begin{aligned}
& a \quad \text{given data} \\
& x := F_1(a) \\
& y := F_2(a,x) \\
(3.1) \qquad & \quad \ldots \\
& z := F_{m-1}(a,x,y,\ldots) \\
& u := F_m(a,x,y,\ldots,z)
\end{aligned}
$$

where all quantities are from appropriate $\mathbb{R}^{n_i}$.

<u>Assume</u> that a high-accuracy algorithm is feasible for each individual computational process F_i.

<u>Design</u> a coupling of the algorithmic processes such that

$u := F(a)$

becomes a high-accuracy algorithm, with a guaranteed (nearly) last bit inclusion of the result u.

An important special case of (S) is the following which arises for just one computational process if the floating-point system in the computation is *not decimal:*

	a	decimal data
$x =$	$F_1(a)$	conversion of data to machine format
$u =$	$F_2(x)$	computation of "$u = F_2(a)$" with converted data x

How is it possible, in an ill-conditioned situation, to obtain $u = F_2(a)$ accurately and reliably, with high-accuracy algorithms for the conversion and for F_2 available?

Note that some of the decimal data at least may well be exact, e.g. constants 10^{-k} or the like.

4. Sequential Defect Correction

To obtain a guideline for an approach to our task (S), we consider once more our analogy from ordinary differential equations. Here, a given approximate solution may be successively improved by *iterative defect correction* (see e.g. [5] and [6]): The interval of integration is traversed several times with basically the same integration method to compute successive corrections from suitably determined defect values. The respective initial data for later steps are not inclusion intervals for the exact solution but the corrected values at the ends of the previous steps ("global connecting strategy", Frank [6], see fig. 4).

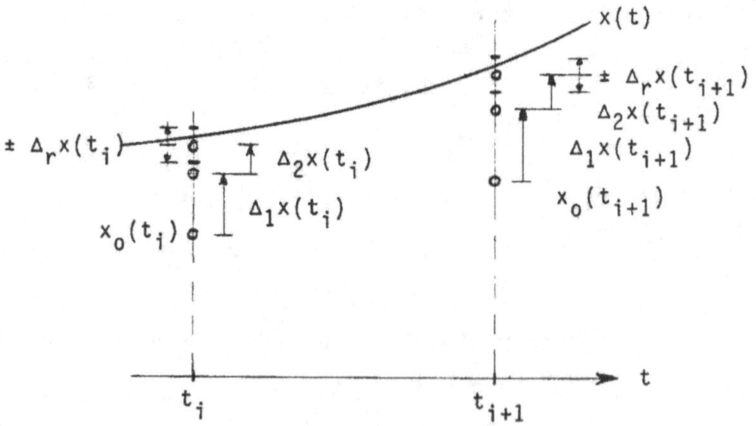

Fig. 4

At any fixed transition point t_i in the interval of integration, the data transmitted in the course of the defect correction iteration consist of the value at t_i of the initial approximation x_0 and of several successive corrections $\Delta_1 x$, $\Delta_2 x, \dots$. If necessary, a final *inclusion* $[-\Delta_r x, +\Delta_r x]$ of the remaining error at t_i of $x_0 + \Delta_1 x + \Delta_2 x + \dots$ may be computed from the defect of that approximation. This would lead to the inclusion for the exact solution x(t):

$$(4.1) \qquad x(t_i) \in x_0(t_i) + \Delta_1 x(t_i) + \Delta_2(t_i) + \dots + [-\Delta_r x(t_i), \Delta_r x(t_i)] \; .$$

In a sequence of *algebraic* operations in a floating-point number system where the only error is round-off it becomes even more mandatory to represent the successive approximations of some exact quantity via an initial approximation and successive corrections. The assertion

$$(4.2) \qquad \begin{array}{l} x \in x_0 + \Delta_1 x + \Delta_2 x + \dots + [-\Delta_r x, +\Delta_r x] \; , \\ \text{with } x_0, \Delta_1 x, \Delta_2 x, \dots, \Delta_r x \in \mathbb{M} \; , \end{array}$$

cannot be replaced by

$$(4.3) \qquad x \in [\underline{x}, \bar{x}] \; , \qquad \text{with } \underline{x}, \bar{x} \in \mathbb{M} \; ,$$

without a considerable loss of information.

<u>Example</u>: Take $\mathbb{M}(10,4)$, i.e. 4-digit decimal floating-point numbers.

$$\text{Let} \qquad \left.\begin{array}{rl} x_0 = & 1.324 \\ \Delta_1 x = & .002\ 189 \\ \Delta_2 x = & .000\ 004\ 641 \\ \Delta_r x = & .000\ 000\ 032\ 86 \end{array}\right\} \in \mathbb{M}(10,4)$$

Then (4.2) represents the assertion

$$(4.4) \qquad x \in [1.326\ 191\ 608\ 14,\ 1.326\ 191\ 673\ 86]$$

while an assertion (4.3) would be restricted to

$$x \in [1.326,\ 1.327] \; .$$

For reasons obvious from our example, we will call the representation of an inclusion interval for some quantity x by a set of floating-point numbers x_o, $\Delta_1 x$, $\Delta_2 x, \ldots, \Delta_r x$ a representation in *"staggered correction format"*. This format obviously permits the transmission of correction information *within a fixed floating-point number system* $\mathbb{M}$ which is equivalent to a dynamic multiple precision in this system, see e.g. (4.4).

But this representation becomes a feasible tool for our purposes only because *data in this representation are acceptable for high-accuracy algorithms*. This is due to two reasons (cf. Appendix):

(i) The staggered correction format is used for the internal representation of the approximate results in any case.

(ii) The computation of the ideal defect values of approximations remains possible when the data are also represented in this form.

To establish the validity of (ii), we note that all defects are in the form of linear residuals (see Appendix, Stage 2). Assume that a, b and x in the linear residual $a^T x - b$ are in staggered correction format:

$$\begin{aligned} a &= a_o + \Delta_1 a + \Delta_2 a + A \,, \\ b &= b_o + \Delta_1 b + \Delta_2 b + B \,, \qquad\qquad a, x \in \mathbb{R}^n \,,\; b \in \mathbb{R} \,, \\ x &= x_o + \Delta_1 x + \Delta_2 x + X \,; \end{aligned}$$

for convenience we have assumed only two scalar corrections in each quantity, A, B, and X are interval corrections. Then

$$\begin{aligned} & a^T x - b = \\ & - b_o \quad - \Delta_1 b \quad - \Delta_2 b \quad - B \\ & + a_o^T x_o + \Delta_1 a^T x_o + \Delta_2 a^T x_o + A^T x_o \\ & + a_o^T \Delta_1 x + \Delta_1 a^T \Delta_1 x + \Delta_2 a^T \Delta_1 x + A^T \Delta_1 x \\ & + a_o^T \Delta_2 x + \Delta_1 a^T \Delta_2 X + \Delta_2 a^T \Delta_2 x + A^T \Delta_2 x \\ & + a_o^T X \quad + \Delta_1 a^T X \quad + \Delta_2 a^T X \quad + A^T X \end{aligned}$$

which is one long scalar product so that the ideal value $\square(a^T x - b)$ is computable.

Naturally, during the iterative defect correction of stage 2 of the algorithm, one will use only the left upper 4 terms for the defect of x_o, the left upper 9 terms for the defect of $x_0 + \Delta_1 x$ etc. while the full interval value is only used in Stage 3.

We have thus gained the following insight: In a sequence (3.1) of dependent high-accuracy algorithms, the results from previous component algorithms must be transmitted in their internal staggered correction format representation for use as data in later component algorithms. The use of such data is feasible in these algorithms.

5. Algorithmic Pattern for Sequential Defect Correction

The decision about the transmission of the information still leaves open the algorithmic strategy for its realization: When should we correct where in the sequence (3.1) and how often? The two extreme strategies are:

(i) Fully sequential strategy: The entire sequence is traversed again and again, with one further correction executed in each component algorithm each time.

(ii) Fully local strategy: The results of the first component algorithm are corrected to an extreme accuracy (but which one?), then the next component algorithm is treated in the same fashion, etc.

These two strategies correspond to the global and local connecting strategies of Frank [6] in our analogy of sections 2 and 3 (see Fig. 2 and 3).

For reasons of intermediate storage, strategy (i) is not attractive; yet strategy (ii) is really not feasible because it is not known a priori which accuracies must be reached in the intermediate component algorithms.

Therefore, at least two passes are needed through the sequence (3.1): During the first pass, sufficient *information about the sensitivity* of the component algorithms with respect to their data (results from previous components) must be collected. This information will permit a *controlled local strategy* in a second pass. A separate third pass may be desirable for the sequential execution of Stage 3 of the high-accuracy algorithms.

This suggests the following algorithmic pattern:

First pass: Local correction to a fixed accuracy

Evaluation of relative condition numbers for the intermediate results with respect to their data

Second pass: Local correction to the necessary accuracy which is now defined through the condition numbers

Third pass: Sequential establishment of inclusions for the intermediate and final results (see Appendix, Stage 3).

Naturally, the second pass may not be necessary in some cases. The third pass may perhaps be combined with the second pass.

Let us look at the first two passes in more detail:

A suitable first pass fixed correction accuracy for "independent" component algorithms would be the following (eps is the relative floating-point accuracy in $\mathbb{M}$).

(6.1) Correct until $\|\Delta x\| < \text{eps} \cdot \|x_0\|$.

After such a correction, the result information may be recombined into two quantities $\overline{x_0}$ and $\overline{\Delta_1 x} \in \mathbb{M}$: E.g., assume $\mathbb{M}(10,4)$ and the following values for x_0 and its iterative corrections until (6.1) holds:

1.456
0.023 45
0.000 9876
0.000 0054 32
―――――――――
1.480 4430 32

Then, $\overline{x_0} = 1.480$ and $\overline{\Delta_1 x} = 4.430 \times 10^{-4}$ will constitute an economic representation of the current approximate result.

In a "dependent" component algorithm which has such results $(\overline{x_0}, \overline{\Delta_1 x})$ amongst its data, we must above all determine the effect of the small perturbation $\overline{\Delta_1 x}$ on the result y_0 obtained as a first approximation with $\overline{x_0}$ only. This is possible because in the scalar product for the defect of y_0 the terms originating from $\overline{\Delta_1 x}$ may be identified and separately accumulated. In view of (6.1), a suitable approximation for the relative condition of y with respect to x would be

$$(6.2) \qquad K_{xy} := \frac{\|\text{effect of } \overline{\Delta_1 x} \text{ upon } \|}{\text{eps } \|y_0\|}$$

The computation of the numerator of (6.2) requires some extra effort (separate treatment of terms which could otherwise be combined), but it is indispensible for a controlled local strategy which should be the most economic overall approach.

In dependent component algorithms, it may then be more reasonable to replace (6.1) by

$$(6.3) \qquad \text{Correct until } \|\Delta y\| < \max(K_{xy}, 1) \text{ eps } \|y_0\| \; ;$$

the recombination at the end will in any case lead to values $\overline{y_0}$ and $\overline{\Delta_1 y}$ with $\|\overline{\Delta_1 y}\| \approx \text{eps } \|\overline{y_0}\|$.

After the first pass, it is possible to determine crude relative condition numbers K_{ie} for the effect of a change in the intermediate result x_i on the end result u by the chain rule, since (6.1) and (6.2) imply

$$(6.4) \qquad \frac{\|\Delta y\|}{\|y\|} \approx K_{xy} \frac{\|\Delta x\|}{\|x\|} \text{ , etc .}$$

The admissible error δ_n in u may then be equidistributed on its sources in the sequence (3.1); this leads to requirements for the local accuracies for the individual component algorithms which should be sufficient for the achievement of an error not larger than δ_n at the end.

If they are not sufficient, e.g. because of strong nonlinear effects, it will become apparent in the third pass; then a repitition of the second pass will be necessary. Naturally, the potential number of such repititions must be limited to a fixed small number; if the third pass has not been successful by that time, an error return must be initiated.

The following example exhibits the necessity of the use of sequential defect correction in a very simple situation. It cannot be treated satisfactorily by a transmission of interval data and interval computations, even with high-accuracy component algorithms.

Example: Compute $u = \frac{a+1}{a^2-2}$ for a = 1.4142, in the floating-point system $\mathbb{M} = \mathbb{M}(16,4)$, i.e. with 4 hexadecimal digits.

We assume that high-accuracy algorithms are available for

conversion from decimal to $\mathbb{M}(16,4)$

evaluation of polynomials } for interval arguments
division }

Our three-pass algorithm will then proceed as follows:

First pass: a) "Conversion" will produce a "double precision" representation $(x_o, \Delta_1 x)$ of a in $\mathbb{M}$.

b) "Evaluation of polynomial" will discover the extremely high relative sensitivity of the denominator y with respect to x, with $K_{xy} > eps^{-1}$.

c) "Division" $\left(u = \frac{x+1}{y}\right)$ will find K_{xu} and $K_{yu} \approx 1$, but produce a poor approximation because of the insufficient accuracy of y.

Second pass: a) $K_{xe} = K_{xy} \cdot K_{yu} + K_{xu} > eps^{-1}$ and $K_{ye} = K_{yu} \approx 1$ request one further correction for the conversion component of the sequence but not for the evaluation and division.

b) Straightforward sequential defect correction yields sufficient accuracy for u.

Third pass: Straightforward inclusion establishment with natural choices for the X, Y and U intervals yields a guaranteed last bit accurate inclusion for u in $\mathbb{M}$:

$$u^* \in [-62\,936.,\ -62\,935.]$$

The exact result is $u^* = -62\,935.349\ldots$

The naive sequencing of the three high-accuracy algorithms, with an interval result from the conversion fed into the following ones etc. would have yielded:

In $\mathbb{M}(16,4)$: no result because $0 \in$ denominator.

In $\mathbb{M}(16,8)$, with the final inclusion rounded into $\mathbb{M}(16,4)$:

$$u^* \in [-62\,943.,\ -62\,918.]\ .$$

6. Conclusions

We have shown that the evolving high-accuracy algorithms for algebraic computations need a non-trivial mechanism for their sequential application if their power is not to be lost. This sequential defect correction process rests strongly on the transmission of results in staggered correction format; this device must be combined with an a posteriori determination of the (local) relative condition numbers. The suggested algorithm realizes - within natural limits - a *dynamic accuracy* feature which makes optimal use of the available floating-point arithmetic.

The implementation of this algorithmic pattern should relieve the user from any undue technical concerns: He should be able to specify blocks of high-accuracy computation (like the one in the example at the end of section 6) in a Fortran context; the necessary, rather complicated infrastructure should be set up automatically by a preprocessor. I hope that some excellent computer scientists will devote themselves to the design of such a computational tool.

If it has been implemented in a sufficiently user oriented fashion, this computational tool will lend itself to many important uses: It will permit a feasible verification of the obtained accuracy even for long algebraic computations by breaking the total computation into a few blocks, with one elementary round-off error committed in each one. More important, it will permit the establishment of the *meaningful accuracy* of computations with inaccurate data. It should also be of use in the debugging of programs. For production programs, this tool may aid in eliminating double and multiple precision where not needed and introducing it at the critical spots which may be found with its help.

Appendix: Structure of High-Accuracy Algorithms

The high-accuracy algorithms referred to in this paper have the following general structure (see, e.g. [2] - [4]):

Stage 1: An approximation x_0 of the exact result x^* is generated by a standard numerical procedure. (If this stage is unsuccessful the algorithm fails; however, a crude approximation is usually sufficient.)

Stage 2: Successive corrections $\Delta_1 x$, $\Delta_2 x, \ldots$ are computed by *iterative defect corrections* (cf. [5]). This leads to an actual improvement in the accuracy of the successive approximations $x_i := x_0 + \Delta_1 x + \ldots + \Delta_i x$ only if the ideal values in $\mathbb{M}$ of the defects of these approximtions can be computed, *irrespective of leading digit cancellation.* Furthermore, the defect of x_i must be computed from the set of values $(x_0, \Delta_1 x, \ldots, \Delta_i x)$ because x_i cannot be formed explicitly in $\mathbb{M}$ without loss of accuracy (see "staggered correction format", section 4).

The requirements of stage 2 are met in the following way:

a) The problem at hand is formally written as a system of linear equations (if not in this form anyway).

b) A technique for the ideal evaluation of a scalar product in the sense of (1.2) is used. (Defects in linear equations are scalar products.)

Ad a): Take the evaluation of a polynomial $p(t) = \sum_{\nu=0}^{n} a_\nu t^\nu$.

This is equivalent to

$$\begin{pmatrix} 1 & & 0 \\ -t & 1 & \\ & \ddots & \\ 0 & -t & 1 \end{pmatrix} \begin{pmatrix} \xi_0 \\ \xi_1 \\ \vdots \\ \xi_n \end{pmatrix} = \begin{pmatrix} a_n \\ a_{n-1} \\ \vdots \\ a_0 \end{pmatrix}, \qquad p(t) = \xi_n .$$

Ad b): The "ideal" value $\square\left(\sum_{j=1}^{m} \xi_j \eta_j\right)$ may be computed recursively, in $\mathbb{M}$, by various algorithms (e.g. Bohlender's algorithm, see e.g. [1]). However, the designation of a section in main storage as "long accumulator" which permits the accumulation of the scalar product in a quasi-fixed-point manner may be more economical in today's computers, see [1].

<u>Stage 3</u>: A (tiny) interval X is computed such that the exact result x^* of the problem satisfies (cf. (4.2)):

$$x^* \in x_o + \Delta_1 x + \Delta_2 x + .. + X \; . \tag{7.1}$$

For the output of a final approximation, (7.1) is then reduced to a *nearest* result $\square x^*$ or to an *inclusion* $[\nabla x^*, \Delta x^*]$, in $\mathbb{M}$.

The computation of X proceeds in the following way:

A) For evaluation problems, the defect of a sufficiently accurate approximation is enclosed in an interval, by interval arithmetic using the ideal scalar product. A corresponding interval correction X is then formed.

B) Solution-of-equation problems are cast into a fixed-point form for the correction Δx of the current approximation $\tilde{x}$:

$$\Delta x = G(\tilde{x}, \Delta x) \; .$$

The computational establishment of

$$G(\tilde{x}, X) \subsetneq X$$

for an interval X implies

$$\exists_1 \; x^* \in \tilde{x} + X$$

by well-known fixed-point theorems (cf. [2]).

Example: Linear System $Ax = b$.

<u>Stage 1</u>: Standard Gauss elimination.

<u>Stage 2</u>: Standard iterative improvement, in staggered correction format, with ideal defects.

<u>Stage 3</u>: Let R be some approximate inverse of A. Multiplication by R and rearrangement transforms

$$A(\tilde{x} + \Delta x) = b \qquad \text{into}$$

$$\Delta x = (I - RA)\Delta x - R(A\tilde{x} - b) =: G(\tilde{x}, \Delta x)$$

For an interval vector X,

$$G(\tilde{x},X) \subset \lozenge\{I - RA\} \mathbin{\lozenge\!\!\!\!\cdot\;} X \mathbin{\lozenge\!\!\!\!+\;} R \mathbin{\lozenge\!\!\!\!\cdot\;} \lozenge\{A\tilde{x} - b\} =: \bar{G}(\tilde{x},X) \; .$$

Here, $\lozenge\{..\} := [\nabla\{..\},\Delta\{..\}]$ (computed via exact scalar product) while $\blacklozenge$ are ordinary interval arithmetic operations.

$\bar{G}(\tilde{x},X) \subsetneqq X$ may be established by the computer and implies:

A is non-singular

$x^* = A^{-1}b \in \tilde{x} + X$

The computational effort is approximately 6 times the Gauss elimination effort.

References

[1] Kulisch, U., Miranker, W.L.: Computer arithmetic in theory and practice, Academic Press 1982.

[2] Kaucher, E., Rump, S.M.: E-methods for fixed point equations f(x)=x, Computing 28 (1982), 31-42.

[3] Rump, S.M.: Solving nonlinear systems with least significant bit accuracy, Computing 29 (1982), 183-200.

[4] Rump, S.M., Böhm, H.: Least significant bit evaluation of arithmetic expressions in single-precision, Computing 30 (1983), 189-199.

[5] Stetter, H.J.: The defect correction principle and discretization methods, Numer. Math. 29 (1978), 425-443.

[6] Frank, R., Oberhuber, C.W.: Iterated defect correction for the efficient solution of stiff systems of ordinary differential equations, BIT 17 (1977), 146-159.

NUMERICAL EXPERIMENTS WITH PARTIALLY SEPARABLE OPTIMIZATION PROBLEMS

by A. Griewank and Ph.L. Toint

Abstract. In this paper, we present some numerical experiments with an algorithm that uses the partial separabilty of an optimization problem. This research is motivated by the very large number of minimization problems in many variables having that particular property. The results discussed in the paper cover both unconstrained and bound constrained cases, as well as numerical estimation of gradient vectors. It is shown that exploiting the present underlying structure can lead to efficient algorithms, especially when the problem dimension is large.

1. Introduction

In the recent years, the solution of large dimensional unconstrained minimization problems has attracted a certain attention in the mathematical programming community. At first, only the nonlinear conjugate gradients algorithm seemed to be competitive in this area, because of its low storage requirements and its sometimes suprisingly good performance on large problems, when used with a suitable restart strategy. However, this algorithm requires a relatively large number of function evaluations, when compared to quasi-Newton Methods on problems with few variables. This is because these latter methods provide an asymptotically correct estimation of the stepsize, and therefore linesearching is reduced to a minimum. This fact led to hope for further progress when quasi-Newton updates were proposed by Toint [18] and Marwil [12], and soon after by Shanno [15], that could exploit any sparsity present in the Hessian matrix of the objective function not only for reducing the storage required by the method but also hopefully to improve on the speed of convergence. Unfortunately, although a sound convergence theory exists for the first of these updates [19], some bad examples where found concerning the numerically more successful "sparse BFGS update" [14], and the general behaviour of these sparse quasi-Newton methods led to some disappointments.

In view of these shortcomings, the authors recently proposed in [8] another type of quasi-Newton algorithm that could incorporate information about the structure of a given minimization problem and possibly increase the efficiency of the resulting procedure. They considered the so called "partially separable" minimization problems, i.e. problems where the objective function has a decomposition of the form

$$f(x) = \sum_{i=1}^{m} f_i(x), \tag{1}$$

where each "element function" $f_i(\cdot)$ has a Hessian matrix of low rank compared to the total number of variables of the problem. This clearly happens in the frequent case of a decomposition of the form (1) and where each $f_i(\cdot)$ only depends on a small number of these variables. Partially separable problems arise naturally in many different fields : finite element calculations, discretized variational problems, transportation networks, spline approximation problems, hydraulics, circuit theory, penalty functions, large dimensional nonlinear least squares, multi-objective decision making, image processing, econometric problems in utility and production theory, to name a few. It is also possible to see [8] that this structure covers, amongst others, all sufficiently differentiable problems with a sparse second derivative matrix.

The idea is then to build approximations to these low rank "element Hessians" separately, since one may expect to obtain them with reasonable accuracy in a number of steps that is much smaller than the total dimension of the problem. Hence the proposal of partitioned updating discussed in [6] : the Hessian matrices of the element functions $f_i(\cdot)$ are updated separately, and only gathered for the definition of the step at the k-th iteration via the relation

$$\sum_{i=1}^{m} B_i^k s^k = -g(x^k), \tag{2}$$

where x^k is k-th approximation to the solution, $g(x^k)$ the gradient vector of f at this point, s^k the k-th step of the algorithm and B_i^k the approximation of the i-th element Hessian at the k-th iteration.

This approach has been theoretically justified in the case where each element function $f_i(\cdot)$ is convex, and local superlinear convergence can be proven in this framework under relatively mild conditions [7]. In particular, it is not necessary to assume that the initial approximations to the element Hessians are sufficiently exact, an unrealistic assumption made in many local convergence proofs. Some preliminary numerical experiments in the "convex decomposition case" were also presented in [6] with rather encouraging results.

The purpose of the present paper is to extend these experiments also to non convex decompositions and to point out some of the advantages that are inherent in a partitioned updating algorithm and could lead to a significant improvement in performance, generality and ease of use of an algorithm for solving partially separable optimization problems. Section 2 presents some of the interesting features that are inherent in the partitioned updating approach and that can be used to some advantage. Section 3 is devoted to the more detailed description of the method tested in this paper. Finally, Section 4 presents the results of the numerical tests, together with some comments and conclusions.

2. Some algorithmic consequences of partitioned updating

The strategy of updating separately the Hessian matrix of each element function, combined with a iterative method for solving the set of equations (2), presents some advantages to the organization of an efficient minimization algorithm that are discussed below, because they are, up to now, unique, and can lead to a significant improvement in performance on some practical problems.

The first observation concerns the speed of convergence of the resulting method. In low dimensions, one expects a decent quasi-Newton algorithm to reach the solution of a given problem in a number of iterations roughly proportional to the number of its variables. A whole theory of "finite termination on quadratics" (see [11], for example) has been devoted to this desirable property. The reason invoked in this kind of theory is that the algorithm will need of the order of n iterations to obtain a reasonable approximation to the n by n Hessian matrix of the objective function f, where n is its number of variables. When considering a partitioned updating method, the prospects improve dramatically : the only dimension relevant with respect to the updating is the maximal dimension of the range of any element Hessian, that may be (and usually is) much smaller than the total number of variables present in the problem. This dimension is typically less than ten, even in problems featuring several thousands of variables. So it is not unreasonable to expect convergence in a number of iterations that is much smaller than the dimension. A similar effect may occur for the conjugate gradient method, where this basic number is then proportional to the number of distinct eigenvalues of the complete Hessian matrix of f. This may also be substantially less than the dimension, although not usually to the extent obtained in partitioned case.

A second important consequence of partitioned updating concerns the possible restarts of the algorithm. In many applications, the optimization part of the problem is itself part of an outer loop, and one wishes to restart the minimization procedure with a perturbed starting point, but also very often with a perturbed objective function. One typically considers a reweighting of the different terms of (1), possibly dropping or adding some of them. Varying penalty or dual parameters, expressing a new goal for a multiple objective decision, varying the discretization widths on a variational problem, considering a reduced or expanded network are some of the circumstances where this type of situation occurs. In this case, the Hessian approximation produced by classical methods at the end of the previous minimization cannot be used to improve the current process, because one does not know how to modify it to reflect the modifications in f. If each of the element Hessian approximation is available separately, this problem disappears, and one is able to build a new starting Hessian approximation very easily. This type of facility can result in a very important gain in computing time : a factor greater than ten is not exceptional for large problems. An example, presented in [6], shows that, for a 3-dimensional finite element problem with 1331 free variables, it is over 20 times faster to use a simple mesh refinement technique with restart, than to solve the problem directly.

In some nonlinear problems, most methods approach the solution along a path that stays more or less in a subspace of the total space where minimization takes place. This is especially true when bounds are imposed on the variables of the problem and when the optimal solution lies on the boundary of the feasible set. The availability of the objective function as a sum of the form (1) allows a partially separable optimization algorithm to take advantage of this situation : only the element functions whose variables are being significantly modified need to be recomputed at a new point. The gains obtained by using this feature may be interesting when the cost of function evaluation is very high. In some cases, the total number of function evaluations required to reach the solution may be smaller than the number of iterations! The fact that f is computed elementwise also allows the use of the knowledge of lower bounds on the $f_i(\cdot)$. The user may detect an unacceptably high function value before all the m element functions are computed at some point, and decide to shorten the step. In the authors' experience, this procedure may save up to 40% of the function evaluations, but is rather delicate to use because it assumes a good knowledge of the problem at hand. The main difficulty is then to find a suitable interpolation procedure based on incomplete function and gradient values. A simple bisection process is always possible, but may be less efficient.

The gradient of a partially separable function is comparatively cheap to approximate by some difference scheme, because the number of additional function values that are required to form these differences is now equal to the maximum number of variables active in any element $f_i(\cdot)$, which is much smaller than the dimension. In most cases, one can expect the algorithm to reach the solution of the problem using less function calls than would be required by a classical algorithm to form the gradient at the initial point. This is at variance with the current belief that optimization without analytical gradients is hopeless for large dimensional problems, because of the estimation costs [5].

A last important point about partitioned updating methods deals with storage requirements. It may be thought that by storing each element Hessian separately one fails to utilize the overlaps between these matrices in an efficient way. However, in many problems (e.g. a number of finite elements calculations), these element Hessians are known to be singular, even w.r.t. the variables that appear explicitly in the corresponding element functions. One may then store these approximations only on their ranges, with the drawback that one therefore needs a projector on each of these ranges. Fortunately, it happens very often that the projector is the same for all elements, up to the assignment of the variables appearing explicitly in the element functions. The storage is then usually reduced, compared to the more conventional sparse methods, because all this range information would be lost by assembling the complete second derivative matrix. A typical example is the minimum surface problem [6] that would require O(5n) storage in the conventional sparse matrix storage scheme, while it only requires O(3n) when storing each element Hessian separately, as just described.

It is also interesting to point out that partitioned updating algorithms may adapt very well to the parallel computing environment. Indeed, the treatment of each element function is done completely independently : updating, function evaluation, and even, to a certain extent, solution of the system (2). Indeed the only operation needed for an iterative solver is the matrix vector product, which can be done element by element and assembled only afterwards. Some numbers will be given below, to illustrate the potential gains due to parallel computing in this framework.

3. A more detailed description of the particular algorithm tested

With all these considerations in mind, a particular algorithm was designed for partially separable minimization problems, and coded into a routine called PSPMIN. It was also decided to incorporate into this algorithm the necessary facilities to treat upper and lower bound constraints on the variable of the problem, because the applicability of the resulting method would be substantially enlarged at a reasonable price in conceptual complexity and coding effort. This algorithm has the following general layout :

1. Choose a starting point, starting approximations to the element Hessians, and evaluate the function and gradient values at this point.

2. Define a step by solving the linear system (2) inexactly inside a predefined trust region around the current point.

3. Ensure a significant reduction in the objective function along the projection of this step on the feasible region, by performing a very weak linesearch.

4. Compute the function value and gradient at the newly found point.

5. Test for termination.

6. Update the approximations to the element Hessians matrices by a suitable quasi-Newton formula, enforcing the condition

$$B_i^{k+1} s^k = g_i(x^{k+1}) - g_i(x^k) \tag{3}$$

for each one of these matrices, where $g_i(\cdot)$ is the gradient of $f_i(\cdot)$.

7. Revise the trust region size, and go back to 2.

Clearly, this outline should be detailed somewhat, if we want to understand the tested algorithm, but, before going into more detail, it seems appropriate to describe first the way the approximations B_i^k to the element Hessians are stored.

3.1. Storing the approximate element Hessian matrices

From what has been said already, it is clear that an element function will not, in general, involve all the variables of the problem. Hence, the Hessian matrix of the i-th element function will be singular with its range included in the span of the variables that appear explicitly in $f_i(\cdot)$. Moreover, it happens quite often that this range is even smaller than the span of its internal variables, for example if some of these variables contribute linearly to the corresponding element function, or, more generally, if this latter function satisfies a relation of the type

$$f_i(x) = f_i(x+aw) \tag{4}$$

for all real a and some vector w in the span of the variables internal to $f_i(\cdot)$. In this case, the considered range is a proper subspace of the above mentioned span, and the matrix B_i (where we dropped the iteration superscript) can be written as

$$B_i = U_i^T C_i U_i, \tag{5}$$

where B_i is an n_i by n_i matrix, U_i is p_i by n_i, C_i is p_i by p_i, and where n_i is the dimension of the span of the variables internal to the i-th element, while p_i is the dimension of the true range. Our algorithm only stores C_i and the user is asked to provide a routine to perform the necessary computations with the matrix U_i. In contrast to [6], we do not assume that the first p_i columns of U_i form the identity matrix, since this would sometimes force the user to renumber the variables internal to some elements, which can be unnatural and very cumbersome. But this supplementary freedom results in the need, for the user, to provide a code for solving the following four equations for z, given v :

$$U_i z = v, \qquad U_i v = z,$$

$$U_i^T z = v, \qquad U_i^T v = z,$$

for any particular i.

The reason why the matrices U_i are not stored explicitly is that, frequently, the same matrix U plays the role of U_i for several, if not all, elements. It would therefore be very inefficient to store O(m) copies of the same matrix. Moreover, these matrices are often very simple to write down, and although the providing of the necessary code may look complex, it is usually very easy.

3.2. Initializing the Hessian approximations

We now turn to a more systematic discussion of the features present in the algorithm outlined at the beginning of this section, and our first remark concerns the way the initial approximation B_i^0 are chosen. In fact, the algorithm provides two options : either choose a suitably scaled multiple of the identity matrix (see [16]), or estimate the initial true second derivative matrix by differences in the gradient vectors. (This last option assumes that the gradients are analytically available). The estimation has some advantages when the user is not sure of the true range of the B_i, because it will automatically preserve the correct subspaces, and it is relatively inexpensive if

$$n_{max} = \max\{ n_i \mid i=1, \ldots, m\} \tag{6}$$

is small compared to n, i.e. if the function f has a small depth, in the sense defined in [8]. The relative efficiency of these two possible choices will be examined in the next Section.

3.3. Defining the step

The next question we consider is how the linear system (2) defining the step direction is solved. In the algorithm tested, a diagonally preconditioned conjugate gradient routine is used to provide an approximation of the exact solution. The accuracy required on the residual for stopping the procedure is proportional to the square root of the gradient norm, therefore increasing as the method approaches a stationary point of the objective function. This inexact solution scheme has the advantage that very little effort is spent to solve (2) accurately in the early stages of the algorithm, when the quadratic approximation to the objective function is still rather poor. If some direction of negative curvature is encountered during the conjugate gradient process, a step is computed to the boundary of the current spherical trust region. On the other hand, if all curvatures are positive along the steps taken by the conjugate gradient procedure, the steplength is then not restricted by the trust region size, allowing therefore the usual full quasi-Newton step. This last feature is useful when minimizing convex functions, because the trust region strategy could, in this case, slow the convergence down in the first iterations of the method.

Alternative procedures have been considered. Elliptic trust region has been tried in conjunction with the considered algorithm, as well as preconditioning the system with a BFGS update of the diagonal. It turned out that both features improved substantially the behaviour of the algorithm on some examples, but were, on the whole, more costly (especially in terms of function evaluation and storage) and less reliable on a larger set of test problems. Therefore, they were not incorporated in the general purpose algorithm. Similarly SSOR preconditioning was not used because, even though it does not require the assembly of the element Hessians, it becomes rather complicated in view of the storage mode of the B_i, and would need additional integer workspace for pointers. However, the authors are very conscious that these alternatives may well pay off in other contexts.

Once the step direction has been determined, the algorithm then tries a stepsize of one, and decides if all element functions have to be recomputed at this new point. If, for example, the current minimization takes place along some boundary of the feasible region, it is likely that some element functions will have their internal variables unchanged, and therefore need not to be evaluated again. The linesearch process itself uses cubic interpolation when gradients are available, and bisection otherwise. The stopping criterion only asks for a sufficient reduction in the value of the objective and an average positive curvature over the step. In many cases, the steplength one is accepted in practice, except during the first iterations. This linesearch can therefore be viewed as a safeguard for the algorithm, enforcing convergence from poor starting points.

The trust region size is then adjusted in the usual way : if the reduction obtained in the objective is not too small compared to the reduction predicted by the local quadratic model, then the next trust region will have a radius twice as large as the current step length; otherwise, this radius will be set to a half of this length.

3.4. Updating the approximated element Hessians

The element Hessian approximations B_i^k are updated using either the BFGS or the rank one quasi-Newton formulae, in contrast to the "curvature shift strategy" studied in [9] for the case where some element functions $f_i(\cdot)$ are not convex. After much experimenting, it was found best to update a particular element Hessian aproximation with the BFGS update, as long as this element function looks convex. If the algorithm detects that the i-th element function is nonconvex, then all subsequent updates to B_i will use the rank one formula. This update is performed as long as the residual is not too orthogonal to the step, so that the rank one formula does not explode numerically. If this simple angle check fails, B_i is left unchanged. The tests that were performed to define this strategy included, amongst other possibilities, BFGS, rank one, DFP, Greenstadt, and PSB updates, and various combinations of them.

3.5. Gradient estimation and bounds handling

When analytical gradient of a particular element function is not available, it is estimated by differences in the values of this element function, with the difference steps determined by an algorithm due to Stewart [17]. Although this method was originally designed for convex functions and in conjunction with the DFP update, it is not too difficult to extend it to the present context : modifications are necessary to be able to deal, for example, with zero or very small gradients or with negative curvature. Most often, only forward differences are used, because the element gradients are generally nonzero at the solution of the overal problem. However, central differences are required when cancellation and rounding errors in the assembly of the complete gradient of f become important.

The upper and lower bound constraints on the variables are dealt with using the projection operator on the feasible region, and linesearch along possibly crooked arcs. This bending of the search direction does not raise any particular difficulty in the organization of the algorithm, although a simple anti-zigzag device due to Bertsekas [1] has been introduced.

3.6. Some programming details

The code written for the tests discussed in the next section, namely the routine PSPMIN, also incorporates a possibility, for the user, to control the whole minimization algorithm in a more efficient way : freedom is left to specify a problem oriented stopping criterion for the method, as well to check the plausibility of each stepsize proposed by the linesearch. This freedom is obtained by dedicating special small well documented routines to these purpose, that can be modified very easily by the user.

The code has also options for checking the analytical element gradients by comparing them to difference estimations, and for keeping certain variables fixed throughout the calculation. This last feature is useful, for example, in problems involving boundary conditions : the element functions need not to be rewritten along the boundary, but the generic element can be used and selected variables are then considered as parameters.

However, none of these special features were exploited in the numerical tests presented below.

4. Numerical results and discussion

All tests were run in double precision on the DEC2060 of the Facultes Universitaires de Namur, a machine with 36 bits words, using the FORTRA fortran compiler. A total of 154 different test problems were used, involving 25 different functions and starting points, with dimensions varying from 2 up to 1000. 52 of these problems involve bounds on the variables. A large number of these test example also feature a decomposition in element functions that are nonconvex. Three possible choices of the parameters of PSPMIN were tested, producing a total of 457 runs of the routine. The complete list of tests together with the detailed results for these runs is available separately as a technical report of the Department of Applied Mathematics of Namur (Belgium) under the reference 83/4. The rest of this section will be devoted to the discussion of these results, as well as to the presentation of some meaningful examples.

The three different parameter settings for the proposed algorithm are defined as follows :

- alg1 will denote the partitioned updating algorithm using analytical gradients and with the initial approximating element Hessian matrices chosen as a suitable multiple of the identity,

- alg2 will denote the same algorithm, but with gradients approximated by differences in the function values,

- alg3 is similar to alg1, but the initial element Hessians are estimated by differences in the gradients.

For these methods, a fractional number of function and gradient evaluations will be given : it is the number of element function calls divided by the number of element functions. It is therefore comparable with the number of function calls for an algorithm that is not using the decomposition of the objective into elements. It also incorporates all element function calls that are used for approximating gradients or initial Hessian matrices, for alg2 and alg3 respectively.

4.1. Comparison with other methods in low dimension

Although some early tests were already promising, we were first concerned with verifying that the proposed partially separable algorithm would still behave reasonnably well when applied on low dimensional problems, involving may be just one element function. In this framework, the routine can be considered just as another implementation of a quasi-Newton procedure for minimization with bound constraints, and can therefore be compared with various methods in the literature. To this aim, some results recently published using well-known test functions were compared with the results obtained by the new algorithm. In all these comparisons, care was taken to ensure that the new routine was run with a stopping criterion at least as asking as those used with the other methods.

In the three following tables, the results are presented according to the following format :

number of iterations / number of function calls, (7)

where a call to the function also incorparates the computation of the gradient, when available. The test functions are defined in the corresponding references.

Table 4-1 presents a comparison of the partitioned updating algorithms alg1 and alg3, together with the results obtained by Shanno in [15] for the "sparse BFGS" (SBFGS), "sparse PSB" (SPSB), BFGS (BFGS) and conjugate gradient (CG) methods.

In Table 4-2, the famous Rosenbrock function and the "singular function" in four variables by Powell are considered, and alg1 and alg3 are compared with a discrete Newton method (DN) as proposed by O'Leary [13], a conjugate gradient (CG), the BFGS method (BFGS) and a modified Newton method (MN) discussed by Gill and Murray in [4], the double dogleg method (DeM) and the optimally conditioned Davidon update (Dav.) tested by Dennis and Mei in [3], and an implementation of the rank one update (RK1) advocated by Cullum and Brayton in [2].

In Table 4-3, alg1 and alg3 are compared with the double dogleg (DeM), Davidon's optimally conditioned method (Dav.), the routines VF02AD by Powell, OPRQP by Bartholomew-Biggs and GRGA by Abadie and Guigou, and with the rank one method of Brayton and Cullum (RK1). The problems involve bounds on the variables, and are

Problems	n	alg1	alg3	SBFGS	SPSB	BFGS	CG
TRIDIA	10	3/ 5.00	2/ 4.00	20/27	22/ 31	33/36	31/ 66
	25	3/ 5.00	3/ 5.00	23/36	25/ 39	44/48	41/ 90
BV	10	16/19.00	11/14.25	38/44	26/ 35	26/29	26/ 74
	20	19/21.00	15/18.11	69/82	88/104	55/58	55/141
	30	19/21.00	18/21.07	30/35	113/137	61/64	240/519
NONDIA	10	23/39.00	34/51.00	31/42	31/ 38	34/41	25/ 72
	20	23/39.00	36/52.00	33/42	36/ 46	30/40	22/ 61
	30	23.39.00	33/53.00	37/46	74/ 76	30/40	24/ 56

Table 4-1: Some problems tested by Shanno

Problems	n	alg1	alg3	DN	CG	BFGS	MN	DeM	Dav.	RK1
Rosenbrock	2	23/ 39.00	35/ 53.00	22/ 67	-	-	-	?/35.8	?/54	?/68
(extended)	50	75/109.00	95/148.00	38/1551	108/201	128/287	62/202	-	-	-
	100	146/245.98	152/247.00	-	191/365	-	-	-	-	-
Powell	4	41/ 47.00	30/ 36.00	11/ 56	-	?/31.6	?/ 26	?/ 87	-	-

Table 4-2: Two famous test functions

extracted from the collection of Hock and Schittkowski [10]. It should be observed that VF02AD, OPRQP and GRGA are routines for solving much more general constrained problems, and are therefore not used in their most efficient context.

All these results imply that the behaviour of the partitioned updating algorithm is at least comparable to that of some good methods for solving unconstrained and bound constrained minimization problems in low dimension.

Problems	n	alg1	alg3	DeM	Dav.	VF02AD	OPRQP	GRGA	RK1
Wood	4	39/47.00	92/133.00	?/52.6	?/50	?/109	?/130	?/127	?/167
Paviani	10	6/ 9.00	6/ 18.00	-	-	?/ 9	?/ 8	?/ 10	-
McCormick	2	6/ 7.00	4/ 10.00	-	-	?/ 8	?/ 9	?/ 36	-

Table 4-3: Problems with bounds

4.2. Results on the test problem collection

We now present and comment on the results that were obtained on the 457 test problems tested with the partially separable algorithm. For clarity, we distinguish two categories of tests : those of "small dimension" (n<100) and those of "large dimension" (n above 100). The first category contains 86 problems, and the second 68. The algorithm was terminated when the euclidean norm of the gradient was less than 10^{-7}, if analytical gradients are available, or less than 10^{-4}, if they are obtained by differences.

	#p	ntot	#it	#fg	#CG
alg1	86	2735	2351	3008.55	17498
alg2	83	2625	1628	7386.91	9629
alg3	84	2635	2196	2955.85	14279

Table 4-4: Counts for the small problems

In Table 4-4, the column headings #p, ntot, #it, #fg and #CG stand respectively for the number of problems successfully solved (i.e. solved within the desired accuracy in less than 500 iterations), the cumulated dimension of these problems, the total number of main iterations of PSPMIN required to solve them, the total number of function calls needed and the total number of internal conjugate gradient iterations performed.

Alternately, we can consider the following statistics presented in Table 4-5, where the column headings av.n, av.it, av.fg, av.CG, f/it and CG/it now mean the average dimension of the successfully solved problems, the average number of main iterations per problem, the average number of function calls, the average number of conjugate gradient iterations, the average number of function calls per iteration and the average number of internal conjugate gradient iterations per main iteration.

	av.n	av.it	av.fg	av.CG	f/it	CG/it
alg1	31.80	27.34	34.98	203.47	1.28	7.44
alg2	31.63	19.61	89.00	116.01	4.54	5.92
alg3	31.37	26.14	35.19	169.99	1.35	6.50

Table 4-5: Statistics for the small problems

If we now turn to the large problems, we obtain Tables 4-6 and 4-7, where the column headings have the same meaning as in Tables 4-4 and 4-5.

	#p	ntot	#it	#fg	#CG
alg1	68	43293	2064	2629.53	39855
alg2	66	42193	1747	8111.03	22782
alg3	63	40133	1652	2417.85	26676

Table 4-6: Counts for large problems

	av.n	av.it	av.fg	av.CG	f/it	CG/it
alg1	636.66	30.35	38.67	586.10	1.27	19.31
alg2	639.29	26.47	122.89	345.18	4.64	13.04
alg3	637.03	26.22	38.38	423.43	1.46	16.15

Table 4-7: Statistics for the large problems

A first observation concerns the performance differences between the small and large test problem sets. These differences are rather small in our opinion : the average number of function calls is about unchanged, as are the average number of iterations per problem and the average number of function evaluations per iteration. Only the number of conjugate gradient iterations increases with dimension, and this even less than proportionally. Hence, the method seems to preserve its efficiency when dealing with large dimensional problems.

We also see, in these tables, that gradient estimation by differences in function values can be rather efficient in the partially separable context, as expected. The solution of some very large problems without derivatives therefore seems possible with this approach. However, the algorithm failed on some problems where the true gradient was difficult to approximate by differences (exponential, ...) : the overal reliability is therefore diminished compared to the situation where analytic gradients can be supplied by the user. But this again is quite natural and has to be expected, independently of the particular decomposition structure of the objective function.

When comparing alg1 and alg3, we observe that alg3 is in general more efficient but again less reliable than alg1. The fact that the initial scaling of the B_i is much more accurate in alg3 results in fast convergence when this scaling remains valid in the sequel of the minimization process. On the other hand, a less accurate scaling may allow lucky initial steps, that can contribute significantly to the performance. An extreme example is the extended Rosenbrock function : alg1 jumps across the valley and converges very fast to the solution, while alg3 takes a shorter initial step and therefore is bound to take the valley turn, with the result of a much slower convergence. But if the user is reasonably confident in the scaling of the problem at the starting point, computing the initial B_i by differences has some attractions.

It is interesting to observe that the partial evaluation of the objective function is often sufficient, even sometimes resulting , as said above, in a number of function evaluations less than the number of iterations. For example, alg1 solves a cubic variant of the Rosenbrock test function in 100 variables in 837 iterations and only 735.56 function calls. This can be very useful indeed, when the cost of one such evaluation is high.

A typical example of the behaviour of the three methods considered as given in Table 4-8, where their performance is compared on the well known tridiagonal Broyden problem. In this table, the format (7) is used again. This is an example where the initial scaling of alg3 is rather advantageous.

	n=10	n=50	n=500	n=1000
alg1	15/21.00	14/21.00	18/25.00	17/21.00
alg2	13/51.25	12/45.50	14/52.06	15/54.03
alg3	10/13.25	11/14.04	10/13.00	10/13.00

Table 4-8: Broyden tridiagonal problem

Finally, the presence of bounds does not seem to deteriorate the behaviour of the algorithms. On problems involving the same objective function, they are typically faster when bounds are present and active than when the problems are unbounded, taking full advantage of the restricted dimensionality to reduce the number of function calls.

4.3. Perpectives of adaptation to parallel computing

One of the interesting facts about partitioned updating algorithms is that they could possibly be rather efficient in a parallel computing environment, as mentioned above. Indeed, their stucture consists of a sequence of loops, each one performing some task completely decoupled from the next one with respect to the loop index :

updating an element Hessian, computing an element function value, computing the product of an element Hessian with a vector, etc. More precisely, we can divide the computing time needed by alg1, say, into three distinct parts :

- a first part corresponding to the computations that are inherently sequential,
- a second part corresponding to the separate calculations involving the element functions (loops from 1 to m),
- and a third part corresponding to the treatment of vectors of dimension n, the number of variables.

Let us call these times tseq, tm and tn respectively. We may then assume an idealized situation where tm and/or tn could be spread between several parallel processors, while tseq would represent the irreducible sequential computing time. If we have some idea of the ratios of tseq, tm and tn to their sum, we may then compute a theoretical speed up of the algorithm in the presence of p parallel processors, say.

These times have been measured for alg1 and alg2 on a set of 12 problems out of our collection and the corresponding ratios rseq, rm and rn computed, with the results shown in Table 4-9. These tests feature element functions that are very cheap to evaluate. If they were expensive, tm would be proportionally larger.

	size	rseq	rm	rn
alg1	small	0.214	0.613	0.173
	large	0.038	0.741	0.221
alg2	small	0.241	0.542	0.217
	large	0.037	0.703	0.260

Table 4-9: Computing time ratios

We see that the most important part of the computing time is spent on operations related to the element functions, while the purely sequential part is rather small. This conclusion is especially true if large problems are considered, even more important if the cost of evaluating the element functions is high. A first substantial speed up could therefore be achieved if the operations concerning the element functions are performed in parallel. A second speed up could then be obtained by sharing the time tn between the processors.

We computed these speed ups with two extreme environments (4 and 1000 parallel processors), and obtained the results appearing in Table 4-10. In this table, the second column gives the amount of time that is shared between the p parallel processors.

	time shared	size	p=4	p=1000
alg1	tm	small	1.685	2.181
		large	2.115	3.591
	tm and tn	small	2.321	4.136
		large	3.601	26.341
alg2	tm	small	1.851	2.580
		large	2.251	3.850
	tm and tn	small	2.436	4.656
		large	3.591	25.666

Table 4-10: Speed up ratios

A main conclusion can be drawn from these (idealized) results. Firstly, if the cost of eveluating the element function is low, an algorithm like alg1 or alg2 is not really suited to an environment with many parallel processors, but its inherent parallelism would be best exploited with less than 10 processors. Furthermore, the gains increase with the dimension of the problems considered. If this cost is high, it then dominates the total cost of the algorithm, and the theoretical speed up ratio is thus nearly equal to the number p of processors available.

5. Conclusion

In this paper, we have presented some of the theoretical advantages of an algorithm for partially separable optimization, especially for large dimensional problems, together with some numerical evidence that these advantages can be obtained in practice. Although it is clear that much additional experience with methods of this type is needed to assert the value of the approach, the preliminary results that are discussed here are encouraging, and lead us to expect much in the future.

6. Contact address for the routine PSPMIN

If you wish to make your own experiments with the algorithm described in this paper, or if you wish to use it to solve some of your numerical optimization problems, please contact :

Ph.L. Toint,
Department of Applied Mathematics,
Facultes Universitaires de Namur,
8, Rempart de la Vierge,
B-5000 Namur (Belgium).

REFERENCES

[1] D.P. Bertsekas.
Projected Newton Methods for Optimization Problems with Simple Constraints.
SIAM Journal of Control and Optimization 20(2):221-246, 1982.

[2] J. Cullum and R.K. Brayton.
Some Remarks on the Symmetric Rank-One Update.
Journal of Optimization Theory and Applications 29(4):493-519, 1979.

[3] J.E. Dennis and H.H.W. Mei.
Two New Unconstrained Optimization Algorithms Which Use Function and Gradient Values.
Journal of Optimization Theory and Applications 28(4):453-482, 1979.

[4] Ph.E. Gill and W. Murray.
Conjugate Gradient Methods for Large Scale Nonlinear Optimization.
Technical Report SOL 79-15, Dept. of Operations Research, Stanford University, Stanford, 1979.

[5] Ph.E. Gill, W. Murray and M.H. Wright.
Practical Optimization.
Academic Press, London, 1981.

[6] A. Griewank and Ph.L. Toint.
Partitioned Variable Metric Updates for Large Structured Optimization Problems.
Numerische Mathematik (39):119-137, 1982.

[7] A. Griewank and Ph.L. Toint.
Local Convergence Analysis for Partitioned Quasi-Newton Updates .
Numerische Mathematik (39):429-448, 1982.

[8] A. Griewank and Ph.L. Toint.
On the Unconstrained Optimization of Partially Separable Functions.
In M.J.D. Powell (editor), Nonlinear Optimization 1981. Academic Press, New-York, 1982.

[9] A. Griewank and Ph.L. Toint.
On the Existence of Convex Decompositions of Partially Separable Functions.
Mathematical Programming to appear, 1983.

[10] W. Hock and K. Schittkowski.
Test Examples for Nonlinear Programming Codes.
Lectures Notes in Economics and Mathematical Systems 187, Springer Verlag, Berlin, 1981.

[11] H.Y. Huang.
Unified Approach to Quadratically Convergent Algorithms For Function Minimization.
Journal of Optimization Theory and Applications 5(6):405-423, 1970.

[12] E. Marwil.
Exploiting Sparsity in Newton-Like Methods.
PhD thesis, Cornell University, Ithaca, New-York, 1978.

[13] D.P. O'Leary.
A Discrete Newton Algorithm For Minimizing A Function of Many Variables.
Mathematical Programming 23:20-33, 1982.

[14] M.J.D. Powell and Ph.L. Toint.
The Shanno-Toint Procedure for Updating Sparse Symmetric Matrices.
I.M.A. Journal of Numerical Analysis 1:403-413, 1981.

[15] D. F. Shanno.
On Variable Metric Methods for Sparse Hessians.
Mathematics of Computation 34:499-514, 1980.

[16] D.F. Shanno and K.H. Phua.
Matrix Conditionning and Nonlinear Optimization.
Mathematical Programming 14, 1978.

[17] G.W. Stewart.
A Modification of Davidon's Minimization Method to Accept Difference Approximations of Derivatives.
Journal of the ACM 14, 1967.

[18] Ph.L. Toint.
On Sparse And Symmetric Matrix Updating Subject To A Linear Equation.
Mathematics of Computation 31:954-961, 1977.

[19] Ph.L. Toint.
On the Superlinear Convergence of an Algorithm for Solving a Sparse Minimization Problem.
SIAM Journal on Numerical Analysis 16:1036-1045, 1979.

THE NUMERICAL SOLUTION OF TOTAL ℓ_p APPROXIMATION PROBLEMS

G A Watson

1. Introduction

Many data fitting problems may be formulated as follows: given an $m \times n$ matrix A, with $m > n$, and a vector $\underset{\sim}{b} \in R^m$, find $\underset{\sim}{x} \in R^n$ to minimize $||\underset{\sim}{r}||$, where

$$\underset{\sim}{r} + \underset{\sim}{b} = A\underset{\sim}{x} . \tag{1.1}$$

It is typically assumed that the expected values of the components of $\underset{\sim}{r}$ are zero, and the particular norm used depends on the distribution of the errors represented by these quantities. For example, if the components r_i are independent and normally distributed, then the maximum likelihood estimation of the data is given by taking the norm to be the least squares norm. The ℓ_p norms $1 \le p < 2$, are also useful in the data fitting context, as less weight is given to isolated gross errors, or wild points, in the data. In this sense, the ℓ_1 norm may be regarded as giving the most robust estimation.

The model equations (1.1) have the underlying assumption that only the components of the vector $\underset{\sim}{b}$ are in error. However, it is often the case that the elements of A are also unreliable, for example if independent, as well as dependent variable values are inexact. One way in which account can be taken of this more general situation is to introduce perturbations also into the elements of A and to solve the following total approximation problem:

$$\text{find } \underset{\sim}{x} \in R^n \quad \text{to minimize} \quad ||E \vdots \underset{\sim}{r}|| \tag{1.2}$$

where

$$\underset{\sim}{r} + \underset{\sim}{b} = (A+E)\underset{\sim}{x}$$

and the norm is now an appropriate matrix norm. A modification of (1.2) is to the situation where some columns of A are known to be exact : for example if one of the basis functions of the linear model is a constant, then the corresponding column of A will not be in error. The more general situation will be considered here where the matrix A is partitioned so that

$$A = [A_1 \vdots A_2]$$

with $A_1 : R^k \to R^m$ (without loss of generality) assumed to be exact. Therefore the problem may be expressed as

$$\begin{aligned} &\text{minimize} && ||E \vdots \underset{\sim}{r}|| \\ &\text{subject to} && \underset{\sim}{r} + \underset{\sim}{b} = (A_1 \vdots A_2 + E)\underset{\sim}{x} \end{aligned} \tag{1.3}$$

where the norm is one on $m \times (n+1-k)$ matrices, and $\underset{\sim}{x} \in R^n$, $\underset{\sim}{r} \in R^m$ and $E : R^{n-k} \to R^m$ are the variables of the problem.

It is clear that (1.3) may be interpreted as an optimization problem with a (generally) non-differentiable objective function, and m nonlinear equality constraints. If a matrix is considered as a particular organization of the elements of an extended vector, then it is possible to make use of standard results from vector analysis in a straightforward manner. In particular, the subdifferential of $\|M\|$, where $M : R^t \to R^m$, is defined by

$$\partial\|M\| = \{G : R^t \to R^m;\ \|S\| \geq \|M\| + \text{trace}((S-M)^T G),$$
$$\forall S : R^t \to R^m\}.$$

It is readily shown that $G \in \partial\|M\|$ is equivalent to the pair of conditions

$$\text{(i)} \quad \|M\| = \text{trace}(G^T M), \tag{1.4}$$

$$\text{(ii)} \quad \|G\|^* \leq 1, \tag{1.5}$$

where

$$\|G\|^* = \max_{\|S\| \leq 1} \text{trace}(S^T G),$$

and $\|.\|^*$ is the norm dual to $\|.\|$.

It is easily verified that the Jacobian matrix of the constraints of (1.3) has full rank, and so multiplier relations may be developed. The appropriate Lagrangian function is

$$L = \|E \vdots \underset{\sim}{r}\| - \underset{\sim}{\lambda}^T(\underset{\sim}{r} + \underset{\sim}{b} - (A_1 \vdots A_2+E)\underset{\sim}{x}),$$

and necessary conditions for a solution to (1.3) are (for example, Hiriart-Urruty [4]) that there exists $G \in \partial\|E \vdots \underset{\sim}{r}\|$ such that

$$(A_1 \vdots A_2 + E)^T \underset{\sim}{\lambda} = \underset{\sim}{0} \tag{1.6}$$

$$(A_1 \vdots A_2 + E)\underset{\sim}{x} = \underset{\sim}{r} + \underset{\sim}{b} \tag{1.7}$$

$$G + \underset{\sim}{\lambda}\underset{\sim}{x}_2^T = 0, \tag{1.8}$$

where $\underset{\sim}{x}_2$ is the vector formed by setting

$$\begin{bmatrix} \underset{\sim}{x} \\ -1 \end{bmatrix} = \begin{bmatrix} \underset{\sim}{x}_1 \\ \underset{\sim}{x}_2 \end{bmatrix} \tag{1.9}$$

with $\underset{\sim}{x}_1 \in R^k$, $\underset{\sim}{x}_2 \in R^{n+1-k}$. Notice that the subgradient matrix G is a rank one matrix.

2. The total ℓ_p problem

Attention will now be focussed on the particular class of matrix norms defined on matrices $M : R^t \to R^m$ by

$$\|M\|_p = \left(\sum_{i=1}^{m} \sum_{j=1}^{t} |m_{ij}|^p\right)^{1/p} \qquad 1 \leq p < \infty \tag{2.1}$$

with m_{ij} the (i,j) component of M. This definition may be extended to the limiting case $p = \infty$ by defining

$$\|M\|_\infty = \max_{\substack{1\le i\le m\\ 1\le j\le t}} |m_{ij}| \ .$$

Perhaps the most commonly occurring member of this class of ℓ_p norms is the Frobenius norm, which corresponds to the case $p = 2$ and gives rise to the total least squares problem. Conventionally the subscript F (or sometimes E for Euclidean) is used to denote this matrix norm, with the subscript 2 being reserved for the matrix norm subordinate to the ℓ_2 vector norm. Other real number subscripts often have the subordinate norm interpretation, but for the sake of notational convenience, this convention will be abandoned.

An effective way in which the total ℓ_p problem may be tackled is to exploit its connection with the following constrained vector norm approximation problem. Define $Z : R^{n+1} \to R^m$ by

$$Z = [A \,\vdots\, \underset{\sim}{b}]$$

and let $\underset{\sim}{v} \in R^{n+1}$ be partitioned so that

$$\underset{\sim}{v} = \begin{bmatrix} \underset{\sim}{v}_1 \\ \underset{\sim}{v}_2 \end{bmatrix}$$

with $\underset{\sim}{v}_1 \in R^k$, $\underset{\sim}{v}_2 \in R^{n+1-k}$, where it is recalled that k is the number of exact columns of A. Then the problem referred to above is

$$\text{minimize} \quad \|Z\underset{\sim}{v}\|_p \quad \text{subject to} \quad \|\underset{\sim}{v}_2\|_q = 1 \ , \tag{2.2}$$

where the subscripts on the norms refer to the usual ℓ_p and ℓ_q vector norms, and where p and q are dual in the sense of satisfying the relationship

$$\frac{1}{p} + \frac{1}{q} = 1 \ .$$

The precise connection between (2.2) and (1.3) will now be explored. Notice first that (2.2) (like (1.3)) is not a convex optimization problem (the feasible region is in fact the outside of the unit ball) so local solutions may occur. Indeed it may only be possible to calculate a point satisfying appropriate necessary conditions for a local solution. It will now be shown that knowledge of such a point can be used directly to give variables satisfying the conditions (1.6)-(1.8); a preliminary lemma is required.

Lemma 1 [9] Let $\underset{\sim}{v} \in R^{n+1}$ solve (2.2). Then there exists $\underset{\sim}{w}_2 \in \partial\|\underset{\sim}{v}_2\|_q$, $\underset{\sim}{g} \in \partial\|Z\underset{\sim}{v}\|_p$ such that

$$Z^T\underset{\sim}{g} = \|Z\underset{\sim}{v}\|_p \begin{bmatrix} \underset{\sim}{0} \\ \underset{\sim}{w}_2 \end{bmatrix} . \tag{2.3}$$

Theorem 1 Let $\underset{\sim}{v} \in R^{n+1}$ with $v_{n+1} \neq 0$, and let $\underset{\sim}{w}_2 \in \partial\|\underset{\sim}{v}_2\|_q$, $\underset{\sim}{g} \in \partial\|Z\underset{\sim}{v}\|_p$ exist such that (2.3) holds. Then (1.6)-(1.8) are satisfied by taking

$$\underset{\sim}{v} = \tau \begin{bmatrix} \underset{\sim}{x} \\ -1 \end{bmatrix} \tag{2.4}$$

$$[E\vdots r] = -Zvw_2^T \tag{2.5}$$

$$\lambda = \tau g \ . \tag{2.6}$$

Proof Since by assumption $v_{n+1} \neq 0$, $x \in R^n$ and τ are uniquely defined by (2.4). Also by assumption $w_2 \in \partial\|v_2\|_q$, $g \in \partial\|Zv\|_p$ exist satisfying (2.3), which may be written

$$[A_1 \vdots A_2 \vdots b]^T\lambda = \tau\|Zv\|_p \begin{bmatrix} 0 \\ w_2 \end{bmatrix}, \tag{2.7}$$

choosing λ to satisfy (2.6). It follows that

$$A_I^T\lambda = 0 \ . \tag{2.8}$$

Let $[E\vdots r]$ be defined by (2.5). Then

$$\begin{aligned}[A_2 + E \vdots b + r]^T\lambda &= [A_2 \vdots b]^T\lambda + [E \vdots r]^T\lambda \ , \\ &= \tau\|Zv\|_p w_2 - v^TZ^T\lambda\, w_2 \qquad \text{using (2.7)} \\ &= 0 \ , \end{aligned} \tag{2.9}$$

where the relationship

$$g^TZv = \|Zv\|_p$$

(the vector analogue of (1.4)) has been used; (1.6) follows from (2.8) and (2.9).

Next

$$\begin{aligned}(A_1 \vdots A_2+E)x - r - b &= Ax + [E \vdots r]x_2 - b \\ &= Z\begin{bmatrix} x_1 \\ x_2 \end{bmatrix} + [E \vdots r]x_2 \\ &= Z\begin{bmatrix} x_1 \\ x_2 \end{bmatrix} - \frac{1}{\tau} Zvw_2^Tv_2 \\ &= 0 \end{aligned}$$

which establishes (1.7). Finally it is necessary to show that $G \in \partial\|E\vdots r\|$ if G is given by (1.8), which will be done by showing that (1.4) and (1.5) are satisfied. Notice that the matrix norm dual to the ℓ_p norm is the ℓ_q norm, where $\frac{1}{p} + \frac{1}{q} = 1$. Then

$$\begin{aligned}\|G\|_q &= \frac{1}{\tau}\|\lambda\|_q\|v_2\|_q \\ &= \|g\|_q \ \|v_2\|_q \\ &\leq 1 \ . \end{aligned}$$

Also

$$\begin{aligned}\text{trace}\,(G^T[E\vdots r]) &= \frac{1}{\tau}\,\text{trace}\,(v_2\lambda^TZvw_2^T) \\ &= \|Zv\|_p\|v_2\|_q \\ &= \|E \vdots r\|_p \ . \end{aligned}$$

Thus $G \in \partial \| [E \vdots \underset{\sim}{r}] \|$, and the proof is concluded. □

Remark A slightly stronger result than Lemma 1 is in fact proved in [9] : it is shown that for every $\underset{\sim}{w}_2 \in \partial \| \underset{\sim}{v}_2 \|_q$ there exists $\underset{\sim}{g} \in \partial \| Z\underset{\sim}{v} \|_p$ such that (2.3) holds. Thus $[E \vdots \underset{\sim}{r}]$ will not normally be uniquely defined by (2.5) unless $\partial \| \underset{\sim}{v}_2 \|_q$ is a singleton.

As indicated previously, in general it will not be possible to guarantee the calculation of a global solution to (2.2). However, if such a solution is obtained, it results in a corresponding global solution to (1.3); it may be helpful to isolate the following result as a preliminary lemma.

Lemma 2 Let $M : R^t \to R^m$, and $\underset{\sim}{s} \in R^t$ be such that $\| \underset{\sim}{s} \|_q = 1$. Then

$$\| M\underset{\sim}{s} \|_p \le \| M \|_p .$$

Proof If $\underset{\sim}{m}_i^T$ denotes the ith row of M,

$$\| M\underset{\sim}{s} \|_p^p = \sum_{i=1}^{m} | \underset{\sim}{m}_i^T \underset{\sim}{s} |^p$$

$$\le \sum_{i=1}^{m} \| \underset{\sim}{m}_i \|_p^p \| \underset{\sim}{s} \|_q^p$$

$$= \| M \|_p^p .$$

□

Theorem 2 Let $\underset{\sim}{v} \in R^{n+1}$ solve (2.2), with $v_{n+1} \ne 0$, and let $\underset{\sim}{w}_2 \in \| \underset{\sim}{v}_2 \|_q$. Then $\underset{\sim}{x}$, $[E \vdots \underset{\sim}{r}]$ given by (2.4), (2.5) solve (1.3).

Proof Let $\bar{\underset{\sim}{x}}$, $[\bar{E} \vdots \bar{\underset{\sim}{r}}]$ satisfy the constraints of (1.3), and let $\bar{\underset{\sim}{v}} = \tau \begin{bmatrix} \bar{\underset{\sim}{x}} \\ -1 \end{bmatrix}$, with τ chosen so that $\| \bar{\underset{\sim}{v}}_2 \|_q = 1$. Then

$$Z\bar{\underset{\sim}{v}} + [\bar{E} \vdots \bar{\underset{\sim}{r}}]\bar{\underset{\sim}{v}}_2 = \underset{\sim}{0}$$

so that

$$\| Z\bar{\underset{\sim}{v}} \|_p = \| [\bar{E} \vdots \bar{\underset{\sim}{r}}]\bar{\underset{\sim}{v}}_2 \|_p$$

$$\le \| \bar{E} \vdots \bar{\underset{\sim}{r}} \|_p \text{ , by Lemma 2.}$$

Since $\| Z\underset{\sim}{v} \|_p = \| E \vdots \underset{\sim}{r} \|_p$, the result follows. □

Remark The condition $v_{n+1} \ne 0$ in these theorems is clearly necessary for $\underset{\sim}{x}$ satisfying (2.4) to exist. In particular, when $\underset{\sim}{v}$ is a global solution to (2.2), it represents an existence condition on solutions to (1.3) : there always exists a matrix minimizing $\| E \vdots \underset{\sim}{r} \|$, but not necessarily a vector $\underset{\sim}{x}$ satisfying the corresponding constraints of (1.3).

3. The weighted total ℓ_p problem

The results of the previous section may be extended to the class of norms defined by

$$\| E \vdots \underset{\sim}{r} \| = \| D[E \vdots \underset{\sim}{r}]T_2 \|_p \tag{3.1}$$

where $D : R^m \to R^m$ and $T_2 : R^{n+1-k} \to R^{n+1-k}$ are positive diagonal matrices. Let

$$T = \begin{bmatrix} T_1 & \vdots & \\ \cdots & \cdots & \cdots \\ & \vdots & T_2 \end{bmatrix}$$

where $T_1 : R^k \to R^k$ is any non-singular matrix (for example I_k). Then the total approximation problem (1.3) with norm given by (3.1) again has a close relationship with a problem of the form (2.2), in this case

$$\text{minimize} \quad \| DZT\underset{\sim}{v} \|_p \quad \text{subject to} \quad \| \underset{\sim}{v}_2 \|_q = 1 \; . \tag{3.2}$$

A key result is the expression for the dual norm. By definition, for any matrix $M : R^{n+1-k} \to R^m$,

$$\begin{aligned} \| M \|^* &= \max_{\| S \| \le 1} \quad \text{trace } (S^T M) \\ &= \max_{\| DST_2 \|_p \le 1} \quad \text{trace } (S^T M) \\ &= \max_{\| DST_2 \|_p \le 1} \quad \text{trace } (T_2^{-1} (DST_2)^T D^{-1} M) \\ &= \max_{\| DST_2 \|_p \le 1} \quad \text{trace } ((DST_2)^T D^{-1} M T_2^{-1}) \\ &= \| D^{-1} M T_2^{-1} \|_q \; . \end{aligned}$$

The following result may be established by a proof which parallels that of Theorem 1.

<u>Theorem 3</u> Let $\underset{\sim}{v} \in R^{n+1}$ with $v_{n+1} \neq 0$, and let $\underset{\sim}{w}_2 \in \partial \| \underset{\sim}{v}_2 \|_q$, $\underset{\sim}{g} \in \partial \| DZT\underset{\sim}{v} \|_p$ exist such that

$$(DZT)^T \underset{\sim}{g} = \| DZT\underset{\sim}{v} \|_p \begin{bmatrix} \underset{\sim}{0} \\ \underset{\sim}{w}_2 \end{bmatrix} .$$

Then (1.6)-(1.8) are satisfied by taking

$$\begin{aligned} T\underset{\sim}{v} &= \tau \begin{bmatrix} \underset{\sim}{x} \\ -1 \end{bmatrix} \\ [E \vdots \underset{\sim}{r}] &= -ZT\underset{\sim}{v}\underset{\sim}{w}_2^T T_2^{-1} \\ \underset{\sim}{\lambda} &= \tau D\underset{\sim}{g} \; . \end{aligned}$$

It does not appear possible to relax the condition that the matrices D and T_2 be diagonal without losing the relationship with a problem of the form of (2.2). As an example, let $k = 0$, and define a norm on matrices $M : R^{n+1} \to R^m$ by

$$\|M\|^2 = \sum_{j=1}^{n+1} \kappa_j(M)^T W_j \kappa_j(M) \tag{3.3}$$

where $\kappa_j(M)$ denotes the jth column of M, and W_j, $j = 1,2,\ldots,n+1$ are given positive definite variance-covariance matrices. The Lagrangian function for the problem (1.2) with this norm may be taken as

$$L = \|E\vdots \underset{\sim}{r}\|^2 - \underset{\sim}{\lambda}^T(\underset{\sim}{r} + \underset{\sim}{b} - (A+E)\underset{\sim}{x}).$$

Then, exploiting the differentiability of L, necessary conditions for a solution to (1.2) are that

$$(A+E)^T\underset{\sim}{\lambda} = \underset{\sim}{0} \tag{3.4}$$

$$(A+E)\underset{\sim}{x} = \underset{\sim}{r} + \underset{\sim}{b} \tag{3.5}$$

$$[E\vdots \underset{\sim}{r}] = \tfrac{1}{2}[x_1 W_1^{-1}\underset{\sim}{\lambda}, \ldots, x_n W_n^{-1}\underset{\sim}{\lambda}, - W_{n+1}^{-1}\underset{\sim}{\lambda}] . \tag{3.6}$$

Notice that $[E\vdots \underset{\sim}{r}]$ does not have rank one unless $W_j = \alpha_j I$, $j = 1,2,\ldots,n+1$. It follows from (3.5) and (3.6) that

$$\underset{\sim}{\lambda} = 2H(\underset{\sim}{x})^{-1}(\underset{\sim}{b} - A\underset{\sim}{x}) \tag{3.7}$$

where

$$H(\underset{\sim}{x}) = \sum_{i=1}^{n} x_i^2 W_i^{-1} + W_{n+1}^{-1} ,$$

and therefore

$$\|E\vdots \underset{\sim}{r}\|^2 = (A\underset{\sim}{x} - \underset{\sim}{b})^T H(\underset{\sim}{x})^{-1}(A\underset{\sim}{x} - \underset{\sim}{b}) . \tag{3.8}$$

Direct calculation shows that the gradient vector of this expression is $(A+E)^T\underset{\sim}{\lambda}$, so that a point $\underset{\sim}{x}$ minimizing (3.8) will result in (3.4) being satisfied. Let

$$\underset{\sim}{v} = \tau\begin{bmatrix}\underset{\sim}{x}\\ -1\end{bmatrix}, \quad \tau > 0 . \tag{3.9}$$

Then (3.8) can be written as

$$f(\underset{\sim}{v}) = \underset{\sim}{v}^T Z^T G(\underset{\sim}{v})^{-1} Z\underset{\sim}{v} \tag{3.10}$$

where

$$G(\underset{\sim}{v}) = \sum_{j=1}^{n+1} v_j^2 W_j^{-1} .$$

If $\underset{\sim}{v}$ minimizes (3.10) with $v_{n+1} \neq 0$, and $\underset{\sim}{x}$ satisfies (3.9), the necessary conditions for a solution to (1.2) will be satisfied. Clearly $f(\underset{\sim}{v})$ is homogeneous of degree 0 in $\underset{\sim}{v}$; however it does not split naturally into a numerator and denominator, and so does not result in a problem of the form (2.2), unless $W_j = \alpha_j I$, $j = 1,2,\ldots,n+1$. There appears to be no advantage in dealing with (3.10) in this case, and the direct minimization of (3.8) using a quasi-Newton method, for example, may be the way to proceed.

4. Solving total ℓ_p problems, $1 < p < \infty$

The assumption will now be made that the total ℓ_p problem can be 'transformed' into a problem of the form (2.2), and so attention will be concentrated on the solution of that particular problem. It will also be assumed that Z has full column rank, so that the optimal value of the objective function in (2.2) is positive.

The case $p = q = 2$ in (2.2) corresponds to the total least squares problem. In particular if $k = 0$, (2.2) can be written in the form

$$\text{minimize} \quad \underset{\sim}{v}^T Z^T Z \underset{\sim}{v} \quad \text{subject to} \quad \underset{\sim}{v}^T \underset{\sim}{v} = 1 , \tag{4.1}$$

and the problem reduces to that of finding the smallest eigenvalue of $Z^T Z$, and corresponding eigenvector, or equivalently the smallest singular value of Z, and corresponding right singular vector. An analysis of this problem, and a method of solution based on the singular value decomposition of Z is given in [3].

When $k > 0$, the eigenvalue problem becomes the generalized eigenvalue problem

$$Z^T Z \underset{\sim}{v} = \mu C \underset{\sim}{v}$$

where $C = \text{diag}\{0,0,\ldots,0,1,1,\ldots,1\}$, there being k zero diagonal elements. Only $n+1-k$ finite eigenvalues exist, and the problem may be reduced to a standard $(n+1-k)$ dimensional eigenvalue problem by well-known techniques (see, for example [6]). A method for this problem based on reduction to a total least squares problem in $(n+1-k)$ variables has been developed by R Byers and the authors of [3], and will be published subsequently.

For all values of p in the range $1 < p < \infty$, (2.2) may be written in the form

$$\text{minimize} \quad \| Z\underset{\sim}{v} \|_p^p \quad \text{subject to} \quad \| \underset{\sim}{v}_2 \|_q^q = 1 \tag{4.2}$$

which has differentiable objective function and constraint. From an algorithmic point of view, it is important to be able to define the following diagonal matrices

$$D(\underset{\sim}{v}) = \text{diag}\{ |(Z\underset{\sim}{v})_i|^{p-2} , \quad i = 1,2,\ldots,m\}$$

$$C(\underset{\sim}{v}) = \text{diag}\{0,0,\ldots,0, |v_{k+1}|^{q-2}, \ldots, |v_{n+1}|^{q-2}\} ,$$

which will be assumed to exist for all $\underset{\sim}{v}$ of interest. Strictly from a data fitting point of view, the important values of p satisfy $p \leq 2$, for which $q \geq 2$ and the existence of $C(\underset{\sim}{v})$ is not an issue. When $p = 1$, the solution to (4.2) is characterized by certain zero components of $Z\underset{\sim}{v}$ (see [5] and Section 5) so it is the case that some elements of $D(\underset{\sim}{v})$ will become increasingly large as p gets closer to 1; however for reasonable values of p, it is possible to work with $D(\underset{\sim}{v})$ except in pathological cases. The problem (4.2) may then be written

$$\text{minimize} \quad \underset{\sim}{v}^T J(\underset{\sim}{v}) \underset{\sim}{v} \quad \text{subject to} \quad \underset{\sim}{v}^T C(\underset{\sim}{v}) \underset{\sim}{v} = 1 \tag{4.3}$$

where

$$J(\underset{\sim}{v}) = Z^T D(\underset{\sim}{v}) Z .$$

The Kuhn-Tucker first order necessary conditions for $\underset{\sim}{v}^*$ to solve (4.3) (corresponding to Lemma 1) are that there exists a scalar μ such that

$$p\,J(\underset{\sim}{v}^*)\underset{\sim}{v}^* - \mu q\; C(\underset{\sim}{v}^*)\underset{\sim}{v}^* = \underset{\sim}{0} \quad . \tag{4.4}$$

Thus $\lambda^* = \mu q/p$ is an eigenvalue of the generalized eigenvalue problem

$$J(\underset{\sim}{v}^*)\underset{\sim}{v} = \lambda C(\underset{\sim}{v}^*)\underset{\sim}{v} \tag{4.5}$$

with eigenvector $\underset{\sim}{v}^*$ (normalized so that $\underset{\sim}{v}^{*T}C(\underset{\sim}{v}^*)\underset{\sim}{v}^* = 1$). Premultiplying by $\underset{\sim}{v}^{*T}$ shows that

$$\lambda^* = \|Z\underset{\sim}{v}^*\|_p^p > 0 \quad \text{by assumption.}$$

Without loss of generality, let the non-zero components of $\underset{\sim}{v}_2^*$ be in the last $s+1$ places. Letting $\ell = n-s$, the problem (4.5) has s other finite eigenvalues, say $\lambda_1, \lambda_2, \ldots, \lambda_s$, and corresponding eigenvectors satisfying

$$J(\underset{\sim}{v}^*)\begin{bmatrix}\underset{\sim}{x}_i\\ \underset{\sim}{y}_i\end{bmatrix} = \lambda_i C(\underset{\sim}{v}^*)\begin{bmatrix}\underset{\sim}{x}_i\\ \underset{\sim}{y}_i\end{bmatrix} \quad , \quad i = 1,2,\ldots,s, \tag{4.6}$$

where $\underset{\sim}{x}_i \in R^\ell$ and $\underset{\sim}{y}_i \in R^{s+1}$, and we may choose the eigenvectors so that

$$[\underset{\sim}{x}_i^T\ \underset{\sim}{y}_i^T]\; C(\underset{\sim}{v}^*)\begin{bmatrix}\underset{\sim}{x}_i\\ \underset{\sim}{y}_i\end{bmatrix} = \delta_{ij} \quad , \quad i,j = 1,2,\ldots,s \tag{4.7}$$

$$[\underset{\sim}{x}_i^T\ \underset{\sim}{y}_i^T]\; C(\underset{\sim}{v}^*)\underset{\sim}{v}^* = 0 \quad , \quad i = 1,2,\ldots,s \;. \tag{4.8}$$

If $J(\underset{\sim}{v}^*)$ is partitioned so that

$$J(\underset{\sim}{v}^*) = \begin{bmatrix} J_{11}^* & \vdots & J_{12}^* \\ \cdots & \cdots & \cdots \\ J_{12}^{*T} & \vdots & J_{22}^* \end{bmatrix} \begin{matrix} \ell \updownarrow \\ \\ (s+1) \updownarrow \end{matrix} \quad \text{(columns: } \leftarrow\ell\rightarrow \ \ \leftarrow(s+1)\rightarrow) \tag{4.9}$$

and X and Y are defined by

$$X = [\underset{\sim}{x}_1\ \underset{\sim}{x}_2\ \ldots\ \underset{\sim}{x}_s]$$

$$Y = [\underset{\sim}{y}_1\ \underset{\sim}{y}_2\ \ldots\ \underset{\sim}{y}_s] \;,$$

it follows using (4.6) and (4.7) that

$$Y^T J_{12}^{*T} X + Y^T J_{22}^* Y = \mathrm{diag}\{\lambda_1, \lambda_2, \ldots, \lambda_s\} \quad . \tag{4.10}$$

The following relationship between eigenvalues of (4.5) is required subsequently.

<u>Lemma 3</u> Let $\underset{\sim}{v}^*$ solve (2.2). Then

$$\frac{\lambda^*}{\lambda_i} \le \frac{p-1}{q-1} \quad , \quad i = 1,2,\ldots,s \;. \tag{4.11}$$

<u>Proof</u> If $\underset{\sim}{v}^*$ solves (2.2), then exploiting the simple form of derivatives of (4.3), second order necessary conditions (see, for example [2]) are

$$\tilde{d}^T((p-1)J(\tilde{v}^*) - \lambda^*(q-1)C(\tilde{v}^*))\tilde{d} \geq 0 \tag{4.12}$$

for all $\tilde{d} : \tilde{d}^T C(\tilde{v}^*)\tilde{v}^* = 0$. It follows using (4.8) that

$$[\tilde{d}_1^T : \tilde{t}^T Y^T] [(p-1)J(\tilde{v}^*) - \lambda^*(q-1)C(\tilde{v}^*)] \begin{bmatrix} \tilde{d}_1 \\ Y\tilde{t} \end{bmatrix} \geq 0$$

for all $\tilde{d}_1 \in R^{\ell}$, $\tilde{t} \in R^s$, or equivalently that the matrix M defined by

$$M = \left[\begin{array}{c:c} J_{11}^* & J_{12}^* Y \\ \hdashline Y J_{12}^{*T} & Y^T J_{22}^* Y - \frac{\lambda^*(q-1)}{p-1} I \end{array}\right] \tag{4.13}$$

is positive semi-definite. If $\tilde{e}_i \in R^s$ denotes the ith co-ordinate vector,

$$M\begin{bmatrix} \tilde{x}_i \\ \tilde{e}_i \end{bmatrix} = \begin{bmatrix} J_{11}^*\tilde{x}_i + J_{12}^*\tilde{y}_i \\ Y^T J_{12}^{*T}\tilde{x}_i + Y^T J_{22}^*\tilde{y}_i - \lambda^* \frac{q-1}{p-1}\tilde{e}_i \end{bmatrix}, \quad i = 1,2,\ldots,s$$

$$= \begin{bmatrix} \tilde{0} \\ (\lambda_i - \lambda^* \frac{q-1}{p-1})\tilde{e}_i \end{bmatrix} \qquad i = 1,2,\ldots,s\,,$$

where (4.10) has been used. Thus the conditions

$$[\tilde{x}_i^T \; \tilde{e}_i^T] M \begin{bmatrix} \tilde{x}_i \\ \tilde{e}_i \end{bmatrix} \geq 0 \quad , \quad i = 1,2,\ldots,s$$

imply that

$$\lambda_i \geq \lambda^* \frac{q-1}{p-1} \quad , \quad i = 1,2,\ldots,s$$

and the result is proved. □

<u>Corollary 1</u> The assumed relationship between p and q gives (4.11) as

$$\frac{\lambda^*}{\lambda_i} \leq (p-1)^2 \quad , \quad i = 1,2,\ldots,s\,.$$

<u>Corollary 2</u> If second order sufficiency conditions hold at $\tilde{v}^*$ (see for example [2]), then strict inequality holds in (4.11).

<u>Corollary 3</u> The generalized eigenvalues λ^*, λ_i, $i = 1,2,\ldots,s$ of (4.5) are positive, and such that $\lambda^* < \lambda_i$, $i = 1,2,\ldots,s$ if $1 < p < 2$.

In order to satisfy (4.4), consider the simple iteration

$$\left.\begin{aligned} J(\tilde{v}^{(i)})\tilde{d} &= C(\tilde{v}^{(i)})\tilde{v}^{(i)} \\ \tilde{v}^{(i+1)} &= \tilde{d}/\|\tilde{d}_2\|_q \end{aligned}\right\} \quad i = 0,1,\ldots \tag{4.14}$$

which may be implemented provided that $\tilde{v}_2^{(0)} \neq \tilde{0}$ and $J(\tilde{v}^{(i)})$ is non-singular for all $i = 0,1,\ldots$: this will be assumed. The vector $\underset{\sim}{d}_2$ is formed from $\underset{\sim}{d}$ by removing the first k components. Numerically, this system of equations may be solved by first forming the QR factors of $D^{\frac{1}{2}}(\underset{\sim}{v}^{(i)})Z$, and then backward and forward substitution. The iteration scheme (4.14) may be regarded as a generalization of inverse iteration, which it is if $p = q = 2$ and $k = 0$.

Theorem 4 Sufficient conditions for (4.14) to converge locally to $\underset{\sim}{v}^*$ satisfying (4.4) are

$$\frac{p-3}{q-1} < \frac{\lambda^*}{\lambda_i} < \frac{p-1}{q-1} \quad , \quad i = 1,2,\ldots,s \tag{4.15}$$

and, if $\ell > 0$, $p < 3$.

Proof It is consistent with (4.6) to write

$$\gamma_i J(\underset{\sim}{v}^*)\begin{bmatrix}\underset{\sim}{x}_i\\ \underset{\sim}{y}_i\end{bmatrix} = C(\underset{\sim}{v}^*)\begin{bmatrix}\underset{\sim}{x}_i\\ \underset{\sim}{y}_i\end{bmatrix} \quad , \quad i = 1,2,\ldots,n+1 \ , \tag{4.16}$$

where

$$\gamma_i = \frac{1}{\lambda_i} \quad , \quad i = 1,2,\ldots,s \ ,$$

$$\gamma_i = 0 \quad , \quad i = s+1,\ldots,n \ ,$$

$$\gamma_{n+1} = \frac{1}{\lambda^*} \ ,$$

$$[\underset{\sim}{x}_{n+1}{}^T \ \underset{\sim}{y}_{n+1}{}^T] = \underset{\sim}{v}^{*T}$$

and the eigenvectors satisfy

$$[\underset{\sim}{x}_i{}^T \ \underset{\sim}{y}_i{}^T]J(\underset{\sim}{v}^*)\begin{bmatrix}\underset{\sim}{x}_i\\ \underset{\sim}{y}_i\end{bmatrix} = 0 \ , \quad i,j = 1,2,\ldots,n+1 \ , \quad i \neq j \ . \tag{4.17}$$

Let
$$\underset{\sim}{v} = \underset{\sim}{v}^* + \sum_{i=1}^{n} \theta_i \begin{bmatrix}\underset{\sim}{x}_i\\ \underset{\sim}{y}_i\end{bmatrix}$$

$$= \underset{\sim}{v}^* + \underset{\sim}{\varepsilon} \quad , \text{ say} \ ,$$

where the numbers θ_i, $i = 1,2,\ldots,n$ are small. Then if $\underset{\sim}{\varepsilon} = \begin{bmatrix}\underset{\sim}{\varepsilon}_1\\ \underset{\sim}{\varepsilon}_2\end{bmatrix}$, with $\underset{\sim}{\varepsilon}_1 \in R^k$,

$$\|\underset{\sim}{v}_2\|_q^q = \|\underset{\sim}{v}_2^* + \underset{\sim}{\varepsilon}_2\|_q^q$$

$$= \|\underset{\sim}{v}_2^*\|_q^q + q\,\underset{\sim}{\varepsilon}^T C(\underset{\sim}{v}^*)\underset{\sim}{v}^* + O(\|\underset{\sim}{\varepsilon}\|^2)$$

$$= 1 + q \sum_{i=1}^{n} \theta_i [\underset{\sim}{x}_i{}^T \ \underset{\sim}{y}_i{}^T] C(\underset{\sim}{v}^*)\underset{\sim}{v}^* + O(\|\underset{\sim}{\varepsilon}\|^2)$$

$$= 1 + O(\|\underset{\sim}{\varepsilon}\|^2), \text{ using (4.8) and (4.17)},$$

showing that $\underset{\sim}{v}_2$ is correctly normalized to first order in $\underset{\sim}{\varepsilon}$.

Now define $\underset{\sim}{\delta}$ by

$$J(\underset{\sim}{v})(\underset{\sim}{v}^* + \underset{\sim}{\delta}) = \alpha\, C(\underset{\sim}{v})(\underset{\sim}{v}^* + \underset{\sim}{\varepsilon}) \tag{4.18}$$

where α is the constant so that

$$\underset{\sim}{\delta} = \sum_{i=1}^{n} \phi_i \begin{bmatrix} \underset{\sim}{x}_i \\ \underset{\sim}{y}_i \end{bmatrix} .$$

As before, if $\underset{\sim}{\delta} = \begin{bmatrix} \underset{\sim}{\delta}_1 \\ \underset{\sim}{\delta}_2 \end{bmatrix}$, with $\underset{\sim}{\delta}_1 \in R^k$, $\underset{\sim}{v}_2^* + \underset{\sim}{\delta}_2$ is correctly normalized to first order in $\underset{\sim}{\delta}$. In addition, $\|\underset{\sim}{\delta}\| \sim \|\underset{\sim}{\varepsilon}\|$ since $J(\underset{\sim}{v}^*)$ is non-singular. Then Taylor expansion of both sides of (4.18) about $\underset{\sim}{v}^*$ gives

$$J(\underset{\sim}{v}^*)(\underset{\sim}{v}^* + \underset{\sim}{\delta}) + (p-2)J(\underset{\sim}{v}^*)\underset{\sim}{\varepsilon} = \alpha\, C(\underset{\sim}{v}^*)(\underset{\sim}{v}^* + \underset{\sim}{\varepsilon})$$
$$+ \alpha(q-2)C(\underset{\sim}{v}^*)\underset{\sim}{\varepsilon} + O(\|\underset{\sim}{\varepsilon}\|^2).$$

Equating zero order terms gives

$$\alpha = \lambda^* ,$$

and equating first order terms gives

$$J(\underset{\sim}{v}^*)\Big(\sum_{i=1}^{n} \phi_i \begin{bmatrix} \underset{\sim}{x}_i \\ \underset{\sim}{y}_i \end{bmatrix} + (p-2) \sum_{i=1}^{n} \theta_i \begin{bmatrix} \underset{\sim}{x}_i \\ \underset{\sim}{y}_i \end{bmatrix} \Big)$$
$$= \lambda^*(q-1)C(\underset{\sim}{v}^*) \sum_{i=1}^{n} \theta_i \begin{bmatrix} \underset{\sim}{x}_i \\ \underset{\sim}{y}_i \end{bmatrix} .$$

Thus

$$J(\underset{\sim}{v}^*)\Big(\phi_i \begin{bmatrix} \underset{\sim}{x}_i \\ \underset{\sim}{y}_i \end{bmatrix} + (p-2)\theta_i \begin{bmatrix} \underset{\sim}{x}_i \\ \underset{\sim}{y}_i \end{bmatrix}\Big) = \lambda^*(q-1)C(\underset{\sim}{v}^*)\theta_i \begin{bmatrix} \underset{\sim}{x}_i \\ \underset{\sim}{y}_i \end{bmatrix} , \quad i = 1,2,\ldots,n .$$

It follows that

$$\lambda_i(\phi_i + (p-2)\theta_i) = \lambda^*(p-1)\theta_i , \quad i = 1,2,\ldots,s$$

or

$$\phi_i/\theta_i = (\lambda^*(q-1) - (p-2)\lambda_i)/\lambda_i , \quad i = 1,2,\ldots,s .$$

Also

$$C(\underset{\sim}{v}^*) \begin{bmatrix} \underset{\sim}{x}_i \\ \underset{\sim}{y}_i \end{bmatrix} = \underset{\sim}{0} , \quad i = s+1,\ldots,n ,$$

so that

$$\phi_i/\theta_i = -(p-2) , \quad i = s+1,\ldots,n .$$

Since local convergence is implied by $|\phi_i/\theta_i| < 1$, $i = 1,2,\ldots,n$, the result follows. Γ

Corollary 1 Let second order sufficiency conditions hold at $\underset{\sim}{v}^*$. Then local convergence is guaranteed for

$$p < 3 + \frac{\lambda^*}{\lambda_i}(q-1) \ , \quad i = 1,2,\ldots,s \ ,$$

or

$$p < 3 \ \text{ if } \ \ell > 0 \ .$$

Corollary 2 Let $\beta = \min\limits_{1\le i\le s} \lambda^*/\lambda_i$, and second order sufficiency conditions hold at $\underset{\sim}{v}^*$. Then (exploiting the relationship between p and q), local convergence is guaranteed for

$$p < 2 + \sqrt{1+\beta}$$

or

$$p < 3 \ \text{ if } \ \ell > 0.$$

It is in fact possible to make the method robust (for any value of p) by exploiting the fact that the direction of progress implicit in the iteration (4.14) is a feasible descent direction for $\|Z\underset{\sim}{v}\|_p^p$ at $\underset{\sim}{v}^{(i)}$ (in the sense that it can lead to a point satisfying the constraints of (4.3) and having a lower function value).

Theorem 5 Given $\underset{\sim}{v}$, $\|\underset{\sim}{v}_2\|_q = 1$ with $J(\underset{\sim}{v})$ non-singular, let $\underset{\sim}{s}$ be defined by

$$J(\underset{\sim}{v})\underset{\sim}{d} = C(\underset{\sim}{v})\underset{\sim}{v}$$

$$\underset{\sim}{s} = \underset{\sim}{d}/\underset{\sim}{d}^T C(\underset{\sim}{v})\underset{\sim}{v} - \underset{\sim}{v} \ . \tag{4.19}$$

Then unless $\underset{\sim}{v}$ satisfies (4.4), $\underset{\sim}{s}$ is a feasible descent direction for $\|Z\underset{\sim}{v}\|_p^p$ at $\underset{\sim}{v}$.

Proof $\underset{\sim}{s}$ is well-defined, since

$$\underset{\sim}{d}^T C(\underset{\sim}{v})\underset{\sim}{v} = \underset{\sim}{d}^T J(\underset{\sim}{v})\underset{\sim}{d} > 0 \ .$$

Taylor expansion gives

$$\|Z(\underset{\sim}{v} + \gamma\underset{\sim}{s})\|_p^p = \|Z\underset{\sim}{v}\|_p^p + \gamma p\, \underset{\sim}{v}^T J(\underset{\sim}{v})\underset{\sim}{s} + O(\gamma^2) \ . \tag{4.20}$$

Further

$$\underset{\sim}{v}^T J(\underset{\sim}{v})\underset{\sim}{s} = \frac{\underset{\sim}{v}^T J(\underset{\sim}{v})\underset{\sim}{d}}{\underset{\sim}{d}^T C(\underset{\sim}{v})\underset{\sim}{v}} - \underset{\sim}{v}^T J(\underset{\sim}{v})\underset{\sim}{v}$$

$$= \frac{1 - (\underset{\sim}{d}^T J(\underset{\sim}{v})\underset{\sim}{d})(\underset{\sim}{v}^T J(\underset{\sim}{v})\underset{\sim}{v})}{\underset{\sim}{d}^T J(\underset{\sim}{v})\underset{\sim}{d}}$$

$$\le 0 \ ,$$

using the Cauchy-Schwartz inequality, with strict inequality unless $\underset{\sim}{d}$ and $\underset{\sim}{v}$ are parallel, when $\underset{\sim}{v}$ satisfies (4.4). Finally, using convexity

$$\|\underset{\sim}{v}_2 + \gamma \underset{\sim}{s}_2\|_q^q \geq \|\underset{\sim}{v}_2\|_q^q + \gamma q \underset{\sim}{s}^T C(\underset{\sim}{v})\underset{\sim}{v} = 1$$

and so renormalizing can only further reduce the objective function value, and the proof is complete. □

This result shows that defining search directions by (4.19) and incorporating a sensible line search procedure can give a method which will converge to a point satisfying first order necessary conditions for a solution to (4.3) from any feasible starting point, for any $p > 1$. When the conditions of Corollary 1 of Theorem 4 hold, it must eventually be possible to take unit steps. In practice, for values of p satisfying $1 < p < 2+\delta$, where δ depends on the problem but may be about 0.6 or 0.7, the method (4.14) appears to converge from any feasible starting point; however from a theoretical point of view, global convergence is an open question when $p \neq 2$.

Most of the results of this section are stated in terms of both p and q, without use being made of the duality relationship. These results are therefore valid for a wider class of problems (2.2), where the assumption is merely that $1 < p, q < \infty$. In particular, the case of orthogonal ℓ_p approximation ($q = 2$) considered in [7], [8] is covered by the analysis given here. The relevance of these more general problems to the total approximation problem (1.3) is illustrated by the following theorem.

Theorem 6 [9] Let p,q be arbitrary with $1 \leq p, q \leq \infty$, let $\underset{\sim}{v}$ solve (2.2) with $v_{n+1} \neq 0$, and let $\underset{\sim}{w}_2 \in \partial\|\underset{\sim}{v}_2\|_q$. Then $\underset{\sim}{x}$, $[E \vdots \underset{\sim}{r}]$ given by (2.4), (2.5) solve the total approximation problem (1.3) with norm given by

$$\|M\| = \max_{\|\underset{\sim}{d}\|_q = 1} \|M\underset{\sim}{d}\|_p .$$

This theorem may appear to be in conflict with what has gone before when $\frac{1}{p} + \frac{1}{q} = 1$. However the apparent inconsistancy is resolved by observing that the minimizing matrix $[E \vdots \underset{\sim}{r}]$ is a rank one matrix, and for such a matrix

$$\begin{aligned} \|\underset{\sim}{s}\underset{\sim}{t}^T\| &= \max_{\|\underset{\sim}{d}\|_q = 1} \|\underset{\sim}{s}\underset{\sim}{t}^T\underset{\sim}{d}\|_p \\ &= \|\underset{\sim}{s}\|_p \max_{\|\underset{\sim}{d}\|_q = 1} |\underset{\sim}{t}^T\underset{\sim}{d}| \\ &= \|\underset{\sim}{s}\|_p \|\underset{\sim}{t}\|_p \\ &= \|\underset{\sim}{s}\underset{\sim}{t}^T\|_p . \end{aligned}$$

5. The total ℓ_1 problem

When $p = 1$, (2.2) becomes

$$\text{minimize } \|Z\underset{\sim}{v}\|_1 \quad \text{subject to} \quad \|\underset{\sim}{v}_2\|_\infty = 1 . \tag{5.1}$$

For this problem it is possible to give necessary and sufficient conditions for a local solution by exploiting the polyhedral nature of the norms. Let σ denote the set

$$\sigma = \{i : (Z\underset{\sim}{v})_i = 0\}$$

and let $\theta_i = \text{sign}((Z\underset{\sim}{v})_i)$, $i \in \sigma^C$.

Theorem 7 Let $\underset{\sim}{v}^* \in R^{n+1}$ be normalized so that $\|\underset{\sim}{v}_2^*\|_\infty = 1$ and attain the norm only in the t-th component, with $\phi = \text{sign}((\underset{\sim}{v}_2^*)_t)$. Then $\underset{\sim}{v}^*$ is a local solution of (5.1) if and only if there exist numbers u_i, $|u_i| \leq 1$, $i \in \sigma$ such that

$$\sum_{i\in\sigma} u_i \underset{\sim}{z}_i^T + \sum_{i\in\sigma^C} \theta_i \underset{\sim}{z}_i^T = \phi \|Z\underset{\sim}{v}^*\|_1 \underset{\sim}{e}_t$$

where $\underset{\sim}{z}_i^T$ denotes the ith row of Z.

Proof This result follows directly from Theorems 3 and 4 of [9] by specializing the subdifferentials. □

A more general result may be given, without the assumption that $\|\underset{\sim}{v}_2^*\|_\infty$ be attained in only one component. This assumption is made for two reasons : firstly, it simplifies the statement of the theorem; secondly it corresponds to the usual (non-degenerate) case. For this case, it follows from (2.5) that

$$[E\vdots\underset{\sim}{r}] = -\phi Z\underset{\sim}{v}^* \underset{\sim}{e}_t^T$$

which shows that at a solution $[E\vdots\underset{\sim}{r}]$ has only one non-zero column.

Theorem 7 may be interpreted in terms of the number of zeros of $Z\underset{\sim}{v}^*$. It is shown in [5] that at a global solution, $Z\underset{\sim}{v}^*$ will have at least n zero components, and at a local solution, $Z\underset{\sim}{v}^*$ will have at least n zero components or there is a nearby point which can be reached without increasing the objective function at which a downhill direction exists.

An algorithm for solving (5.1) in the absence of degeneracy is given in [5]. A feasible descent step is calculated by the use of a reduced gradient technique, and the objective function is minimized in this direction. Finally, the current approximation is renormalized. In the absence of degeneracy, the algorithm is finite, and may be interpreted as eventually being a vertex to vertex descent process, where a vertex corresponds to a feasible point with $Z\underset{\sim}{v}$ having exactly n zero components.

6. Some numerical results

To illustrate the performance of the simple iteration method (4.14), some results are displayed for the calculation of total ℓ_p approximations for the matrix $Z : R^5 \to R^{21}$ given in Table 1 : this is the stack loss data set considered in Daniel and Wood [1]. For different values of $p < 3$ the method was run from different starting points, for different values of k. Table 2 shows the number of iterations

i	A				b
1	1	80	27	89	42
2	1	80	27	88	37
3	1	75	25	90	37
4	1	62	24	87	28
5	1	62	22	87	18
6	1	62	23	87	18
7	1	62	24	93	19
8	1	62	24	93	20
9	1	58	23	87	15
10	1	58	18	80	14
11	1	58	18	89	14
12	1	58	17	88	13
13	1	58	18	82	11
14	1	58	19	93	12
15	1	50	18	89	8
16	1	50	18	86	7
17	1	50	19	72	8
18	1	50	19	79	8
19	1	50	20	80	9
20	1	56	20	82	15
21	1	70	20	91	15

Table 1
The stack loss data set

required from $\underset{\sim}{v}^{(0)} = \underset{\sim}{e}_5$ (which is always feasible) to obtain a difference in objective function values for (4.3) of less than 0.0001. The method was terminated if the objective function value was increased at any step, and this is what happened for $p = 2.8$, $k = 2$. The results show that the number of iterations tends to increase as p moves away from 2, and this is typical behaviour.

In Table 3 are shown the computed vectors $\underset{\sim}{v}^*$, and corresponding objective function values λ^*. For comparison, the solution $\underset{\sim}{v}^*$, λ^* for $p = 1$ is also given, as obtained by the method developed in [5].

p \ k	0	1	2	3	4
2.8	7	19	-	28	29
2.5	5	8	9	11	10
2.0	3	7	5	8	2
1.8	4	9	7	8	5
1.5	7	11	11	9	9
1.2	13	16	16	21	21
1.1	19	15	15	21	21
1.05	21	11	11	14	14

Table 2

Numbers of iterations

p	$\underset{\sim}{v}^*$					λ^*
2.8	.99745	-.00779	-.01949	-.00381	.01243	.00498
2.5	.99872	-.00646	-.01798	-.00483	.01095	.01097
2.0	.99979	-.00592	-.01571	-.00560	.00998	.04470
1.8	.99994	-.00595	-.01473	-.00580	.00978	.07880
1.5	1	-.00623	-.01265	-.00606	.00944	.18546
1.2	1	-.00717	-.01120	-.00582	.00965	.44001
1.1	1	-.00779	-.01090	-.00555	.00992	.58818
1.05	1	-.00827	-.01098	-.00529	.01030	.67984
1	1	-.00845	-.01079	-.00524	.01031	.7840

Table 3

Final coefficient values, k = 0

References

1. Daniel, C. and F.S. Wood. Fitting Equations to Data, Wiley, New York (1971).
2. Fletcher, R.. Practical Methods of Optimization, Vol. 2 Constrained Optimization, Wiley, Chichester (1981).
3. Golub, G.H. and C.F. van Loan. An analysis of the total least squares problem, SIAM J. Num. Anal. 17 (1980), pp. 883-893.
4. Hiriart-Urruty, J.B.. Tangent cones, generalized gradients and mathematical programming in Banach spaces, Math. of O.R. 4 (1979), pp. 79-97.
5. Osborne, M.R. and G.A. Watson. An analysis of the total approximation problem in separable norms, and an algorithm for the total ℓ_1 problem, preprint.
6. Peters, G. and J.H. Wilkinson. $A\underset{\sim}{x} = \lambda B\underset{\sim}{x}$ and the generalized eigenproblem, SIAM J. Num. Anal. 7 (1970), pp. 479-492.
7. Späth, H.. On discrete linear orthogonal L_p approximation, Z. Angew. Math. Mech. 62 (1982), pp. 354-355.
8. Watson, G.A.. Numerical methods for linear orthogonal L_p approximation, IMA J. Num. Anal. 2 (1982), pp. 275-287.
9. Watson, G.A.. The total approximation problem, in Approximation Theory IV, ed. L.L. Schumaker, Academic Press (to appear).

Postscript If $J(\underset{\sim}{v}^{(i)})$ is nonsingular for all i, it may be shown that the iteration (4.14) is globally convergent from any feasible initial approximation provided that $1 < p \leq 2$.

Postpostscript Reference 5 will appear in SIAM J. Sci. Stat. Comput.

AN IMPLICIT DIFFUSIVE NUMERICAL PROCEDURE FOR A SLIGHTLY COMPRESSIBLE MISCIBLE DISPLACEMENT PROBLEM IN POROUS MEDIA

Thom C. Potempa and Mary Fanett Wheeler

ABSTRACT

An implicit diffusive numerical procedure is formulated for computing the mole fraction distribution in a slightly compressible multicomponent single phase miscible displacement problem in a porous medium. This scheme has several desirable features including a substantial reduction of the "grid orientation" effect often observed with other methods. Theoretical convergence analyses for rectangular regions and computational results for this method are presented.

1. INTRODUCTION

In this paper we shall consider the flow of two slightly compressible miscible fluids in a porous medium. Although the assumptions made in this study are unrealistic for a physically interesting system, the qualitative aspects of this model can adequately represent some of the features encountered in the displacement of oil in a petroleum reservoir. A fluid, designated as the resident fluid, is assumed to be originally present in the reservoir. A second fluid, designated as the invading fluid, is injected at wells that are called injection wells in an attempt to produce the resident fluid at wells that are designated as production wells. The simulated reservoir is assumed to be homogeneous in the vertical direction, and hence is modeled over a two dimensional domain Ω. No fluid flow is assumed to occur through the boundary of the domain $\partial\Omega$.

The resident and invading fluids are assumed to mix in all proportions and to only form a single phase upon mixing. The mixing is assumed to obey an ideal mixing rule, i.e., the mixing of n volumes of the resident component and m volumes of the invading component results in $n+m$ volumes of mixture. The fluids are assumed to be slightly compressible. Under these assumptions, the density ρ of the mixture is related to the pressure P and the mass fraction X of the invading component by

$$\rho = \rho(X,P) = \rho_0(X)[1+\alpha(X)P]$$

The density of the mixture at standard conditions ρ_0 is written as a linear combination of the density of the resident fluid ρ_r and the invading fluid ρ_i as

$$\rho_0 = [1 - X]\rho_r + X\rho_i .$$

The compressibilty α of the mixture is written as a linear combination of the compressibilities of the individual components as

$$\alpha = [1 - X]\alpha_r + X\alpha_i .$$

Another fluid property which is dependent upon the fluid composition that is used in the equations stated below is the viscosity μ. A power law mixing rule that is commonly used in investigations of this type [8, 14, 18] is employed. Letting μ_r denote the viscosity of the resident fluid and μ_i denote the viscosity of the invading fluid, the viscosity rule is given by

$$\mu = ([1 - X]\mu_r^{\frac{1}{4}} + X\mu_i^{\frac{1}{4}})^4 .$$

The kinematic viscosity of the mixture ν is the quotient of the viscosity and the density of the mixture; i.e.

$$\nu = \frac{\mu}{\rho} .$$

Let the total density of the invading component be given by

$$\hat{\rho} = X\rho .$$

In a porous media, the volumetric flux v, is related to the pressure distribution by an empirical relationship that is known as Darcy's law [1, 11]. In this investigation, this law is used to relate the mass flux distribution u to the pressure distribution, the absolute permeability K, and the kinematic viscosity. Let

$$u = -\frac{K}{\nu}\nabla P .$$

The rational for introducing the mass flux vector is beyond the scope of this work and is found in [15]. The utilization of this quantity is inconsequential when modeling a displacement problem where every component is assumed to partition into only one phase, and is included only for the sake of completeness. The mass flux vector, however, is the basis for a model of multiple phase mass transfer [15] that performs significantly better than the standard fractional flow model [6] in the simulation of displacements where a component partitions into both a liquid and a vapor phase.

The differential system describing displacement under the above assumptions can be derived by standard material balance arguments of transport phenomenon [4]. The governing equations with initial and boundary conditions are given by

$$(1.1)\quad (a)\quad \phi\frac{\partial\rho}{\partial t} = -\nabla\cdot u + q,\ (x,t)\in\Omega\times[0,T],$$

$$(b)\quad P(x,0) = P_0(x),\ x\in\Omega,$$

$$(c)\quad u\cdot\upsilon = 0 \quad \text{on} \quad \partial\Omega,$$

$$(1.2)\quad (a)\quad \phi\frac{\partial\hat{\rho}}{\partial t} = \phi\frac{\partial\rho X}{\partial t} = -\nabla\cdot Xu + \nabla\cdot D\nabla X + q\tilde{X},\ (x,t)\in\Omega\times[0,T],$$

$$(b)\quad X(x,0) = X_0(x),\ x\in\Omega,$$

$$(c)\quad D\nabla X\cdot\upsilon = 0 \quad \text{on} \quad \partial\Omega,$$

where υ is the outer normal to the boundary $\partial\Omega$. Here q, the mass source term, is a linear combination of Dirac measures and is defined to be positive at injection wells. The coefficient ϕ denotes the porosity of the media, a dimensionless variable which is a measure of the amount of volume available for fluid flow. We are assuming that ϕ is independent of pressure P. The coefficient D is the mass diffusivity governed by Fick's Law for diffusion; we assume $D = D_0$, a positive constant. The effective mass fraction $\tilde{X}$ is set to one at injection wells, $q > 0$, and equals X at production wells, $q < 0$.

A development of slightly different equations describing single phase compressible miscible displacement in a porous media can be found in Douglas and Roberts [9]. For the slightly compressible case they introduce and analyze under certain restrictions, namely q smooth, finite element methods for approximating the concentration of one of the fluids and the pressure of the mixture.

The purpose of this paper is to formulate and analyze under certain restrictions a computationally efficient numerical scheme for approximating the solution of (1.1) and (1.2) which exhibits minimal "grid orientation". Briefly, the grid orientation effect refers to a severe dependence in certain model problems of the solution on the spatial orientation of the computational grid. This computational anomaly is important in practical reservoir engineering simulations since very different numerical results may lead to contradictory conclusions; namely, whether a given oil field is economically feasible to produce.

In §2 a continuous time diffusive (upstream weighting) numerical procedure is defined for a general polygonal domain. A convergence analysis with a number of restrictions is given in §3 for a fully discrete scheme. For this analysis the most important restriction is that the solution is smooth; i.e. q is smoothly distributed and the coefficients are smooth. We also assume that the domain Ω is a rectangle and that the approximating spaces are bilinear. In §4 computational results are presented for rectangular and triangular domains Ω.

2. FORMULATION OF IMPLICIT DIFFUSIVE SCHEME

One of the authors, Potempa, in his 1982 Ph.D. dissertation [13] introduced a new numerical method for solving the equations describing multicomponent single phase flow in a porous medium. He has successfully used this procedure in modeling compressible miscible displacement problems [13, 14] and steamflooding problems [16]. These implementations have been restricted to rectangular domains and have used bilinear approximating spaces. One can view these schemes as nine point finite difference methods as was done by Bell, Shubin, and one of the authors, Wheeler, [3] for the incompressible miscible displacement problem. In this section, Potempa's scheme is extended to general polygonal domain for the compressible miscible displacement problem.

Let $T_h = \{\tau_1, \tau_2, \ldots, \tau_{s_h}\}$ be a quasi-regular triangulation of Ω with τ_i a polygon. Let M_h be a C^0 piecewise polynomial space defined on T_h and assume that the w_i form a nodal basis for M_h; i.e

$$M_h = \text{span}\{w_1, w_2, \ldots, w_{N_h}\}$$

with $w_i(n_k) = \delta_{ik}$, $1 \le i, k \le N_h$, where $\{n_1, n_2, \ldots, n_{N_h}\} \subset \bar{\Omega}$ and δ denotes the Kronecker δ. Let Ω_i, which shall be referred to as computational molecule, be the region under the surface in $\mathbb{R}^3$ defined by w_i.

The continuous time implicit diffusive approximation to the solution of (1.1) - (1.2), (P,X), is defined by $(P_h, X_h)\colon [0, T] \to M_h \times M_h$ which satisfies the mass balance equations

$$V_i \frac{\partial \rho_{hi}}{\partial t} = \sum_{j=1}^{N_h} \Gamma_{ij} + Q_{hi}, \tag{2.1}$$

$$(2.2)\quad V_i \frac{\partial \hat{\rho}_{hi}}{\partial t} = \sum_{j=1}^{N_h} \{ (\frac{\Gamma_{ij} + |\Gamma_{ij}|}{2}) X_{hj} + (\frac{\Gamma_{ij} - |\Gamma_{ij}|}{2}) X_{hi} - D_{ij} X_{hj} \}$$

$$+ Q_{hi} \tilde{X}_{hi}, \quad i = 1, 2, \ldots, N_h.$$

Here $X_h(t) = \sum_{i=1}^{N_h} X_{hi}(t) w_i$, $P_h(t) = \sum_{i=1}^{N_h} P_{hi}(t) w_i$, $\rho_{hi} = \rho(X_{hi}, P_{hi})$, $\hat{\rho}_{hi} = X_{hi}\, \rho_{hi}$, and $V_i = \int_\Omega \phi w_i \,dx\,dy$, the pore volume of the cell. The diffusion coefficient D_{ij} is given by $D_{ij} = \int_\Omega D \nabla w_i \cdot \nabla w_j \,dx\,dy$. The coefficient Γ_{ij} represents a mass transfer rate between the comtational molecules Ω_i and Ω_j and

$$\Gamma_{ij} = \int_\Omega U_h (w_j \nabla w_i - w_i \nabla w_j)\,dx\,dy,$$

where U_h, the approximation to the mass flux, is defined by

$$U_h = -\frac{K(x,y)}{\mu(X_h)} \rho(X_h, P_h) \nabla P_h.$$

In practice one can easily compute Γ_{ij} by approximating the coefficient functions K, μ and ρ at the center of mass of each τ_k.

Let $\{x^1, x^2, \ldots, x^{NWI}\}$ and $\{x^{NWI+1}, \ldots, x^{NW}\}$ denote the locations of the injection and production wells respectively, i.e., the point sources and sinks. We assume that $x^k = n_\ell$ for some ℓ. Following Peaceman [12] the approximation Q_h to the source term q in (1.1a) and (1.2a) is defined by

$$Q_{hi} = \sum_{k=1}^{NWI} q_k(t)\, \delta_{ki} + \sum_{k=NWI+1}^{NW} \frac{K\theta_k}{\mu(X_h)} \rho(X_h, P_h)(P_{bh} - P_h)\,\delta_{ki},$$

where $q_k(t)$ is the mass flow rate at the kth well, P_{bh} is the wellbore pressure, and θ_k is a constant dependent upon the triangulation. The calculation of θ_k is based upon Peaceman's procedure [12] and will not be discussed here.

Various timestepping procedures can be used in solving the system of ordinary differential equations (2.1) - (2.2). Generally first order Euler schemes are coded such as fully implicit backward differences, sequential Euler, or IMPES, an implicit pressure and explicit mass fraction Euler scheme [1, 2, 11]. The fully implicit Euler scheme involves solving the coupled system at each time step. The sequential and IMPES formulations involve a decoupling of the system.

In §4 the computer codes used were written using C^0 piecewise linear subspaces and an IMPES timestepping scheme.

3. THEORETICAL RESULTS

Let Ω be a rectangle in $\mathbb{R}^2$. In this section we define and analyze a fully discrete version of (2.1) and (2.2) based on backward differencing in time and M_h a tensor product C^0 piecewise linear subspace. We assume that the δ_{x^k} are smoothly distributed and that $\rho = \rho(P) = \rho_0(1 + \alpha_0 P)$, ρ_0 and α_0 constants.

Let $(u,v) = \int_\Omega u\, v\, dx\, dy$ and $\|u\|^2 = (u,u)$. Let $W_s^k(\Omega)$ be the Sobolev space on Ω norm

$$\|\phi\|_{W_s^k} = \left\{ \sum_{|\alpha| \le k} \left\| \frac{\partial^\alpha \phi}{\partial x^\alpha} \right\|_{L^s(\Omega)} \right\}^{1/s}$$

with the usual modification for $s = \infty$. When $s = 2$, let $\|\phi\|_{W_s^k} = \|\phi\|_{H^k} \equiv \|\phi\|_k$.

For simplicity we assume Ω is the rectangle $(0,XL) \times (0,XL)$ which is divided into squares of size $\Delta x = \Delta y = h$ and that $[0.T]$ is divided into uniform intervals with $t^n = n\Delta t$. Define

$$M_h = \{v \in C^0(0,L) \,|\, v \in P^1(I_i),\ i = 1, \ldots, N\},$$

where $I_i = (x_{i-1}, x_i)$, $x_i = (i-1)h$ and $P^1(I_i)$ is the set of linear functions defined on I_i. Set

$$\mathcal{M}_h = M_h \otimes M_h .$$

Defining

$$v_i(x) = \begin{cases} \dfrac{x - x_{i-1}}{h}, & x \in I_i, \\ \dfrac{x_{i+1} - x}{h}, & x \in I_{i+1}, \\ 0, & x \in I_i \cup I_{i+1}, \end{cases}$$

we have that

$$w_{ij}(x,y) = v_i(x) v_j(y),\ 1 \le i,\ j \le N,$$

forms a basis for $\mathcal{M}_h$.

Let $\psi_{ij}^{n} = \psi^{n}(x_i,y_j) = \psi(x_i,y_j,t^n)$ and $d_t\,\psi_{ij}^{n} = (\psi_{ij}^{n+1} - \psi_{ij}^{n})/\Delta t$. Denote the approximation of P by P_h: $\{0 = t^0, t^1, \ldots, t^M = T\} \to M_h$ and the approximation of X by X_h: $\{t^0, t^1, \ldots, t^M\} \to M_h$. Assuming that P_h^n and X_h^n are known we determine P_h^{n+1} and X_h^{n+1} as follows:

$$(3.1)\quad V_{ij}\, d_t\, \rho_{hij}^{n} = (a(P_h^n,X_h^n)\nabla P_h^{n+1},\nabla w_{ij}) + (Q_h^n,w_{ij})$$

$$(3.2)\quad V_{ij}(d_t\hat{\rho}_{hij}^{n}) = \sum_{k,\ell=-1,0,1} \Big(\Gamma_{ij}^{k\ell}(U^{n+1})\Big(\frac{X_{hi+k,j+\ell}^{n+1}+X_{hij}^{n+1}}{2}\Big) + |\Gamma_{ij}^{k\ell}(U^{n+1})|\Big(\frac{X_{hi+k,j+\ell}^{n+1}-X_{hij}^{n+1}}{2}\Big)\Big) - (D\nabla X_h^{n+1},\nabla w_{ij}) + (Q_h^n\tilde{X}_h^{n+1},w_{ij})$$

where

$$(3.3)\quad \begin{cases} V_{ij} = \int_{\Omega}\phi w_{ij}\,dx\,dy, \\ \rho_h^n = \rho(P_h^n) = \rho_0(1+\alpha_0 P_h^n), \\ a(P_h^n,X_h^n) = \dfrac{K(x,y)}{\mu(X_h^n)}\rho(P_h^n), \\ Q_h^n = q(\rho_h^n,X_h^n,P_h^{n+1}), \\ \hat{\rho}_h^n = X_h^n\rho_h^n, \\ U^{n+1} = -a(P_h^n,X_h^n)\nabla P_h^{n+1}, \end{cases}$$

and

$$\Gamma_{ij}^{k\ell}(U^{n+1}) = \iint_{\gamma_{ij}^{k\ell}} \Big(U_1^{n+1}\frac{\partial w_{ij}}{\partial x}(1-\delta_{k0})v_{j+\ell}(y) + U_2^{n+1}\frac{\partial w_{ij}}{\partial y}(1-\delta_{\ell 0})v_{i+k}(x)\Big)dx\,dy,$$

with $\gamma_{ij}^{k\ell} = \operatorname{supp} w_{ij} \cap \operatorname{supp} v_{i+k}(x) \cap v_{j+\ell}(y)$. As mentioned in §2 $\Gamma_{ij}^{k\ell}$ represents flow from $\Omega_{i+k,j+\ell}$ into Ω_{ij} where Ω_{ij} is the region under the surface in $\mathbb{R}^3$ defined by w_{ij}.

We remark that the system (3.1) and (3.2) has been linearized and is sequential; i.e. one solves for P_h^{n+1} in (3.1) and then for X_h^{n+1} in (3.2)

The procedure (3.1) and (3.2) was analyzed in [3] under the assumption $\rho \equiv$ constant. In particular they noted that the first term on the right hand side of (3.2) represents a discretization of the convection term of

(1.2). The second term represents a numerical diffusion - dispersion term that depends on the velocity field and the mesh size.

In our analysis we shall fequently use the norm equivalence of the L^2 norm and a discrete L^2 norm on M_h; i.e. there exist positive constants K_1 and K_2 such that

$$(3.4) \qquad K_1 \|\psi\| \le |||\psi||| \le K_2\|\psi\|, \ \psi \in M_h,$$

where $|||\psi|||^2 = \sum_{ij} V_{ij}\psi_{ij}^2$.

For convenience we shall also assume that there exist positive constants ρ_*, ρ^*, a_* and a^* such that

$$(3.4') \qquad \begin{aligned} \rho_* &\le \rho_h^n \le \rho^*, \\ a_* &\le a(P_h^n, X_h^n) \le a^*, \end{aligned}$$

and that $|\frac{\partial a}{\partial P}|$ and $|\frac{\partial a}{\partial X}|$ are bounded by K^* over $\mathbb{R}^2$. These assumptions are unnecessary since one can use an induction argument to bound P_h^n, $n \le M$, from which (3.4′) follows.

We prove a maximum principle for X_h^n for the special case related to petroleum reservoir computations (the δ_{x^i} are not smoothly distributed). A more general result can be formulated for the smooth case but will not be discussed here.

<u>THEOREM 3.1</u>

If $0 \le X_h^0 \le 1$, then the X_{hij}^n computed using (3.1) and (3.2) satisfy

$$0 \le X_{hij}^n \le 1.$$

<u>PROOF</u>.

This result can be seen by examining the structure of the matrix problem associated with (3.1) and (3.2) and applying an induction argument.

The matrix problem can be written in the form

$$(D + M + A + Z)X_h^{n+1} = DX_h^n + Ze ,$$

where D is a positive diagonal matrix, Z is a nonnegative diagonal matrix, $M = (m)_{k\ell}$ is a diagonally dominant matrix such that $m_{k\ell} \ge 0$ if $\ell = k$ and $m_{k\ell} \le 0$ otherwise, A is a diagonally dominant positive semidefinite matrix, and the vector e is a vector consisting of all ones. Since the matrix $(D + M + A + Z)$ is monotone, $X_h^{n+1} \geq 0$. One obtains $X_h^{n+1} \leq 1$ by noting that $(D + M + A + Z)e = (D + Z)e$.//

We derive estimates of $\xi^n = P_h^n - \tilde{P}^n$ and $\upsilon^n = X_h^n - \hat{X}^n$ where $\tilde{P}^n$ is an elliptic projection of P^n and satisfies

$$(3.5') \qquad (a(P^n,X^n)\nabla(\tilde{P}^n - P^n), \nabla w_{ij}) + \lambda((\tilde{P}^n - P^n), w_{ij}) = 0,$$

with λ a positive constant, and $\hat{X}^n \in M_h$ is the bilinear interpolant of X^n. Set $\eta^n = \tilde{P}^n - P^n$ and $\beta^n = \hat{X}^n - X^n$. From approximation theory [5 and elliptic finite element results [10, 17]

$$(3.5) \qquad \|\eta^n\|_{W_p^0} + h\,\|\nabla\eta^n\|_{W_p^0} \leqslant K_3 h^2 (\log h^{-1})^{\bar{r}}\, \|P^n\|_{W_p^2},$$

with $\bar{r} = 0$ if $p = 2$ and $\bar{r} = \frac{1}{2}$ if $2 < p \leq \infty$, and

$$(3.6) \qquad \|\beta^n\|_{W_p^0} + h\|\nabla\beta^n\|_{W_p^0} \leqslant K_4 h^2 \Big(\Big\|\frac{\partial^2 X^n}{\partial x^2}\Big\|_{L^p(\Omega)} + \Big\|\frac{\partial^2 X^n}{\partial y^2}\Big\|_{L^p(\Omega)}\Big), \; 2 \leq p \leq \infty.$$

Multiplying (1.1) by w_{ij} and integrating by parts, we see that

$$(3.7) \qquad (\Phi\rho_t, w_{ij}) + (a(P,X)\nabla P, \nabla w_{ij}) = (q, w_{ij}),$$

where $a(P,X) = K\rho(P)/\mu(X)$. Combining (3.7) and (3.5') we obtain

$$(3.8) \qquad V_{ij}\, d_t\, \rho_{ij}^n + (a(P^{n+1},X^{n+1})\nabla\tilde{P}^{n+1}, \nabla w_{ij})$$
$$= (q^{n+1}, w_{ij}) - \lambda(\eta^{n+1}, w_{ij}) - (\Phi\rho_t^{n+1}, w_{ij}) + V_{ij}\, d_t\, \rho_{ij}^n$$

Subtracting (3.8) from (3.1), we note that

$$V_{ij}\, d_t\, (\rho_h - \rho)_{ij}^n + (a(P_h^n, X_h^n)\nabla\xi^{n+1}, \nabla w_{ij})$$
$$(3.9) \quad = (Q_h^n - q^{n+1}, w_{ij}) - ((a(P_h^n,X_h^n) - a(P^{n+1},X^{n+1}))\nabla\tilde{P}^{n+1}, \nabla w_{ij})$$
$$+ ((\Phi\rho_t^{n+1}, w_{ij}) - V_{ij}\, d_t\, \rho_{ij}^n + \lambda(\eta^{n+1}, w_{ij}).$$

Multiplying (3.9) by ξ_{ij}^{n+1} and summing on i and j and noting that $d_t(\rho_h - \rho)_{ij}^n = \rho_0\alpha_0(d_t\xi_{ij}^n + d_t\eta_{ij}^n)$, we obtain

$$\frac{\rho_0\alpha_0}{\Delta t}(|||\xi^{n+1}|||^2 - |||\xi^n|||^2) + (a(P_h^n,X_h^n)\nabla\xi^{n+1}, \nabla\xi^{n+1})$$
$$(3.10) \quad \leqslant (Q_h^n - q^{n+1}, \xi^{n+1}) - ((a(P_h^n,X_h^n) - a(P^{n+1},X^{n+1}))\nabla\tilde{p}^{n+1}, \nabla\xi^{n+1})$$
$$+ (\lambda\eta^{n+1}, \xi^{n+1}) + (\Phi\rho_t^{n+1}, \xi^{n+1}) - \langle d_t\rho^n + \rho_0\alpha_0\, d_t\,\eta^n, \xi^{n+1}\rangle,$$

where

$$(3.10') \qquad \langle\phi, \psi\rangle = \sum_{ij} V_{ij}\phi_{ij}\psi_{ij}.$$

Applying standard norm inequalities to (3.10) and using (3.4′) and (3.4) we deduce that

$$
\begin{aligned}
&\frac{\rho_0\alpha_0}{\Delta t}(|||\xi^{n+1}|||^2 - |||\xi^n|||^2) + a_*\|\nabla\xi^{n+1}\|^2 \\
&\leq \frac{a_*}{2}\|\nabla\xi^{n+1}\|^2 + K_5(K^*,a_*,a^*)(1+\|\nabla\tilde{P}\|_{L^\infty})(\|\upsilon^n\|^2 \\
(3.11)\qquad &+ \|\xi^n\|^2 + \|\beta^n\|^2 + \|\eta^n\|^2 + \Delta t\|P_t\|^2_{L^2((t^n,t^{n+1});L^2)} \\
&+ \Delta t\|X_t\|^2_{L^2((t^n,t^{n+1});L^2)}) + K_6(|||\xi^{n+1}|||^2 \\
&+ |||d_t\eta^n|||^2 + \Delta t\|P_{tt}\|^2_{L^2((t^n,t^{n+1});L^2)} \\
&+ \frac{h^2}{\Delta t}\|P_t\|^2_{L^2((t^n,t^{n+1});H^1)}).
\end{aligned}
$$

Multiplying (3.11) by Δt and summing on n, $n = 0, 1, \ldots, M$, and using the approximations (3.5) and (3.6), we obtain

$$
\begin{aligned}
&\rho_0\alpha_0(|||\xi^M|||^2 - |||\xi^0|||^2) + \sum_{n=0}^{M-1}\frac{a_*}{2}\|\nabla\xi^{n+1}\|^2\Delta t \\
(3.12)\qquad &\leq K_7(h^2 + (\Delta t)^2) + K_8(\sum_{n=0}^{M-1}(\|\xi^n\|^2 + \|\upsilon^n\|^2)\Delta t \\
&+ K_9|||\xi^M|||^2\Delta t.
\end{aligned}
$$

Setting $P_h^0 = \tilde{P}^0$ and applying the discrete Gronwall lemma to (3.12), we conclude that

$$
\begin{aligned}
&\rho_0\alpha_0|||\xi^M|||^2 + \sum_{n=0}^{M-1} a_*\|\nabla\xi^{n+1}\|^2\Delta t \\
(3.13)\qquad &\leq K_{10}(h^2 + (\Delta t)^2) + K_{11}\sum_{n=0}^{M-1}\|\upsilon^n\|^2\Delta t,
\end{aligned}
$$

for Δt sufficiently small.

The bilinear interplant $\hat{X}^{n+1}$ of X^{n+1} satisfies the equation

$$
\begin{aligned}
&V_{ij}\, d_t(\rho\hat{X})_{ij}^n + (D_0\nabla\hat{X}^{n+1},\nabla w_{ij}) \\
(3.14)\qquad &= (D_0\nabla\beta^{n+1},\nabla w_{ij}) + (q\tilde{X}^{n+1},w_{ij}) \\
&+ (u^{n+1}X^{n+1},\nabla w_{ij}) - T_{ij}^{1,n},
\end{aligned}
$$

where

$$
(3.15)\qquad T_{ij}^{1,n} = (\phi(\rho X)_t^{n+1},w_{ij}) - V_{ij}d_t(\rho X)_{ij}^n .
$$

Subtracting (3.14) from (3.2) and adding a mesh dependent diffusion ter $h\mathcal{D}^*(U^n;X_h^{n+1},w_{ij})$ to each side of the resulting equation, we obtain

$$
\begin{aligned}
&V_{ij}(d_t(\rho_h X_h)_{ij}^n - d_t(\rho\hat{X})_{ij}^n) + (D_0\nabla\upsilon^{n+1},\nabla w_{ij}) \\
(3.16)\qquad &+ h\mathcal{D}^*(U^n;\upsilon^{n+1},w_{ij}) \\
&= T_{ij}^{1,n} + T_{ij}^{2,n} + T_{ij}^{3,n},
\end{aligned}
$$

where

$$
\begin{aligned}
T_{ij}^{2,n} &= -(D_0\nabla\beta^{n+1},\nabla w_{ij}) - (u^{n+1}\chi^{n+1},\nabla w_{ij}) \\
(3.17)\qquad &+ (U^n X_h^{n+1},\nabla w_{ij}) - h\mathcal{D}^*(U^n;\hat{X}^{n+1},w_{ij}) \\
&+ (Q_h^n\tilde{X}_h^{n+1} - q^{n+1}\tilde{X}^{n+1},w_{ij})
\end{aligned}
$$

and

$$
\begin{aligned}
T_{ij}^{3,n} &= \sum_{k,\ell=-1,0,1} \Gamma_{ij}^{k\ell}(U^n)\frac{X_{hi+k,j+\ell}^{n+1}+X_{kij}^{n+1}}{2} \\
(3.18)\qquad &+ |\Gamma_{ij}^{k\ell}(U^n)|\frac{X_{hi+k,j+\ell}^{n+1} - X_{hij}^{n+1}}{2} \\
&- (U^n X_h^{n+1},\nabla w_{ij}) + h\mathcal{D}^*(U^n;X_h^{n+1},w_{ij}) .
\end{aligned}
$$

The mesh dependent diffusion term is defined as

$$
(3.19)\qquad \mathcal{D}^*(u;z,w_{ij}) = (A(u)\nabla z,\nabla w_{ij}) + \frac{h^2}{36}(F(u)z_{xy},(w_{ij})_{xy})
$$

with $u^t = (u_1,u_2)$ and

$$
A(u) = \frac{1}{12}\begin{pmatrix} 4|u_1| + |u_1+u_2| + |u_1-u_2| & |u_1+u_2| - |u_1-u_2| \\ |u_1+u_2| + |u_1-u_2| & 4|u_2| + |u_1+u_2| + |u_1-u_2| \end{pmatrix}
$$

and

$$
F(u) = 2|u_1| + 2|u_2| - \tfrac{1}{2}(|u_1+u_2| + |u_1-u_2|).
$$

The following result was derived in [3].

<u>LEMMA 3.1</u>

If $z \in M_h$ and $\bar{u} = (\bar{u}_1,\bar{u}_2)^t$ with $\bar{u}_1$ and $\bar{u}_2$ constants, then

$$\sum_{k,\ell=-1,0,1} \Gamma_{ij}^{k\ell}(\bar{u})\left(\frac{z_{i+k,j+\ell}+z_{ij}}{2}\right) + |\Gamma_{ij}^{k\ell}(\bar{u})|\left(\frac{z_{i+k,j+\ell}-z_{ij}}{2}\right)$$
$$= (\bar{u}z,\nabla w_{ij}) - hD^*(\bar{u};z,w_{ij}).//$$

In estimating bounds for υ we shall use similar arguments to those given in [3] for the case $\rho \equiv$ constant; i.e., multiply (3.16) by υ_{ij}^{n+1} and sum on i, j, followed by multiplication by Δt, and summation n. The major change is in the treatment of the accumulation terms.

Using the notation (3.10′), we have

$$\begin{aligned} &\langle d_t(\rho_h X_h)^n - d_t(\rho\hat{X})^n, \upsilon^{n+1}\rangle \\ (3.20)\qquad &= \langle d_t(\rho\upsilon)^n + d_t(\theta X_h)^n, \upsilon^{n+1}\rangle , \end{aligned}$$

where $\theta = \rho_h - \rho$. The first term on the right hand side of (3.20) can be expressed as

$$\begin{aligned} \langle d_t(\rho\upsilon)^n, \upsilon^{n+1}\rangle &= \langle d_t(\rho\upsilon^2)^n, 1\rangle + \frac{\Delta t}{2}\langle \rho^n, (d_t\upsilon^n)^2\rangle \\ (3.21)\qquad &\quad - \tfrac{1}{2}\langle \rho^n, d_t(\upsilon^2)^n\rangle . \end{aligned}$$

The second term we rewrite in the form

$$\begin{aligned} \langle d_t(\theta X_h)^n, \upsilon^{n+1}\rangle &= \langle d_t\theta^n, (\upsilon^{n+1})^2\rangle + \langle d_t\theta^n, \hat{X}^{n+1}\upsilon^{n+1}\rangle \\ (3.22)\qquad &\quad + \langle \theta^n, (d_t\upsilon^n)\upsilon^{n+1}\rangle + \langle \theta^n, (d_t\hat{X}^n)\upsilon^{n+1}\rangle . \end{aligned}$$

Now

$$\begin{aligned} \langle \theta^n, (d_t\upsilon^n)\upsilon^{n+1}\rangle &= \tfrac{1}{2}(\langle d_t(\theta\upsilon^2)^n, 1\rangle - \langle d_t\theta^n, (\upsilon^{n+1})^2\rangle \\ (3.23)\qquad &\quad + \Delta t\langle \theta^n, (d_t\upsilon^n)^2\rangle). \end{aligned}$$

Combining (3.20) - (3.23), we obtain

$$\begin{aligned} &\langle d_t(\rho_h X_h)^n - d_t(\rho\hat{X})^n, \upsilon^{n+1}\rangle \\ &= \langle d_t(\rho\upsilon^2)^n, 1\rangle + \tfrac{1}{2}(\Delta t\langle \rho_h^n, (d_t\upsilon^n)^2\rangle \\ (3.24)\qquad &\quad - \langle \rho^n, (d_t\upsilon^2)^n\rangle) + \langle d_t\theta^n, (\tfrac{1}{2}(\upsilon^{n+1})^2 + \hat{X}^{n+1}\upsilon^{n+1}\rangle \\ &\quad + \tfrac{1}{2}\langle d_t(\theta\upsilon^2)^n, 1\rangle + \langle \theta^n, (d_t\hat{X}^n)\upsilon^{n+1}\rangle . \end{aligned}$$

Multiplying (3.24) by Δt and summing on n, $n = 0, \ldots .\ M-1$, we see that

$$\sum_{n=0}^{M-1}(\langle d_t(\rho_h X_h)^n - d_t(\rho\hat{X})^n, \upsilon^{n+1}\rangle)\Delta t$$

(3.25)
$$= \langle \tfrac{1}{2}\rho_h^M \upsilon^M, \upsilon^M\rangle + \sum_{n=0}^{M-1}(\tfrac{1}{2}\langle \rho_h^n, \Delta t(d_t\upsilon^n)^2\rangle + \tfrac{1}{2}\langle d_t\rho^n, (\upsilon^{n+1})^2\rangle + \langle \theta^n, (d_t\hat{X}^n)\upsilon^{n+1}\rangle + \langle d_t\theta^n, \tfrac{1}{2}(\upsilon^{n+1})^2 + \hat{X}^{n+1}\upsilon^{n+1}\rangle)\Delta t.$$

The above result involved assuming $X_h^0 = \hat{X}^0$ and summing by parts the term $-\frac{1}{2}\sum_{n=0}^{M-1}\langle \rho^n, (d_t\upsilon^2)^n\rangle\Delta t$.

Multiplying (3.16) by $\upsilon_{ij}^{n+1}\Delta t$, and summing on i, j and n, and using (3.25), we obtain

(3.26)
$$\tfrac{1}{2}\langle \rho_h^M \upsilon^M, \upsilon^M\rangle + \sum_{n=0}^{M-1}(\tfrac{1}{2}\langle \rho_h^n, \Delta t(d_t\upsilon^n)^2\rangle + (D_0\nabla\upsilon^{n+1}, \nabla\upsilon^{n+1}) + h\mathcal{D}^*(U^n;\upsilon^{n+1},\upsilon^{n+1}))\Delta t$$
$$= \sum_{n=0}^{M-1}\{-(\tfrac{1}{2}\langle d_t\rho^n, (\upsilon^{n+1})^2\rangle + \langle \theta^n, (d_t\hat{X}^n)\upsilon^{n+1}\rangle + \langle d_t\theta^n, (\tfrac{1}{2}(\upsilon^{n+1})^2 + \hat{X}^{n+1}\upsilon^{n+1})\rangle) + \sum_{\ell=1}^{3}\sum_{ij} T_{ij}^{\ell,n}\upsilon_{ij}^{n+1}\}\Delta t.$$

Observe that

(3.27)
$$-\tfrac{1}{2}\langle d_t\rho^n, (\upsilon^{n+1})^2\rangle + \langle \theta^n, (d_t\hat{X}^n)\upsilon^{n+1}\rangle \leq K_{12}'(\|d_t\rho\|_{L^\infty} + \|d_tX\|_{L^\infty})(|||\upsilon^{n+1}||| + |||\xi^n||| + |||\eta^n|||)\,|||\upsilon^{n+1}|||,$$
$$\leq K_{12}(\|\upsilon^{n+1}\|^2 + \|\xi^n\|^2 + \|\eta^n\|^2).$$

Let $\psi_h^{n+1} \in M_h$ be the bilinear interpolant of $\psi^{n+1} = \frac{1}{2}(\upsilon^{n+1})^2 + \hat{X}^{n+1}\upsilon^{n+1}$. Multiplying (3.9) by $(\psi_h^{n+1})_{ij}$ and summing on i and j, we see that

(3.28)
$$\langle d_t\theta^n, \psi^{n+1}\rangle = \langle d_t(\rho_h - \rho)^n, \psi_h^{n+1}\rangle$$
$$\leq K_{13}\{\|\nabla\xi^{n+1}\|\,\|\nabla\psi_h^{n+1}\| + \|\nabla\tilde{p}^{n+1}\|_{L^\infty}(\|\xi^n\| + \|\upsilon^n\| + h^2 + (\Delta t)^2)\|\nabla\psi_h^{n+1}\| + (\|\xi^n\| + \|\upsilon^n\| + h^2 + (\Delta t)^2)\|\psi_h^{n+1}\|\}.$$

By the triangle inequality,

$$\|\nabla\psi_h^{n+1}\| \leq \|\nabla(\psi - \psi_h)^{n+1}\| + \|\nabla\psi^{n+1}\|.$$

Using bilinear interpolation results [5], we note that on $R_{ij} = (x_{i-1},x_i) \times (y_{j-1},y_j)$

$$\|\nabla(\psi-\psi_h)^{n+1}\|^2_{L^2(R_{ij})} \leqslant \text{Const}\, h^2(\|\psi^{n+1}_{xx}\|^2_{L^2(R_{ij})} + \|\psi^{n+1}_{yy}\|^2_{L^2(R_{ij})})$$

$$(3.29)\quad \leqslant \text{Const}\, h^2\Big[\int_{R_{ij}} ((\upsilon^{n+1}_x)^4 + (\upsilon^{n+1}_y)^4 + (\hat{X}^{n+1}_x\upsilon^{n+1}_x)^2 + (\hat{X}^{n+1}_y\upsilon^{n+1}_y)^2)\,dxdy\Big]$$

$$\leqslant \text{Const}\ \|\nabla\upsilon^{n+1}\|^2_{L^2(R_{ij})}[\|\upsilon^{n+1}\|^2_{L^\infty} + \|\hat{X}^{n+1}\|^2_{L^\infty}].$$

Similarly,

$$\|\nabla\psi^{n+1}\|^2 \leqslant \text{Const}\|\nabla\upsilon^{n+1}\|^2(\|X^{n+1}_h\|_{L^\infty} + \|\hat{X}^{n+1}\|_{L^\infty}) .$$

Thus, by the maximum principle

$$(3.30)\qquad \|\psi^{n+1}_h\|_1 \leqslant \text{Const}\|\upsilon^{n+1}\|_1 .$$

From (3.28) - (3.30), we deduce that

$$\sum_{n=0}^{M-1} < d_t\theta^n,\psi^{n+1} > \Delta t$$

$$(3.31)\quad \leqslant \sum_{n=0}^{M-1}\{(D_0/6)\|\nabla\upsilon^{n+1}\|^2 + K_{14}(\|\xi^n\|^2 + \|\upsilon^n\|^2 + \|\upsilon^{n+1}\|^2$$

$$+\|\nabla\xi^{n+1}\|^2 + h^2 + (\Delta t)^2)\}\Delta t.$$

We have

$$(3.32)\qquad \sum_{n=0}^{M-1}\ \sum_{ij} T^{1,n}_{ij}\upsilon^{n+1}_{ij} \leqslant K_{15}(\|\upsilon^{n+1}\|^2 + h^2 + (\Delta t)^2)$$

and

$$\sum_{ij} T^{2,n}_{ij}\upsilon^{n+1}_{ij} \leqslant K_{16}(h\|\nabla\upsilon^{n+1}\| + \|u^{n+1}-U^n\|\,\|X^{n+1}_h\|_{L^\infty}\|\nabla\upsilon^{n+1}\|$$

$$+ \|u^{n+1}\|_{L^\infty}(\|\upsilon^{n+1}\| + h^2)\|\nabla\upsilon^{n+1}\|$$

$$+ h(\|U^n - u^{n+1}\|\,\|\nabla\hat{X}^{n+1}\|_{L^\infty}\|\nabla\upsilon^{n+1}\|$$

$$(3.33)\quad + \|u^{n+1}\|_{L^\infty}\|\nabla\hat{X}^{n+1}\|\,\|\nabla\upsilon^{n+1}\|)$$

$$+ (\|\upsilon^n\| + \|\xi^n\| + h^2+(\Delta t)^2)\|\upsilon^{n+1}\|$$

$$\leqslant D_0/6\|\nabla\upsilon^{n+1}\|^2 + K_{17}(\|\xi^n\|^2_1 + h^2 + \|\upsilon^n\|^2 + (\Delta t)^2 + \|\upsilon^{n+1}\|^2) .$$

In [3] it was shown that

$$|T^{3,n}_{ij}| \leqslant \text{Const}\|U^n - \bar{u}^{n+1}\|_{L^2(\hat{R}_{ij})}\|\nabla X^{n+1}_h\|_{L^2(\hat{R}_{ij})} ,$$

where $\bar{u}^{n+1} = u(x_i, y_j, t^{n+1})$ and $\hat{R}_{ij}$ is the square $(x_{i-1}, x_{i+1}) \times (y_{j-1}, y_{j+1})$. This result uses Lemma 3.1.

Thus,

$$
\begin{aligned}
\sum_{ij} |T_{ij}^{3,n} \upsilon_{ij}^{n+1}| &\leq K_{18}\Big(\sum_{ij} \|U^n - \bar{u}^{n+1}\|_{L^2(\hat{R}_{ij})}(\|\nabla\upsilon^{n+1}\|_{L^2(\hat{R}_{ij})} \\
&\quad + h\|\nabla\hat{X}\|_{L^\infty(\hat{R}_{ij})})|\upsilon_{ij}^{n+1}| \\
(3.34) \qquad &\leq K_{19}(\|U^n - u^{n+1}\|^2 + h^2 + \|\upsilon^{n+1}\|^2) + \frac{D_0}{6}\|\nabla\upsilon^{n+1}\|^2 \\
&\leq K_{20}(\|\xi^n\|_1^2 + \|\upsilon^n\|^2 + h^2 + \|\upsilon^{n+1}\|^2) + \frac{D_0}{6}\|\nabla\upsilon^{n+1}\|^2 .
\end{aligned}
$$

Substituting the bounds (3.27), (3.31), (3.32), (3.33), and (3.34) into (3.26), we obtain

$$
\begin{aligned}
&\langle \rho_h^M \upsilon^M, \upsilon^M \rangle + \sum_{n=1}^{M} (D_0 \nabla\upsilon^n, \nabla\upsilon^n) \\
(3.35) \qquad &\leq K_{21}\Delta t\|\upsilon^M\|^2 + K_{22}\sum_{n=0}^{M-1}(\|\upsilon^n\| + \|\xi^n\|_1^2)\Delta t \\
&\quad + K_{23}\|\nabla\xi^M\|^2\Delta t + K_{24}(h^2 + (\Delta t)^2)
\end{aligned}
$$

Using (3.13) to estimate $\sum_{n=0}^{M} \|\xi^n\|_1^2 \Delta t$ in (3.35), we have

$$
\begin{aligned}
&\langle \rho_h^M \upsilon^M \ \upsilon^M \rangle + D_0 \sum_{n=1}^{M} \|\nabla\upsilon^n\|^2 \Delta t \\
&\leq \mathrm{Const}(h^2 + (\Delta t)^2 + \sum_{n=1}^{M} \|\upsilon^n\|^2 \Delta t)
\end{aligned}
$$

Using (3.4') and (3.4) and applying a discrete Gronwall lemma to the result, we conclude that for Δt sufficiently small

$$
(3.36) \qquad \|\upsilon^M\|^2 + D_0 \sum_{n=1}^{M} \|\nabla\upsilon^n\|^2 \Delta t \leq \mathrm{Const}(h^2 + (\Delta t)^2)
$$

The estimates (3.13) and (3.36) and the approximation results (3.5) and (3.6) imply the following result.

THEOREM 2.

Assume that inequalities (3.4') hold and the δ_{x^k} are smoothly distributed. Assume that $P_h^n \epsilon M_h$ and $X_h^n \epsilon M_h$ are computed from (3.1) - (3.3) with $P_h^0 = \tilde{P}^0$ and $X_h^0 = \hat{X}^0$ where $\tilde{P}^n$ and $\hat{X}^n$ are an elliptic projection of P^n and the bilinear interpolant of X^n respectively. If the

X_h^n's satisfy a maximum principle, P and X are sufficiently smooth, and Δt is sufficiently small, then

$$\|P_h^M - P^M\| + \|X_h^M - X^M\| \leq \text{Const}(h + \Delta t). //$$

4 COMPUTATIONAL RESULTS

Computational results obtained from an implementation of numerical procedure described in §2 for the compressible miscible displacement problem are discussed below. Three specific questions are addressed in this study. The sensitivity of areal sweep efficiency to the fluid compressibility is examined for displacements exhibiting an unfavorable mobility ratio. A comparison of the areal sweep efficiency of both the repeated five spot and the inverted seven spot patterns (Figures 1 and 2) is performed for unfavorable mobility displacements. Finally, the numerical dispersion present in the procedure is illustrated by simulation of a displacement by a small slug. Computational results concerning the "grid orientation" effect can be found in [13, 14, 16] and demonstrate the insensitivity of this procedure to the orientation of the grid over rectangular domains for miscible displacement problems (incompressible and slightly compressible) and the steamflooding problem.

The base case data that is used in these computational experiments is presented in Table 1. Deviations from this set of data are indicated for each computational experiment. A pore volume of fluid, which is abbreviated as pv, is defined to be the amount of fluid which would occupy one volume of the simulated reservoir at the initial pressure. An $N \times N$ uniform mesh is used for all experiments, and the choice of N varies for each experiment. Elements of symmetry such as one-eighth of the five spot pattern are used to minimize the computational cost. A discussion of these reduced elements of symmetry is found in many sources, including [7].

Under the assumptions of uniform mesh, the well constant θ_k can be expressed as a function of the number N. The calculation of the well constant differs for simulated five spot patterns and simulated seven spot patterns, since the triangulations are based upon squares and equilateral triangles, respectively. To account for the element of symmetry used in the simulation, the well constant θ_k must be multiplied by an appropriate geometric factor. For example, for an eighth of a five spot pattern, the well constant must be multiplied by one-eighth.

The sensitivity of the areal sweep efficiency to the compressibility is studied. For this study, α_r and α_i are varied over the range $\alpha_r = \alpha_i = 0, 10^{-4}, 10^{-3}$, and 10^{-2}. A compressibility of 10^{-2} indicates that the displacing fluid behaves like a gas, and the lower compressibilities are representative of the behavior of highly compressible liquids. These experiments are performed on an eighth of a five spot pattern with $N = 28$.

Figure 3 illustrates the results of these computations. From these results one can infer that a compressibility of the order of 10^{-6}, which is common for most liquids, has very little effect upon the ultimate recovery of the resident fluid. The major effect of a high fluid compressibility is to retard the time of breakthrough of the invading fluid at the production wells. The percentage of resident fluid recovered at the time of breakthrough is quite insensitive to the value of the compressibility.

A comparison of the sweep efficiency at an unfavorable mobility ratio is undertaken for both the inverted seven spot pattern and the repeated five spot pattern. Tensor product linear basis functions that are modified to take into account the elements of symmetry are used to model the five spot pattern. Modified linear basis functions defined over equilateral triangles are used to simulate the inverted seven spot pattern. For the five spot pattern the experiment is performed on an eighth of the pattern with $N = 42$. A twelfth of the pattern is used for the inverted seven spot pattern with a $N = 48$. The mass diffusivity is set to 0 for these calculations.

Figure 4 illustrates the calculated sweep efficiency for the two pattern floods. The recovery efficiency is clearly about ten percent greater for the inverted seven spot pattern over the five spot pattern. This conclusion does not consider variations in reservoir properties, such as the permeability, which can severly effect the sweep efficiency. Use of the inverted seven spot pattern also helps to minimize the effect of these variations. Compositional profiles for the elements of symmetry used in this study at one pore volume of fluid injected are illustrated in Figure 5 for the repeated five spot pattern and in Figure 6 for the inverted seven spot pattern.

The effect of numerical dispersion is studied by simulating the displacement of the resident fluid by a slug of invading fluid of size 3×10^{-1} pore volumes, which is followed by a chaser fluid with fluid properties that are identical to the resident fluid. The compressibility of the resident fluid is set to $\alpha_r = 3 \times 10^{-6}$ psi^{-1}. Two different displacing fluids are used in this investigation. The first displacing fluid has similar properties to the resident fluid, with the exception of the viscosity. This fluid behaves like an incompressible liquid for all practical purposes, and is labeled as such. The second displacing fluid behaves like a gas, and is labeled in that manner. The stock tank density is set to $\rho_i = 1$ lb/cu ft, and the compressibility is set to $\alpha_i = 10^{-2}$ psi^{-1}. A quarter of a five spot pattern is used in this study, with $N = 28$.

The compositional profiles of the invading fluid are illustrated in Figures 7-12 at times 300 days, 350 days, and 400 days for the gas and the liquid slugs. The profiles at 300 days corresponds to the end of the injection of the invading fluid. The profiles at this time do not exhibit a significant amount of numerical dispersion. This is expected due to the continuity of the invading bank. The sweep of the slug corres ponding to liquid injection is much larger than the sweep corresponding gas slug, due to the high compressibility of the gas. By 350 days, the numerical dispersion has taken its toll on the gas slug. The invading bank is mixed with the resident fluid in large proportions. The highest mass fraction of the gaseous component is only 2×10^{-1}. The liquid slug, on the other hand, maintains much of its continuity. The gas slug has completely disappeared by 400 days. By this time, however, the numerical dispersion has not strongly affected the integrity of the liquid slug.

These results indicate that this procedure is inadequate for the simulation of reservoir processes involving a small bank of less than one half of a pore volume injected of the invading fluid in a process involving gas injection. Many realistic processes do involve small banks of primarily gaseous material, particularly carbon dioxide floods. Many other interesting processes, such as continuous drive steam injection and continuous drive miscible displacement, involve large slugs. Such large slug injections are adequately modeled using this procedure.

ACKNOWLEDGEMENT

The authors wish to thank IBM Corporation for their support and assistance in this research.

Table 1

Base Case Data

Pattern Size (acres)	
Five Spot	91.82
Seven Spot	59.64
Fluid Properties	
Stock Tank Density (lbs/cu ft)	
ρ_r	60
ρ_i	60
Compressibility (psi^{-1})	
α_r	0
α_i	0
Viscosity (cp)	
μ_r	1
μ_i	.01
Mass Diffusivity (lbs/ft/day)	60
Rock Properties	
Permeability (md)	100
Porosity	.1
Initial Conditions	
Pressure (psi)	100
Mass fraction	0.
Well Data	
Injection Rate (pv/day)	.001
Bottom Hole Pressure (psi)	100
Well Constant θ_k	$1/\ln(C(\tau)/n^2)$
Five Spot	$C(\tau) = 239600$
Seven Spot	$C(\tau) = 190000$

REFERENCES

1. Aziz, K., and Settari, A., Petroleum Reservoir Simulation, Applied Science Publishers Ltd., London (1979).

2. Bavly, D., Harris, C. D., and Sheldon, J. W., "A method for general reservoir behavior simulation on digital computers, SPE, Paper 1521G presented at the 35th SPE Annual Fall Meeting, Denver, Colorado Oct. 2 - 5 (1960).

3. Bell, J. B., Shubin, and Wheeler, M. F., "Analysis of a new method for computing the flow of miscible fluids in a porous medium (to appear).

4. Bird, R. B., Lightfoot, W. E., and Stewart, E. N., Transport Phenomenon, John Wiley and Sons, New York (1960).

5. Bramble, J. and Hilbert, S., "Estimation of linear functionals on Sobolev space with application to Fourier transforms and spline interpolation, SIAM, J. Numer. Anal., 7, No. 1, 112-124 (1970).

6. Buckley, S. E. and Leverett, M. C., "Mechanism of fluid displacement in sands", Trans. AIME, 146, 107-110 (1942).

7. Coats, K. H. and Ramesh, A. B., "Effect of grid type and difference scheme on pattern steamflood simulation results", SPE, Paper 11079 presented at the Sixth SPE Symposium on Reservoir Simulation, New Orleans, La. (1982).

8. Darlow, B. D., Ewing, R., and Wheeler, M. F., "Mixed finite element method for miscible displacement problems in porous media", SPE, Paper 10501 presented at the Sixth SPE Symposium on Reservoir Simulation, New Orleans, La. (1982).

9. Douglas, J. Jr. and Roberts, J. E., "Numerical methods for a model for compressible miscible displacement in porous media" (to appear).

10. Nitsche, J. A.,"L_∞ - convergence of finite element approximation", Proc. of the Second Conference on Finite Elements, Rennes France (1975).

11. Peaceman, D. W., Fundamentals of Numerical Reservoir Simulation, Elservier Scientific Publishing Co., New York (1977).

12. Peaceman, "Interpretation of well block pressures in numerical reservoir simulations with nonsquare grid blocks and anesotropic permeabilitys", SPE, Paper 10528 presented at the Sixth Symposium on Reservoir Simulation, New Orleans, La. (1982).

13. Potempa, T. C., Finite Element Methods for Convection Dominated Transport Problems, Ph. D. Thesis, Rice University (1982).

14. Potempa, T. C., "A numerical model of two dimensional, two component, single phase flow in a porous media" (to appear).

15. Potempa, T. C., "The effect of the definition of fractional flow upon grid effects in a numerical model of thermal processes" (to appear).

16. Potempa, T. C., "Three dimensional simulation of steam displacement processes with minimal grid orientation", SPE, Paper 11726 presented at the SPE California Regional Meeting, Ventura Calif. (1983).

17. Scott, R., "Optimal L_∞ estimates for the finite element method on irregular meshes", Math Comp. 30, 681-697 (1976).

18. Young, L., "A finite element method for reservoir simulation", SPE Journal, Trans. AIME, 21, 115-128 (1981).

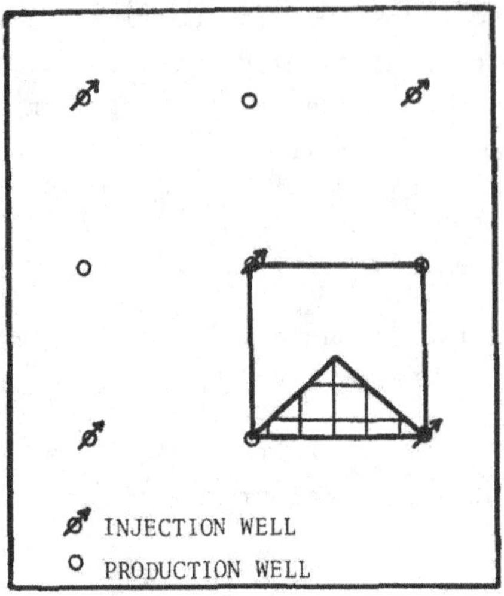

FIGURE 1

Discretization of the Repeated Five Spot

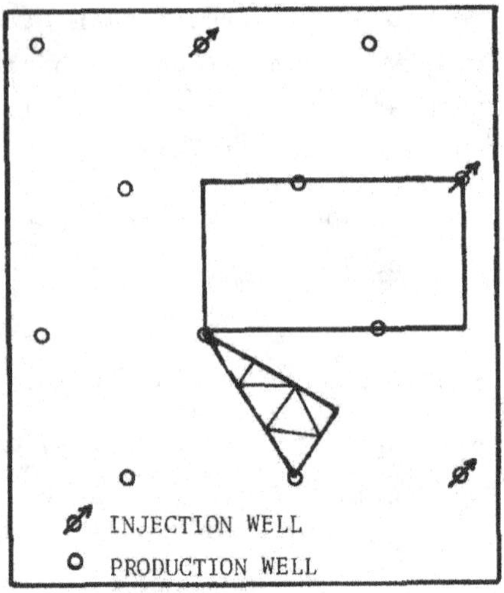

FIGURE 2

Discretization of the Inverted Seven Spot

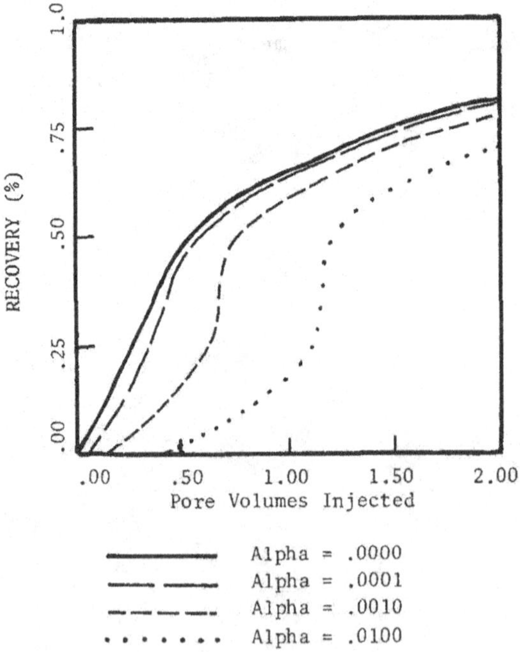

FIGURE 3

The Effect of Fluid Compressibility Upon Areal Sweep Efficiency

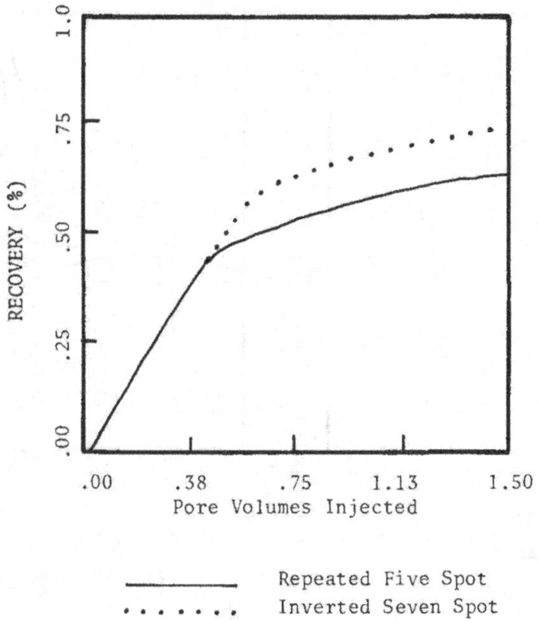

FIGURE 4

Simulated Recovery
Using Different Well
Configurations

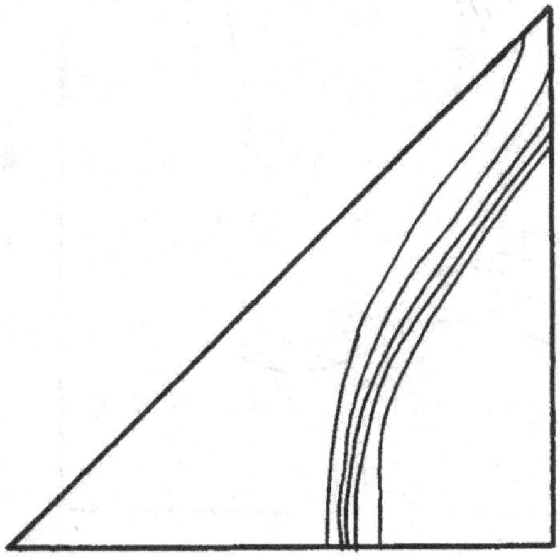

FIGURE 5

Compositional Profile
at 1 PVI, Repeated Five
Spot Pattern

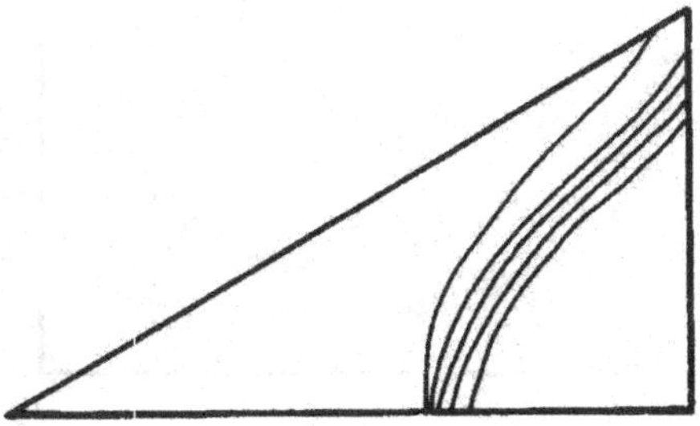

FIGURE 6

Compositional Profile
at 1 PVI, Inverted Seven
Spot Pattern

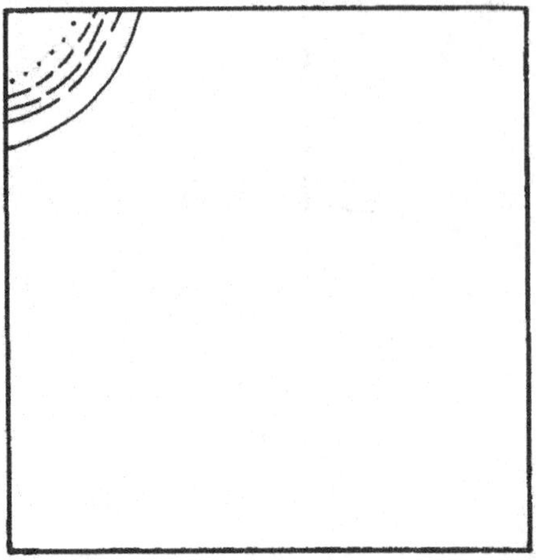

FIGURE 7

Compositional Profiles
at 300 Days, Gas Slug

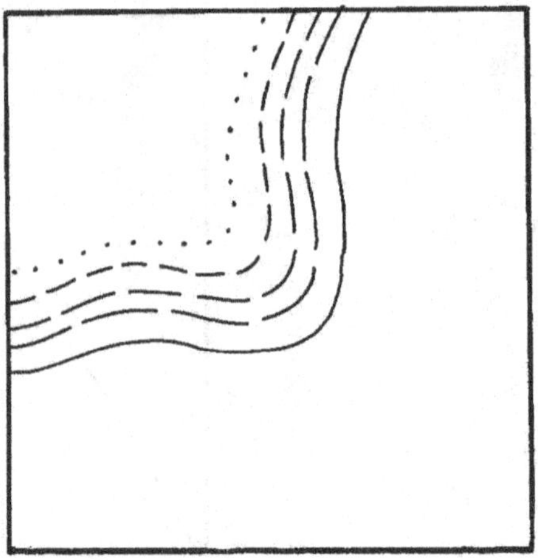

FIGURE 8

Compositional Profiles
at 300 Days, Liquid Slug

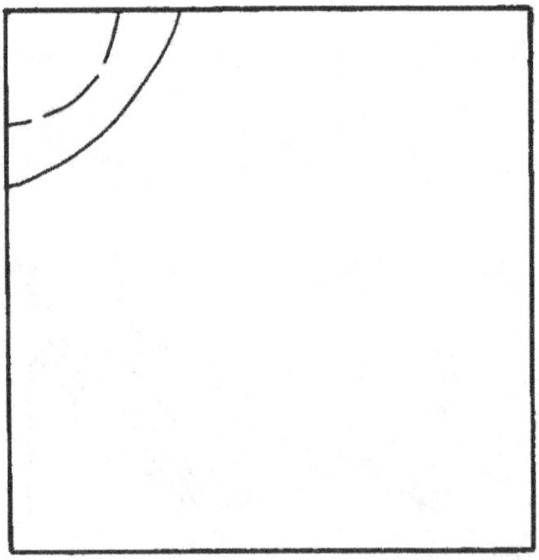

FIGURE 9

Compositional Profiles
at 350 Days, Gas Slug

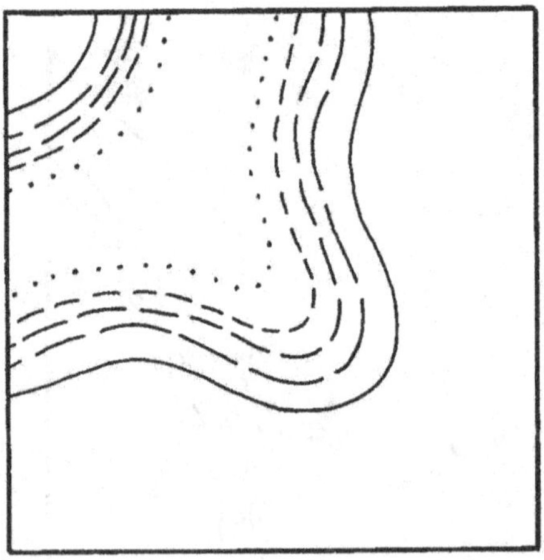

FIGURE 10

Compositional Profiles
at 350 Days, Liquid Slug

FIGURE 11

Compositional Profiles
at 400 Days, Gas Slug

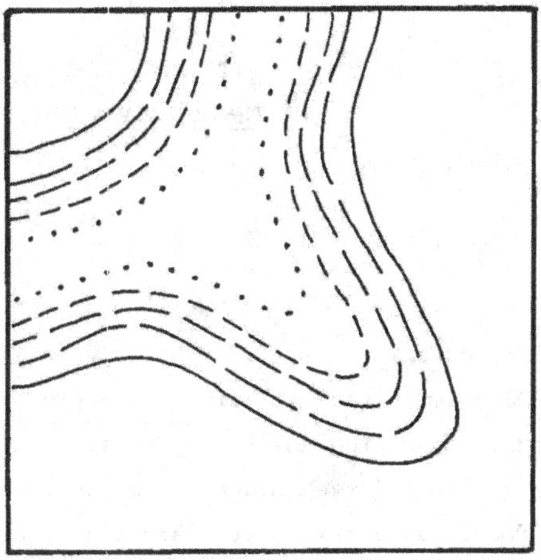

FIGURE 12

Compositional Profiles
at 400 Days, Liquid Slug

SINGULARITIES IN THREE-DIMENSIONAL ELLIPTIC PROBLEMS AND THEIR TREATMENT WITH FINITE ELEMENT METHODS

J.R. Whiteman

1. Introduction

The importance of being able to use finite element methods to treat effectively bounday value problems involving singularities has been recognised in recent years. As a result a large amount of work has been undertaken by mathematicians and engineers to produce variants of standard finite element methods suited to problems of this type.

The work by mathematicians has been mainly in the context of two-dimensional Poisson problems, and has involved not only suitable finite element methods but also associated error analysis, see e.g. Babuska and Rosenzweig [4], Babuska and Rheinboldt [3] and Schatz and Wahlbin [18]. These theoretical techniques rely heavily on analytical results concerning the forms of singularities and regularity properties of the solutions of singular problems, see e.g. Kondrat'ev [10], Lehman [11] and Maz'ja and Plamenevskii [12]. The availability of many such results for two dimensional Poisson problems explains why mathematicians have concentrated on this type of problem.

Simultaneously engineers have created special elements and methods which produce accurate approximations to the solutions of more complicated three-dimensional problems containing boundary singularities. One such important area involving singular problems is that of fracture mechanics, particularly linear elastic fracture. In this case it has been found that the special finite element methods which have been devised are numerically effective for applications in which a linear elastic fracture model of material behaviour is appropriate. However, little is known theoretically about the performance of such methods.

In the context of singularities there thus exists a considerable gulf between the area for which theoretical finite element analysis currently exists and those areas for which the finite element method is used effectively in engineering practice. The focus of this paper is therefore on the gap which lies between the above roughly defined boundaries of theory and practice. Since knowledge of the forms of singularities is fundamental to the successful finite element treatment of the above types of singular problem, the forms of singularities in some three-dimensional Poisson and linear elasticity problems are first discussed. The effects of these singularities on the rates of convergence with decreasing mesh size (h-convergence) of finite element methods are then described. It is not the intention here to discuss the relative merits of the many singularity variants of finite element methods. Instead one such method is chosen and in Section 3 is discussed with reference to theoretical error analysis. Numerical results indicating its performance for a three-dimensional test problem are then presented.

Finally in the fracture mechanics context a short discussion is given in Section 5 which has the aim of stimulating thought on the actual material behaviour in the neighbourhood of a crack. The possibility of the onset of *plastic* material response is accepted and some remarks concerning the effect of admitting this type of nonlinear behaviour are made. The question of whether it is sensible in practice to seek more and more accurate numerical solutions to possibly inadequate mathematical models of material behaviour is raised. The introduction of this topic is aimed at illustrating still further the gulf between theory and practice.

2. Boundary Value Problems Containing Singularities

Let $\Omega \subset \mathbb{R}^3$ be a simply connected polyhedral domain with boundary $\partial\Omega$, where $\partial\Omega \equiv \partial\Omega_1 \cup \partial\Omega_2$, with $\partial\Omega_1 \cap \partial\Omega_2 = \emptyset$. Singularities will be considered in the context of two classes of boundary value problems, one from potential theory and one from linear elasticity.

In the first class of problems the scalar *potential* function $u(\underline{x})$ satisfies Poisson's equation in Ω, together with appropriate boundary conditions on $\partial\Omega$. Thus for $\underline{x} \equiv (x_1,x_2,x_3)^T$

$$\left.\begin{aligned} -\Delta\, u(\underline{x}) &= f(\underline{x}) \;, & \underline{x} &\in \Omega \;, \\ u(\underline{x}) &= 0 \;, & \underline{x} &\in \partial\Omega_1 \;, \\ \frac{\partial u(\underline{x})}{\partial n} &= g(\underline{x}) \;, & \underline{x} &\in \partial\Omega_2 \;, \end{aligned}\right\} \tag{2.1}$$

where $f \in L_2(\Omega)$, $g \in L_2(\partial\Omega_2)$ and $\partial/\partial n$ is the derivative in the direction of the outward normal to $\partial\Omega_2$.

A weak form of (2.1) is defined in the setting of the Sobolev space $H^1(\Omega)$. Let

$$V \equiv \left\{ v : v \in H^1(\Omega) \text{ and } v\big|_{\partial\Omega_1} = 0 \right\} , \tag{2.2}$$

then the weak solution $u \in V$ of (2.1) satisfies

$$a(u,v) = F(v) \quad \forall \; v \in V \;, \tag{2.3}$$

where

$$a(u,v) \equiv \int_\Omega \nabla u \, \nabla v \, d\underline{x} \;, \qquad u,v \in V \tag{2.4}$$

and

$$F(v) \equiv \int_\Omega f\, v\, d\underline{x} + \int_{\partial\Omega_2} g\, v\, ds \;, \qquad v \in V \;. \tag{2.5}$$

The second class consists of problems from linear elasticity in which the vector $\underline{u}(\underline{x}) \equiv \{u_1(\underline{x}),\, u_2(\underline{x}),\, u_3(\underline{x})\}^T$ is the *displacement* at the point $\underline{x} \equiv \{x_1,x_2,x_3\}^T \in \Omega$ and satisfies the Navier equations of elasticity in Ω together with appropriate boundary consitions on $\partial\Omega$. Thus

$$\left.\begin{aligned} -\mu\,\Delta\underline{u}(\underline{x}) - (\lambda+\mu)\,\mathrm{grad}\,\mathrm{div}\,\underline{u}(\underline{x}) &= \underline{f}(\underline{x})\,, & \underline{x} &\in \Omega\,, \\ \underline{u}(\underline{x}) &= 0\,, & \underline{x} &\in \partial\Omega_1 \\ \sum_{j=1}^{3} \sigma_{ij}(\underline{u}(\underline{x}))n_j &= g_i(\underline{x})\,, & \underline{x} &\in \partial\Omega_2\,, \\ & & 1 \leqq i &\leqq 3\,, \end{aligned}\right\} \tag{2.6}$$

where $\underline{f} \in (L^2(\Omega))^3$, $g \equiv (g_1,g_2,g_3) \in (L_2(\partial\Omega_2))^3$ are boundary tractions, n_j is the jth component of the unit outward normal $\underline{n} \equiv (n_1,n_2,n_3)$ to $\partial\Omega_2$ at $\underline{x}$, the stress tensor $\sigma_{ij}(\underline{u}(\underline{x}))$ is, for $\underline{x} \in \Omega$, defined as

$$\sigma_{ij}(\underline{u}) = \sigma_{ji}(\underline{u}) \equiv \lambda\left(\sum_{k=1}^{3} \varepsilon_{kk}(\underline{u})\right)\delta_{ij} + 2\mu\varepsilon_{ij}(\underline{u})\,, \qquad 1 \leqq i,j \leqq 3\,, \tag{2.7}$$

and in (2.6) and (2.7) $\lambda > 0$, $\mu > 0$ are the Lamé constants which are defined, see e.g. Nečas and Hlávaček [14], in terms of the Young's modulus E and Poisson's ratio ν as

$$\lambda = \frac{E\nu}{(1+\nu)(1-2\nu)} \quad \text{and} \quad \mu = \frac{E}{2(1+\nu)}$$

The strain tensor $\varepsilon_{ij}(\underline{u})$ is defined similarly in terms of displacements as

$$\varepsilon_{ij}(\underline{u}) = \varepsilon_{ji}(\underline{u}) = \frac{1}{2}\left(\frac{\partial u_i}{\partial x_j} + \frac{\partial u_j}{\partial x_i}\right)\,, \quad 1 \leqq i,j < \leqq 3\,. \tag{2.8}$$

Similarly to (2.3), but now in the space $\widetilde{V}$, where

$$\widetilde{V} \equiv \left\{\underline{v} : \underline{v} \in (H^1(\Omega))^3\,,\ v_i\big|_{\partial\Omega_1} = 0 \quad 1 \leqq i \leqq 3\right\}, \tag{2.9}$$

a weak form of (2.6) is defined in which $\underline{u} \in \widetilde{V}$ satisfies

$$\widetilde{a}(\underline{u},\underline{v}) = \widetilde{F}(\underline{v}) \qquad \underline{v} \in \widetilde{V}\,, \tag{2.10}$$

where

$$\begin{aligned} \widetilde{a}(\underline{u},\underline{v}) &\equiv \int_\Omega \sum_{i,j=1}^{3} \sigma_{ij}(\underline{u})\,\varepsilon_{ij}(\underline{v})\,d\underline{x} \\ &= \int_\Omega \left\{\lambda\,\mathrm{div}\,\underline{u}\,\mathrm{div}\,\underline{v} + 2\mu \sum_{i,j=1}^{3} \varepsilon_{ij}(\underline{u})\,\varepsilon_{ij}(\underline{v})\right\}d\underline{x} \qquad \underline{u},\underline{v} \in \widetilde{V} \end{aligned} \tag{2.11}$$

and

$$\widetilde{F}(\underline{v}) \equiv \int_\Omega \underline{f}.\underline{v}\,d\underline{x} + \int_{\partial\Omega_2} \underline{g}.\underline{v}\,ds\,, \qquad v \in \widetilde{V}\,. \tag{2.12}$$

Problems (2.1) and (2.6) are, as is usual, now treated by considering their respective weak forms (2.3) and (2.10). Important features of these weak forms, particularly relevant to finite element error analysis, are the continuity and V-ellipticity ($\widetilde{V}$-ellipticity) of the symmetric bilinear forms a(u,v) and $\widetilde{a}$(u,v). Continuity is clear; the V-ellipticity of a(u,v) is considered by Ciarlet [5] and Grisvard [8]; the $\widetilde{V}$-ellipticity of $\widetilde{a}$(u,v) has been proved by Fichera [6], see also [5].

Our interest here lies in problems of the above type in which the boundaries contain re-entrant vertices and edges, and which as a consequence contain point singularities at the vertices and line singularities along the edges. Problems of type (2.3) are considered first. As has been remarked earlier, whilst much has been written about singularities in two-dimensional forms of these problems, see e.g. Whiteman [23] and the large number of references cited therein, significantly less is known about singularities in three-dimensional problems. For regions Ω containing re-entrant vertices and edges as in Fig. 1, with notation as indicated,

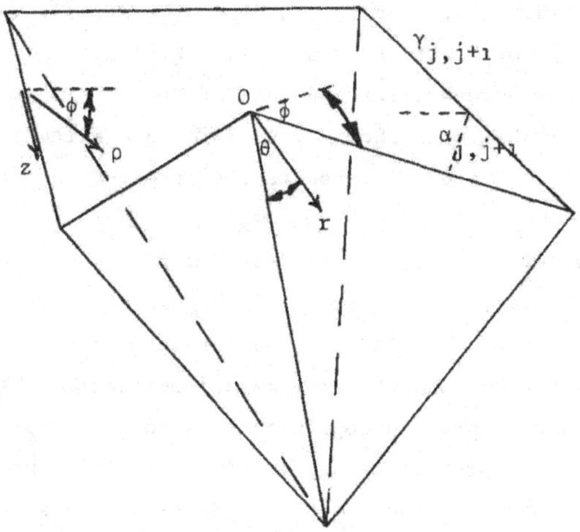

Fig. 1

Grisvard [9] and Stephan [19] have considered (2.3) for the case $g(\underline{x}) = 0$, $\underline{x} \in \partial\Omega_2$, and it has been shown that $u(\underline{x})$ can be written in the form

$$u = \sum_{\text{vertices}} a_i \chi_i(r_i,\theta_i,\phi_i) u_i(r_i,\theta_i,\phi_i) + \sum_{\text{edges}} b_j(z_j)\Xi_j(\rho_j,z_j,\phi_j) v_j(\rho_j,\phi_j) + w , \qquad (2.13)$$

where $w \in H^2(\Omega)$, (r_i,θ_i,ϕ_i) and (ρ_j,z_j,ϕ_j) are respectively spherical and cylindrical polar coordinates local to the ith vertex and j^{th} edge, u_i and v_j are (singular) functions appropriate to the relevant corner or edge, a_i are unknown coefficients, $b_j(z_j) \in H^2(\gamma_{j,j+1})$ are unknown (edge) functions and χ_i and Ξ_j are *cut-off* functions. The size of the interior angle at a vertex or an edge determines whether or not the u_i or v_j is a singular function and, when this is so, these functions represent the dominant part of the singularity which occurs. For a point on an edge the form of the v_j in the plane through the point orthogonal to the edge is that of the corresponding two-dimensional Poisson problem in that plane. The form of the (singular) function u_i at the ith vertex will in general be $r_i^{\mu_i} Q_i(\theta_i,\phi_i)$ where, as has been shown by Grisvard [9], the μ_i is related to the smallest eigenvalue of the Laplace-Beltrami equation defined on that part of the unit sphere centred on the corner interior to Ω. Thus in order to calculate u_i it is necessary to solve an associated eigenvalue problem. This has to date been done for two simple cases by Stephan and Whiteman [20], whilst for a more complicated case Fichera [7] has obtained upper and lower bounds for the corresponding eigenvalue.

We turn now to problems of the type (2.6). The most common occurrence of these is in linear elastic fracture in which a re-entrant edge (see Fig. 1) is the apex of a flat crack, re-entrant angle $\alpha_{j,j+1} = 2\pi$, and along the crack front there is a line singularity. For such a crack at a point internal to the front the displacement $\underline{u}$ has a $\rho^{\frac{1}{2}}$ form so that from (2.8) the strain, and hence the stress, has a $\rho^{-\frac{1}{2}}$ form. At the end point of this type of front, such as O in Fig. 1, the form of the singularity is less well understood. However, for the case of Mode I linear elastic fracture under symmetrical loading, it has been shown by Thompson and Whiteman [22] that, near the vertex point where the crack front intersects the stress-free surface, the stresses have an $r^{-\frac{1}{2}}$ form.

In linear elastic fracture the *stress intensity factor* is a quantity of significance. This factor is essentially the magnitude of the singularity at a crack tip and corresponds to the coefficient of the leading singular term in the displacement near the crack tip. In an engineering context it is more important to obtain a calculated value for this quantity, than to produce approximations to the displacements; which are the primary unknowns in the problem being treated. There is thus a post-processing (recovery/retrieval) situation. For the example in Section 4 an approximation to the stress intensity factor is produced via the J-integral of Rice [17] using the differential stiffness method of Parks [16].

The main interest here is of course the singularity which occurs at a vertex or edge. In the context of problems of types (2.3) and (2.10) the effect of the above singularities is to produce solutions of the problems concerned which have low regularity. This has important consequences in the finite element context, as it causes a reduction of the theoretical rate of convergence with decreasing mesh size of the finite element solution to the weak solution. This reduction is now indicated.

3. Finite Elements Methods

When the Galerkin method is applied to the problem (2.3), the region Ω is partitioned into elements in the usual way. If conforming trial and test functions are employed, the solution $u \in V$ is approximated by $u_h \in S^h$, where $S^h \subset V$ is a finite dimensional space of piecewise polynomial functions of degree p, and u_h satisfies

$$a(u_h, v_h) = F(v_h) \quad \forall \quad v_h \in S^h . \tag{3.1}$$

Use of the V-ellipticity and continuity of the bilinear form a(u,v) enables the error $\| u - u_h \|$ to be bounded in terms of $\| u - v_h \| \quad \forall \quad v_h \in S^h$ and approximation theory results can then be used to produce error bounds of the form

$$\| u - u_h \| \leq K \, h^{\gamma} |u|_k , \tag{3.2}$$

where K is an unknown constant depending on p, $|u|_k$ is the kth order seminorm of u and

γ depends on p, k and the choice of norm on the left hand side of the inequality. The rate of convergence is particularly dependent on the regularity of u, which has been shown in Section 2 to depend on the shape of the boundary and on the boundary conditions.

For the case of the L_∞-norm, for a *two-dimensional* problem of the type (2.3) in which the solution $u \in H^2(\Omega)$, it has been shown by Natterer [13] that on a quasi-uniform mesh the convergence rate can be $2-\varepsilon$ for arbitrary ε. For two dimensional problems of type (2.3) with homogeneous Dirichlet conditions over $\partial\Omega$, but containing vertex singularities, Schatz and Wahlbin [18] have derived bounds which give γ in terms of the re-entrant angles at the vertices. They show that the L_∞ rate of convergence is lower local to a re-entrant vertex than it is away from the vertex. The case of a general (non-singular) Poisson problem with $\Omega \subset \mathbb{R}^N$ has been considered by Nitsche [15] who derives L_∞ results but with restrictions on the degree of polynomial in the piecewise test and trial functions.

It has long been known *experimentally* that "standard" finite element techniques are less accurate near point and line singularities. Although the above *theoretical* results have been proved only for simple two-dimensional problems containing singularities, they do for this case quantify the decrease in convergence rate. It should particularly also be noted that, when the criteria of the relevant theorems are fulfilled, convergence of u_h to u is guaranteed, albeit at a slower rate than for problems not involving singularities.

All the remarks to date in this Section have been directed towards the scalar function $u(\underline{x})$ which is the solution of problem (2.3). It is clear that with appropriate change of notation, error bounds comparable to (3.2) can be derived for the corresponding approximations to the displacement $\underline{u}(\underline{x})$ of the elasticity problem (2.10). The situation regarding slower rates of convergence in the presence of boundary singularities is also relevant to this problem.

The effects, theoretical and experimental, outlined above have motivated attention to singularity treatment and many special finite element methods have been proposed. Examples of these are local mesh refinement, use of special elements and augmentation with singular functions; they have been extensively surveyed by Atluri [1] and Whiteman [24]. It is not the intention here to discuss these methods in general, but rather to consider how one particular technique, using the "$\frac{1}{4}$-point" element, improves accuracy and to illustrate its use.

We consider again (2.3) and the Galerkin approximation (3.1). In the application of the Galerkin technique with conforming functions (3.1) is written as

$$\sum_e a(u_h, v_h)\Big|_e = \sum_e F(v_h)_e \tag{3.3}$$

where the summation is over the elements e, the totality of which partitions Ω. When an isoparametric method is used, each element in physical $\underline{x}$-space is mapped in turn onto a *standard* element in local $\underline{\xi}$-space. Under the transformation the global derivatives of u_h and v_h are transformed into local derivatives so that in the element

$$\{\partial(u_h(\underline{x})_{glob}\}_e = J^{-1}\{\partial\hat{u}_h(\underline{\xi})_{loc}\}_e \qquad (3.4)$$

where J is the Jacobian of the transformation. For an element involving a point of singularity the global derivatives of the approximating function $u_h(\underline{x})$ should possess the same form as those of $u(\underline{x})$ at the singular point. One way of achieving this is to introduce the singularity at a corner of the element via the Jacobian of the transformation. In the $\frac{1}{4}$-point element this is achieved by using quadratic basis functions and by moving the nodes at the mid-points of the sides meeting at the singularity to positions on the sides one quarter of the length of the side from the singular point. Subject to certain constraints on the shape of the elements used, this technique can produce the required ρ-behaviour in u_h about the singular point local to the element, thus improving the accuracy of the approximation.

The technique has been found experimentally to be effective in improving accuracy near a point of singularity, particularly for linear elastic fracture where $\rho^{\frac{1}{2}}$ behaviour of $\underline{u}$ has to be approximated. Although proposed originally for two-dimensional elements, the generalisation to three-dimensions for line singularities is straightforward, see [23]. It is clearly desirable to relate experimental evidence to theoretical error inequalities.

In the context of (3.2) the use of $\frac{1}{4}$-point elements cannot affect the value of k used in the bound, as this is dependent on the smoothness of u. Neither is the degree of the approximating polynomial in the element changed. This suggests that γ remains unchanged and that improvement in accuracy locally must occur due to a reduction in the size of the constant K.

4. Test Problem - Numerical Results (G.M. Thompson)

In order to demonstrate the use of $\frac{1}{4}$-point elements a three-dimensional problem from linear elastic fracture is considered. The finite element calculations for this were computed by G.M. Thompson [21].

The problem is of type (2.6) in which a homogeneous isotropic linear elastic solid $E = 26.9 \times 10^6$, $\nu = 0.3$, of brick shape and containing a flat edge crack is subjected to symmetric loading of 10^5psi on its end faces in the Mode I manner. On account of symmetry only the half region Ω of Fig. 2 need be considered in which PAEQ is the edge crack with front PQ, AB = AD = 0.1, AE = 0.02 and AP/AD = 0.1. By virtue of the symmetry $u_1 = 0$ on DPHQ; the loading imposes the condition $\sigma_{11} = 10^5$ on BCGF, whilst the remaining faces are stress free. Body forces are neglected.

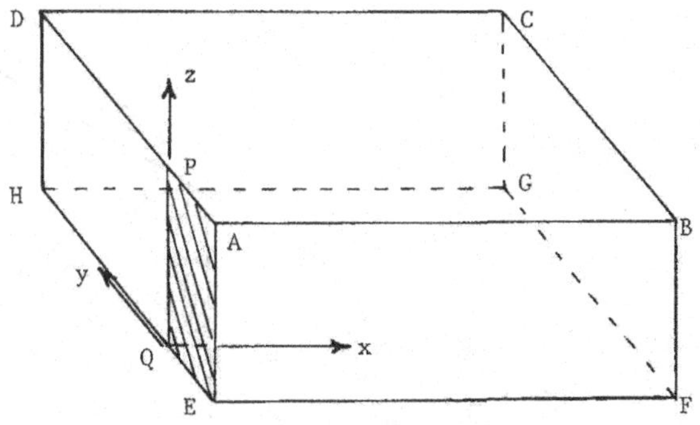

Fig. 2

The finite element method is applied to this problem via the formulation (2.10) for the mesh as shown in Fig. 3. In this standard twenty node quadratic isoparametric brick elements are used throughout, except for elements involving the crack front where degenerated fifteen node isoparametric triangular prismatic elements are employed. The method is used for the two cases in which the prismatic elements are (a) standard (b) $\frac{1}{4}$-point elements. Values for the stress intensity factor for both cases are given in Fig. 4 through the thickness of the solid at nine points along the crack front. It can be seen that the values for K_I calculated with $\frac{1}{4}$-point elements are greater than the corresponding values calculated with standard elements. This can be expected because the $\frac{1}{4}$-point elements model the $\rho^{-\frac{1}{2}}$ strain behaviour better near the crack front, giving there higher values for the strain approximation and hence for K_I.

5. Remarks

As was stated in the introduction the intention has been here to consider finite element methods in the context of three dimensional linear singularities, both theoretically and practically. The importance of knowing the forms of the singularities in the problems considered has been stressed and, using knowledge of these forms, the improvement of the approximations with $\frac{1}{4}$-point elements has been discussed theoretically and indicated experimentally. The error analysis of Section 3 has been presented in terms of the difference between the *primary* variable and its approximation. Often, as illustrated in the problem of Section 4, the quantity of most interest is one which is derived from the primary variable by post-processing. There is thus clearly a great need for error analysis which covers post-processing both in the fracture context and more generally. Recent work by Babuska and Miller [2] addresses certain post-processing approaches for stress-intensity factors.

The theoretical derivation of the forms of singularities in problems of potential theory and linear elasticity is mathematically elegant and, in the context of such

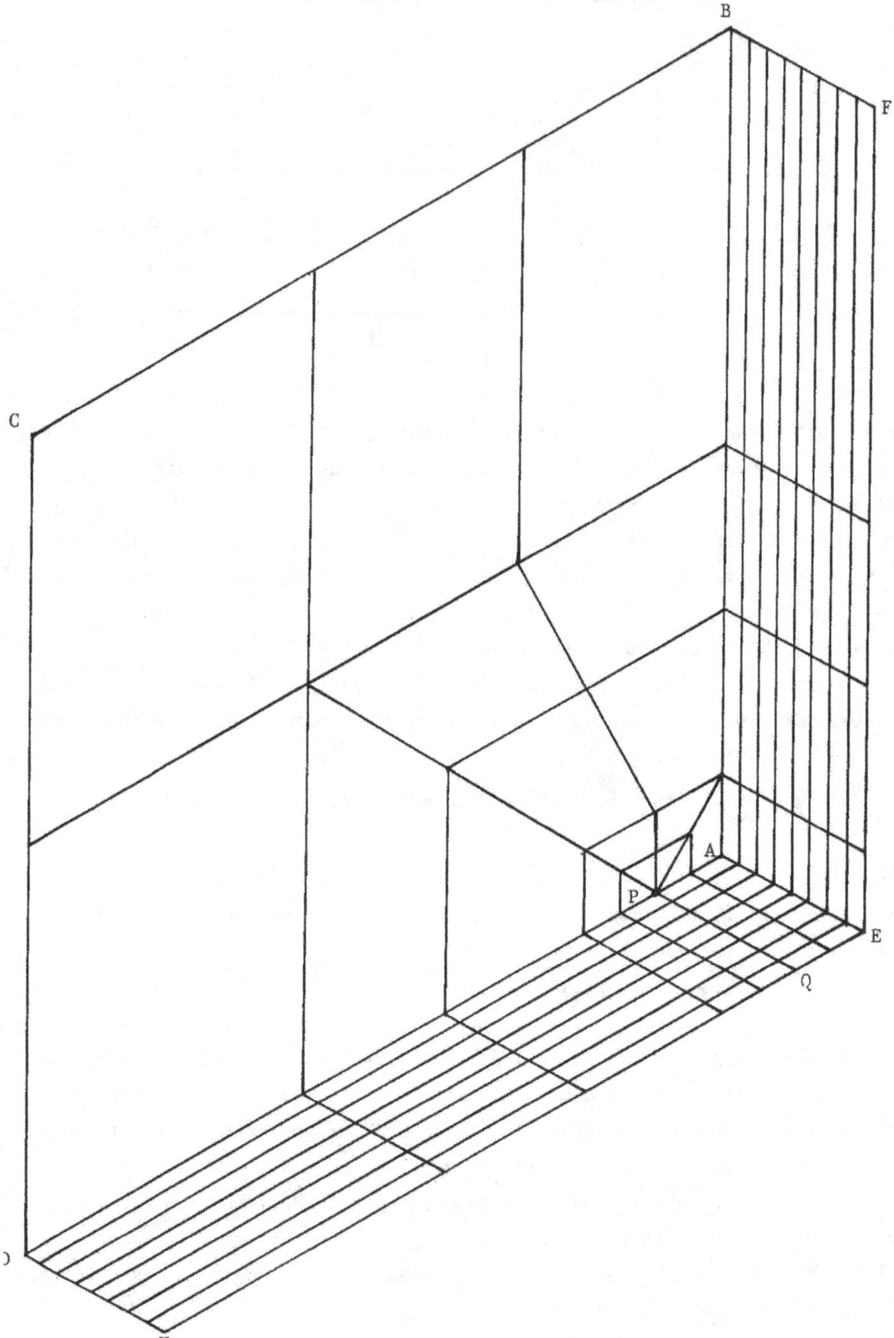

Fig. 3

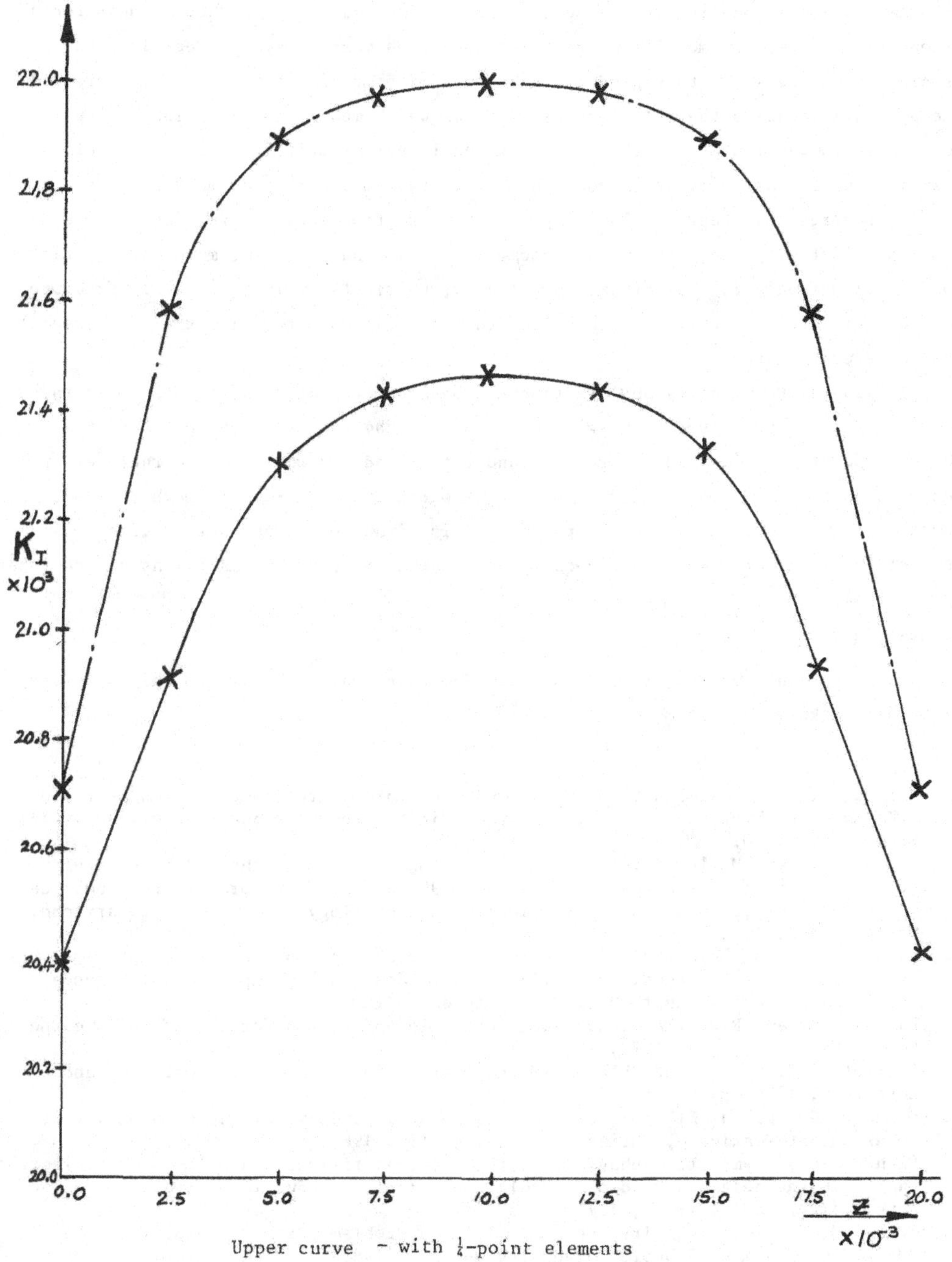

Upper curve - with ¼-point elements

Lower curve - with standard prismatic elements

Fig. 4

problems, computationally advantageous. However, the adequacy of these linear formulations for the task of modelling practical phenomena must always be kept in mind. It is clearly not sensible to expend large amounts of computing time to obtain very accurate approximations to the solutions of problems which model behaviour incompletely. An example in the context of this paper is the linear elastic fracture model which assumes the fracturing material to be brittle. Many materials become ductile at certain levels of stress, so that an elastic-plastic deformation model is required. With the onset of plasticity the problem has become nonlinear and, clearly, much more difficult to treat. Not only is the mathematical theory, particularly in the case of fracture, an order of magnitude more difficult, but the computational techniques are necessarily much more complicated.

The gap between theory and practice has been demonstrated in this paper in the context of linear problems. The gulf between the subclass of linear problems for which computational techniques are well understood and for which considerable error analysis exists and these nonlinear problems, where the situation is much less satisfactory, is even more strikingly apparent. This situation is perhaps inevitable in a subject where research is simultaneously undertaken by both mathematicians and engineers

Acknowledgement

The contributions of D. Harrison, G.M. Thompson and T.J.W.Ward to this work are gratefully acknowledge.

References

1. Atluri, S.N., Higher-order, special, and singular finite elements. Chapter 4 of A.K. Noor and W. Pilkey (eds.), Survey of Finite Element Methods. American Society Mech. Eng., 1980.
2. Babuska, I. and Miller A., The post-processing approach in the finite element method Parts 1, 2 and 3. Technical Notes BN-992, 993 and 1007, Laboratory for Numerical Analysis, Institute for Physical Science and Technology, University of Maryland, College Park, 1982-83.
3. Babuska, I. and Rheinboldt, W., Reliable error estimation and mesh adaptation for the finite element method. pp.67-108 of J.T. Oden (ed.), Computational Methods in Nonlinear Mechanics. North Holland, Amsterdam, 1980.
4. Babuska, I. and Rosenzweig, M.B., A finite element scheme for domains with corners. Numer. Math. 20, 1-21, 1972.
5. Ciarlet, P.G., The Finite Element Method for Elliptic Problems. North Holland, Amsterdam, 1978.
6. Fichera, G., Linear Elliptic Differential Systems and Eigenvalue Problems. Lecture Notes in Mathematics 8, Springer-Verlag, Berlin, 1965.
7. Fichera, G., Asymptotic behaviour of the electric field and density of the electric charge in the neighbourhood of singular points of a conducting surface. Russian Math. Surveys 30, 107-127, 1975.
8. Grisvard, P., Numerical Treatment of Elliptic Problems with Singularities. Lecture Notes, ICPAM Summer School, University of Nice, 1982.
9. Grisvard, P., Behaviour of the solutions of an elliptic boundary value problem in a polygonal or polyhedral domain. pp.207-274 of B. Hubbard (ed.). Numerical Solution of Partial Differential Equations III, SYNSPADE 1975. Accademic Press, New York, 1976.
10. Kondrat'ev, V.A., Boundary problems for elliptic equations in domains with conical or angular points. Trans. Moscow Math. Soc. 16, 227-313, 1967.

11. Lehman, R.S., Development at an analytic corner of solutions of elliptic partial differential equations. J. Maths. Mech. 8, 727-760, 1959.
12. Maz'ja, V.G. and Plamenevskii, B.A., On boundary value problems for a second order elliptic equation in a domain with edges. Vestnik Leningrad Univ. Mathemat. 8, 99-106, 1980.
13. Natterer, F., Uber die punktweise Konvergenz finiter Elemente. Numer. Math. 25, 67-77, 1975.
14. Nečas, J. and Hlávaček, I., Mathematical Theory of Elastic and Elastic-Plastic Bodies. Elsevier, Amsterdam, 1981.
15. Nitsche, J., L_∞-convergence of finite element approximation. Proc. 2nd Conference on Finite Elements, Rennes, 1975.
16. Parks, D., A stiffness derivative finite element technique for determination of crack tip stress intensity factors. Int. J. Fracture 10, 487-502, 1974.
17. Rice, J.R., A path independent integral and the approximate analysis of strain concentration by notches and cracks. J. Appl. Mech. 35, 379-386, 1968.
18. Schatz, A. and Wahlbin, L., Maximum norm estimates in the finite element method on plane polygonal domains, Parts I and II. Math. Comp. 32, 73-109, 1978 and Math. Comp. 33, 465-492, 1979.
19. Stephan, E., A modified Fix method for the mixed boundary value problem of the Laplacian in a polyhedral domain. Technical Report 538, Fachbereich Mathematik, Technische Hochschule Darmstadt, 1980.
20. Stephan, E. and Whiteman, J.R., Singularities in the Laplacian at corners and edges of three dimensional domains and their treatment with finite element methods. Technical Report BICOM 81/1, Institute of Computational Mathematics, Brunel University, 1981.
21. Thompson, G., The Finite Element Solution of Fracture Problems in Two- and Three-Dimensions. Ph.D. Thesis, Department of Mathematics and Statistics, Brunel University, 1983.
22. Thompson, G.M. and Whiteman, J.R., The notch problem in three dimensional linear elasticity. Technical Report, Institute of Computational Mathematics, Brunel University. (to appear)
23. Whiteman, J.R., Finite elements for singularities in two- and three-dimensions. pp.37-55 of J.R. Whiteman (ed.), The Mathematics of Finite Elements and Applications IV, MAFELAP 1981. Academic Press, London, 1982.
24. Whiteman, J.R., Problems with singularities. Sections II.6.0 and II.6.1. of H. Kardestuncer (ed.) Finite Element Handbook. (to appear)

Vol. 926: Geometric Techniques in Gauge Theories. Proceedings, 1981. Edited by R. Martini and E.M.de Jager. IX, 219 pages. 1982.

Vol. 927: Y. Z. Flicker, The Trace Formula and Base Change for GL (3). XII, 204 pages. 1982.

Vol. 928: Probability Measures on Groups. Proceedings 1981. Edited by H. Heyer. X, 477 pages. 1982.

Vol. 929: Ecole d'Eté de Probabilités de Saint-Flour X - 1980. Proceedings, 1980. Edited by P.L. Hennequin. X, 313 pages. 1982.

Vol. 930: P. Berthelot, L. Breen, et W. Messing, Théorie de Dieudonné Cristalline II. XI, 261 pages. 1982.

Vol. 931: D.M. Arnold, Finite Rank Torsion Free Abelian Groups and Rings. VII, 191 pages. 1982.

Vol. 932: Analytic Theory of Continued Fractions. Proceedings, 1981. Edited by W.B. Jones, W.J. Thron, and H. Waadeland. VI, 240 pages. 1982.

Vol. 933: Lie Algebras and Related Topics. Proceedings, 1981. Edited by D. Winter. VI, 236 pages. 1982.

Vol. 934: M. Sakai, Quadrature Domains. IV, 133 pages. 1982.

Vol. 935: R. Sot, Simple Morphisms in Algebraic Geometry. IV, 146 pages. 1982.

Vol. 936: S.M. Khaleelulla, Counterexamples in Topological Vector Spaces. XXI, 179 pages. 1982.

Vol. 937: E. Combet, Intégrales Exponentielles. VIII, 114 pages. 1982.

Vol. 938: Number Theory. Proceedings, 1981. Edited by K. Alladi. IX, 177 pages. 1982.

Vol. 939: Martingale Theory in Harmonic Analysis and Banach Spaces. Proceedings, 1981. Edited by J.-A. Chao and W.A. Woyczyński. VIII, 225 pages. 1982.

Vol. 940: S. Shelah, Proper Forcing. XXIX, 496 pages. 1982.

Vol. 941: A. Legrand, Homotopie des Espaces de Sections. VII, 132 pages. 1982.

Vol. 942: Theory and Applications of Singular Perturbations. Proceedings, 1981. Edited by W. Eckhaus and E.M. de Jager. V, 363 pages. 1982.

Vol. 943: V. Ancona, G. Tomassini, Modifications Analytiques. IV, 120 pages. 1982.

Vol. 944: Representations of Algebras. Workshop Proceedings, 1980. Edited by M. Auslander and E. Lluis. V, 258 pages. 1982.

Vol. 945: Measure Theory. Oberwolfach 1981, Proceedings. Edited by D. Kölzow and D. Maharam-Stone. XV, 431 pages. 1982.

Vol. 946: N. Spaltenstein, Classes Unipotentes et Sous-groupes de Borel. IX, 259 pages. 1982.

Vol. 947: Algebraic Threefolds. Proceedings, 1981. Edited by A. Conte. VII, 315 pages. 1982.

Vol. 948: Functional Analysis. Proceedings, 1981. Edited by D. Butković, H. Kraljević, and S. Kurepa. X, 239 pages. 1982.

Vol. 949: Harmonic Maps. Proceedings, 1980. Edited by R.J. Knill, M. Kalka and H.C.J. Sealey. V, 158 pages. 1982.

Vol. 950: Complex Analysis. Proceedings, 1980. Edited by J. Eells. IV, 428 pages. 1982.

Vol. 951: Advances in Non-Commutative Ring Theory. Proceedings, 1981. Edited by P.J. Fleury. V, 142 pages. 1982.

Vol. 952: Combinatorial Mathematics IX. Proceedings, 1981. Edited by E. Billington, S. Oates-Williams, and A.P. Street. XI, 443 pages. 1982.

Vol. 953: Iterative Solution of Nonlinear Systems of Equations. Proceedings, 1982. Edited by R. Ansorge, Th. Meis, and W. Törnig. VII, 202 pages. 1982.

Vol. 954: S.G. Pandit, S.G. Deo, Differential Systems Involving Impulses. VII, 102 pages. 1982.

Vol. 955: G. Gierz, Bundles of Topological Vector Spaces and Their Duality. IV, 296 pages. 1982.

Vol. 956: Group Actions and Vector Fields. Proceedings, 1981. Edited by J.B. Carrell. V, 144 pages. 1982.

Vol. 957: Differential Equations. Proceedings, 1981. Edited by D.G. de Figueiredo. VIII, 301 pages. 1982.

Vol. 958: F.R. Beyl, J. Tappe, Group Extensions, Representations, and the Schur Multiplicator. IV, 278 pages. 1982.

Vol. 959: Géométrie Algébrique Réelle et Formes Quadratiques, Proceedings, 1981. Edité par J.-L. Colliot-Thélène, M. Coste, L. Mahé, et M.-F. Roy. X, 458 pages. 1982.

Vol. 960: Multigrid Methods. Proceedings, 1981. Edited by W. Hackbusch and U. Trottenberg. VII, 652 pages. 1982.

Vol. 961: Algebraic Geometry. Proceedings, 1981. Edited by J.M. Aroca, R. Buchweitz, M. Giusti, and M. Merle. X, 500 pages. 1982.

Vol. 962: Category Theory. Proceedings, 1981. Edited by K.H. Kamps, D. Pumplün, and W. Tholen, XV, 322 pages. 1982.

Vol. 963: R. Nottrot, Optimal Processes on Manifolds. VI, 124 pages. 1982.

Vol. 964: Ordinary and Partial Differential Equations. Proceedings, 1982. Edited by W.N. Everitt and B.D. Sleeman. XVIII, 726 pages. 1982.

Vol. 965: Topics in Numerical Analysis. Proceedings, 1981. Edited by P.R. Turner. IX, 202 pages. 1982.

Vol. 966: Algebraic K-Theory. Proceedings, 1980, Part I. Edited by R.K. Dennis. VIII, 407 pages. 1982.

Vol. 967: Algebraic K-Theory. Proceedings, 1980. Part II. VIII, 409 pages. 1982.

Vol. 968: Numerical Integration of Differential Equations and Large Linear Systems. Proceedings, 1980. Edited by J. Hinze. VI, 412 pages. 1982.

Vol. 969: Combinatorial Theory. Proceedings, 1982. Edited by D. Jungnickel and K. Vedder. V, 326 pages. 1982.

Vol. 970: Twistor Geometry and Non-Linear Systems. Proceedings, 1980. Edited by H.-D. Doebner and T.D. Palev. V, 216 pages. 1982.

Vol. 971: Kleinian Groups and Related Topics. Proceedings, 1981. Edited by D.M. Gallo and R.M. Porter. V, 117 pages. 1983.

Vol. 972: Nonlinear Filtering and Stochastic Control. Proceedings, 1981. Edited by S.K. Mitter and A. Moro. VIII, 297 pages. 1983.

Vol. 973: Matrix Pencils. Proceedings, 1982. Edited by B. Kågström and A. Ruhe. XI, 293 pages. 1983.

Vol. 974: A. Draux, Polynômes Orthogonaux Formels - Applications. VI, 625 pages. 1983.

Vol. 975: Radical Banach Algebras and Automatic Continuity. Proceedings, 1981. Edited by J.M. Bachar, W.G. Bade, P.C. Curtis Jr., H.G. Dales and M.P. Thomas. VIII, 470 pages. 1983.

Vol. 976: X. Fernique, P.W. Millar, D.W. Stroock, M. Weber, Ecole d'Eté de Probabilités de Saint-Flour XI - 1981. Edited by P.L. Hennequin. XI, 465 pages. 1983.

Vol. 977: T. Parthasarathy, On Global Univalence Theorems. VIII, 106 pages. 1983.

Vol. 978: J. Ławrynowicz, J. Krzyż, Quasiconformal Mappings in the Plane. VI, 177 pages. 1983.

Vol. 979: Mathematical Theories of Optimization. Proceedings, 1981. Edited by J.P. Cecconi and T. Zolezzi. V, 268 pages. 1983.

Vol. 980: L. Breen. Fonctions thêta et théorème du cube. XIII, 115 pages. 1983.

Vol. 981: Value Distribution Theory. Proceedings, 1981. Edited by I. Laine and S. Rickman. VIII, 245 pages. 1983.

Vol. 982: Stability Problems for Stochastic Models. Proceedings, 1982. Edited by V. V. Kalashnikov and V. M. Zolotarev. XVII, 295 pages. 1983.

Vol. 983: Nonstandard Analysis-Recent Developments. Edited by A. E. Hurd. V, 213 pages. 1983.

Vol. 984: A. Bove, J. E. Lewis, C. Parenti, Propagation of Singularities for Fuchsian Operators. IV, 161 pages. 1983.

Vol. 985: Asymptotic Analysis II. Edited by F. Verhulst. VI, 497 pages. 1983.

Vol. 986: Séminaire de Probabilités XVII 1981/82. Proceedings. Edited by J. Azéma and M. Yor. V, 512 pages. 1983.

Vol. 987: C. J. Bushnell, A. Fröhlich, Gauss Sums and p-adic Division Algebras. XI, 187 pages. 1983.

Vol. 988: J. Schwermer, Kohomologie arithmetisch definierter Gruppen und Eisensteinreihen. III, 170 pages. 1983.

Vol. 989: A. B. Mingarelli, Volterra-Stieltjes Integral Equations and Generalized Ordinary Differential Expressions. XIV, 318 pages. 1983.

Vol. 990: Probability in Banach Spaces IV. Proceedings, 1982. Edited by A. Beck and K. Jacobs. V, 234 pages. 1983.

Vol. 991: Banach Space Theory and its Applications. Proceedings, 1981. Edited by A. Pietsch, N. Popa and I. Singer. X, 302 pages. 1983.

Vol. 992: Harmonic Analysis, Proceedings, 1982. Edited by G. Mauceri, F. Ricci and G. Weiss. X, 449 pages. 1983.

Vol. 993: R. D. Bourgin, Geometric Aspects of Convex Sets with the Radon-Nikodým Property. XII, 474 pages. 1983.

Vol. 994: J.-L. Journé, Calderón-Zygmund Operators, Pseudo-Differential Operators and the Cauchy Integral of Calderón. VI, 129 pages. 1983.

Vol. 995: Banach Spaces, Harmonic Analysis, and Probability Theory. Proceedings, 1980–1981. Edited by R.C. Blei and S.J. Sidney. V, 173 pages. 1983.

Vol. 996: Invariant Theory. Proceedings, 1982. Edited by F. Gherardelli. V, 159 pages. 1983.

Vol. 997: Algebraic Geometry – Open Problems. Edited by C. Ciliberto, F. Ghione and F. Orecchia. VIII, 411 pages. 1983.

Vol. 998: Recent Developments in the Algebraic, Analytical, and Topological Theory of Semigroups. Proceedings, 1981. Edited by K.H. Hofmann, H. Jürgensen and H. J. Weinert. VI, 486 pages. 1983.

Vol. 999: C. Preston, Iterates of Maps on an Interval. VII, 205 pages. 1983.

Vol. 1000: H. Hopf, Differential Geometry in the Large. VII, 184 pages. 1983.

Vol. 1001: D.A. Hejhal, The Selberg Trace Formula for PSL(2, ℝ). Volume 2. VIII, 806 pages. 1983.

Vol. 1002: A. Edrei, E.B. Saff, R.S. Varga, Zeros of Sections of Power Series. VIII, 115 pages. 1983.

Vol. 1003: J. Schmets, Spaces of Vector-Valued Continuous Functions. VI, 117 pages. 1983.

Vol. 1004: Universal Algebra and Lattice Theory. Proceedings, 1982. Edited by R. S. Freese and O. C. Garcia. VI, 308 pages. 1983.

Vol. 1005: Numerical Methods. Proceedings, 1982. Edited by V. Pereyra and A. Reinoza. V, 296 pages. 1983.

Vol. 1006: Abelian Group Theory. Proceedings, 1982/83. Edited by R. Göbel, L. Lady and A. Mader. XVI, 771 pages. 1983.

Vol. 1007: Geometric Dynamics. Proceedings, 1981. Edited by J. Palis Jr. IX, 827 pages. 1983.

Vol. 1008: Algebraic Geometry. Proceedings, 1981. Edited by J. Dolgachev. V, 138 pages. 1983.

Vol. 1009: T. A. Chapman, Controlled Simple Homotopy Theory and Applications. III, 94 pages. 1983.

Vol. 1010: J.-E. Dies, Chaînes de Markov sur les permutations. IX, 226 pages. 1983.

Vol. 1011: J.M. Sigal. Scattering Theory for Many-Body Quantum Mechanical Systems. IV, 132 pages. 1983.

Vol. 1012: S. Kantorovitz, Spectral Theory of Banach Space Operators. V, 179 pages. 1983.

Vol. 1013: Complex Analysis – Fifth Romanian-Finnish Seminar. Part 1. Proceedings, 1981. Edited by C. Andreian Cazacu, N. Boboc, M. Jurchescu and I. Suciu. XX, 393 pages. 1983.

Vol. 1014: Complex Analysis – Fifth Romanian-Finnish Seminar. Part 2. Proceedings, 1981. Edited by C. Andreian Cazacu, N. Boboc, M. Jurchescu and I. Suciu. XX, 334 pages. 1983.

Vol. 1015: Equations différentielles et systèmes de Pfaff dans le champ complexe – II. Seminar. Edited by R. Gérard et J.P. Ramis. V, 411 pages. 1983.

Vol. 1016: Algebraic Geometry. Proceedings, 1982. Edited by M. Raynaud and T. Shioda. VIII, 528 pages. 1983.

Vol. 1017: Equadiff 82. Proceedings, 1982. Edited by H. W. Knobloch and K. Schmitt. XXIII, 666 pages. 1983.

Vol. 1018: Graph Theory, Łagów 1981. Proceedings, 1981. Edited by M. Borowiecki, J. W. Kennedy and M.M. Sysło. X, 289 pages. 1983.

Vol. 1019: Cabal Seminar 79–81. Proceedings, 1979–81. Edited by A. S. Kechris, D. A. Martin and Y. N. Moschovakis. V, 284 pages. 1983.

Vol. 1020: Non Commutative Harmonic Analysis and Lie Groups. Proceedings, 1982. Edited by J. Carmona and M. Vergne. V, 187 pages. 1983.

Vol. 1021: Probability Theory and Mathematical Statistics. Proceedings, 1982. Edited by K. Itô and J.V. Prokhorov. VIII, 747 pages. 1983.

Vol. 1022: G. Gentili, S. Salamon and J.-P. Vigué. Geometry Seminar "Luigi Bianchi", 1982. Edited by E. Vesentini. VI, 177 pages. 1983.

Vol. 1023: S. McAdam, Asymptotic Prime Divisors. IX, 118 pages. 1983.

Vol. 1024: Lie Group Representations I. Proceedings, 1982–1983. Edited by R. Herb, R. Lipsman and J. Rosenberg. IX, 369 pages. 1983.

Vol. 1025: D. Tanré, Homotopie Rationnelle: Modèles de Chen, Quillen, Sullivan. X, 211 pages. 1983.

Vol. 1026: W. Plesken, Group Rings of Finite Groups Over p-adic Integers. V, 151 pages. 1983.

Vol. 1027: M. Hasumi, Hardy Classes on Infinitely Connected Riemann Surfaces. XII, 280 pages. 1983.

Vol. 1028: Séminaire d'Analyse P. Lelong – P. Dolbeault – H. Skoda. Années 1981/1983. Edité par P. Lelong, P. Dolbeault et H. Skoda. VIII, 328 pages. 1983.

Vol. 1029: Séminaire d'Algèbre Paul Dubreil et Marie-Paule Malliavin. Proceedings, 1982. Edité par M.-P. Malliavin. V, 339 pages. 1983.

Vol. 1030: U. Christian, Selberg's Zeta-, L-, and Eisensteinseries. XII, 196 pages. 1983.

Vol. 1031: Dynamics and Processes. Proceedings, 1981. Edited by Ph. Blanchard and L. Streit. IX, 213 pages. 1983.

Vol. 1032: Ordinary Differential Equations and Operators. Proceedings, 1982. Edited by W. N. Everitt and R. T. Lewis. XV, 521 pages. 1983.

Vol. 1033: Measure Theory and its Applications. Proceedings, 1982. Edited by J. M. Belley, J. Dubois and P. Morales. XV, 317 pages. 1983.

Vol. 1034: J. Musielak, Orlicz Spaces and Modular Spaces. V, 222 pages. 1983.